the technological bluff

JACQUES ELLUL

The Technological Bluff

Translated by

Geoffrey W. Bromiley

WILLIAM B. EERDMANS PUBLISHING COMPANY
GRAND RAPIDS, MICHIGAN

Copyright © 1990 by Wm. B. Eerdmans Publishing Co.
255 Jefferson S.E., Grand Rapids, Mich. 49503
All rights reserved

Printed in the United States of America

Library of Congress Cataloging-in-Publication Data

Ellul, Jacques.
    [Bluff technologique. English]
    The technological bluff / Jacques Ellul; translated by
Geoffrey W. Bromiley.
       p.   cm.
    Translation of: Le bluff technologique.
    Includes bibliographical references.
    ISBN 0-8028-3678-X
  I. Title.
T14.E54513   1990
303.48'3—dc20   90-40134
                                   CIP

CONTENTS

PART III • THE TRIUMPH OF THE ABSURD

PART IV ● FASCINATED PEOPLE

TRANSLATOR'S PREFACE

The area of technique, technology, and the technological system is one in which Jacques Ellul undoubtedly speaks with a prophetic voice. Already events have justified many of his earlier warnings or even shown them to be too restrained. In this latest work, which rests on solid research and covers a wide and varied range of topics, his message comes through with unusual urgency and force. He might be mistaken on some of the details but this ought not to blind us to the truth in the general thesis. Conversely, even if we find the thesis implausible, this ought not to prevent us from hearing the challenge of the details. As Ellul sees it, either technology as a whole or individual technologies are hurrying us into a situation of catastrophe. Lulled by the bluff of technological discourse, we sleepily fail to perceive the perils that confront us. Ellul's purpose, therefore, is to jolt us awake before we pass the point of no return. A special concern, perhaps, is to awaken the timid churches that are paralyzed by their fear of being derided as old-fashioned and outdated.

Since most of the specific examples that Ellul adduces are taken from his native France, readers should be advised that in a few instances where explanatory footnotes might be needed the details have been omitted. The context will show what is meant in the case of some systems such as Minitel for which the United States has no existing equivalent yet. Established usage in English makes it difficult to retain the distinction that Ellul himself always makes and emphasizes between *la technique* (technique) and *la technologie* (technology). It should be remembered, however, that in the title Ellul has the stricter sense in view, that is, technology as discourse, study, or system. The same applies, of course, to his discussion of technological terrorism.

Whether the powerful tocsin of Ellul can arrest the march of events may well be doubted, but in this regard Ellul has greater hopes, perhaps, from the New World than the Old. With the same sober optimism we present his exposure of the bluff to the English-speaking public.

Pasadena, Pentecost 1989   GEOFFREY W. BROMILEY

FOREWORD

On seeing the title of this book some people might exclaim: "What! Another book on technology!" They would not be wrong. In 1950 I wrote a comprehensive study of "The Technical Society." This was meant to be the title, but the publisher, a sociologist, would not let me use it, stating that he himself was writing a work with this title, although in fact he never finished it. Hence my own book became *La technique ou l'enjeu du siècle* (1954; translated into English as *The Technological Society*). But at that time two important publishers had also rejected the manuscript on very similar grounds. What kind of a subject was this? they asked. Whom did I expect to interest in a subject that was no subject? Yet even at that time there were studies of the industrial mechanism (a model being the work of Georges Friedmann) and literary works on society that carried references to technique (e.g., G. Duhamel's unjustly forgotten *Les scènes de la vie future*). There were also works (e.g., Huxley's *Brave New World*) at a level much higher than the dreadful and stupid books and films, no less vulgar than misleading, that go by the name "science fiction." But there was nothing on technique itself, or on the society about to be assimilated by it.[1] My book needed the influence of M. Duverger with a fourth publisher to

---

1. In this book we shall retain the distinction in French between *la technique*, which Ellul defines as "the totality of methods rationally arrived at and having absolute efficiency in every field of human activity," and *la technologie*, which is the study of technique. As in *The Technological Society* (for the most part, except for the title!), we shall translate the former as "technique" and the latter as "technology." For Ellul's elaboration of this distinction, see, e.g., *The Technological Society*, tr. John Wilkinson (New York: Knopf, 1964), pp. xxv-xxvi; idem, *Perspectives on Our Age*, ed. William H. Vanderburg, tr. Joachim Neugroschel (New York: Seabury, 1981), pp. 32-33.—TRANS.

appear finally in 1954. Today the wheel has turned full circle and there is a multiplicity of writings on the subject. Hardly a week passes without a work coming out on these questions in France.

Among such works I find three categories. First there are popular descriptive accounts of technique itself, usually of high quality, and dealing with specific subjects such as the computer or the laser and their application (e.g., the computer in schools or offices). Then there are lighter works consisting of essays on modern society under the impact of techniques and their rapid development (the consumer society, the computer society, the wasteful society), the upshot being that as technique achieves domination the society becomes one of networks rather than groups; we shall return to this idea. Finally, there are philosophical works. Since Heidegger and Habermas technique has become a primary theme in philosophy. Many philosophers are trying to understand the phenomenon (e.g., Cérézuelle, Janicaud, Neirynck, Hottois) or to see what its influence is on the world (e.g., Baudrillard, Morin, Brun). I refer only to books in French! Thus some think that this book of mine cannot hope to say anything new. Nevertheless, I see some important things that have not yet been brought to light.

I am not going to repeat what I said in my two earlier works.[2] But I will obviously begin with the conclusions I then reached and the phenomena I discerned. Only a few weeks ago an American sociologist told me how very relevant my first work on the subject still is. With no false modesty I can say that social, economic, and technical developments have confirmed in its entirety what I said thirty years ago. I have no need to correct or modify anything. Unfortunately, these analyses have been largely ignored in France, due to labeling. When my first book came out it was described as anti-technique, as hostile to progress, as reactionary. There was thus no point in reading my other books; their contents were known in advance. The Protestant world in particular has paid no attention to what I have written in this area. The label adequately described the product. It is true that many writers, without knowing their origin, have used analyses and conclusions of mine that have passed into common discourse (e.g., the neutrality of technique, or the ineluctable character of its development that escapes our grasp, or its universality). But the label remains. Although I was never against technique, and stated expressly that this would be absurd, and that I did not want a return to the Middle Ages, the label resulted from the rigor of my analysis, which gave to readers with a serene and superficial

---

2. *The Technological Society* and *The Technological System*, tr. Joachim Neugroschel (New York: Continuum, 1980).

view of technique the impression of being in the presence of a strange, aggressive, and constrictive world. Obviously, someone who wrote in this way had to be an enemy of technique. They preferred the soothing language of advertising which argues that technique brings freedom.

A recent example of the game of labeling, which leads to two false conclusions, may be found in a work by A. Bressand and C. Distler,[3] who say that those who are most opposed to technology are now reexamining it. Thus Ellul now grants it the benefit of ambivalence. (A first mistake here is that I already granted it ambivalence in 1950, and this had nothing whatever to do with the fact that it offers us very useful and satisfying products, which I have never denied.) The first false conclusion is that of opposition to technique. That is just as absurd as opposition to an avalanche or to cancer. It is childish to say that someone is against technique. The second false conclusion concerns my work *Changer de révolution*, which the authors quote but totally misunderstand. They say that I changed my view of technique because some new techniques serve the cause of decentralization and offer more free time. But I never denied this. The trouble is that they ignore half my argument. Some techniques can have positive effects so long as there is at the same time a change in society: the coming to power of a revolutionary socialism of liberty, which has nothing whatever to do with modern socialism or communism but demands a return to the ideas of Proudhon and Bakunin and a serious consideration of those of Castoriadis. Also demanded is a basic socioeconomic upheaval (e.g., in remuneration and distribution). Previously, there was no way to do this, but new techniques make it possible, though naturally they do not bring it about automatically. In saying this I was not reexamining any of my previous analyses. I was simply pointing out that change is possible if a politico-economic about-face goes hand in hand with the new techniques. I was also pointing out that the time was short in which to do this, perhaps only months, at most only a few years. Those years have now passed. It is now too late to change the course of technique. We have lost a decisive opportunity in human history.

I have adduced the example of Bressand and Distler to show how slight and superficial is the way in which most authors read my books. I am not referring, of course, to those who really read them. My warning today is the same as in 1954, when I wanted to alert people to the future potential of technique and to the risks entailed by its growth so that they might be able to react and to master it, lest otherwise it escape their control. I began with a warning, for one person warned is worth

3. *Le Prochain Monde* (Paris: Seuil, 1985).

two who are not. But people did not read the warning, and the book found its way onto library shelves and among the quiet studies of slightly outdated intellectuals. The only ones to take it seriously belonged to a society in which it was already too late to do anything—the USA. There both intellectuals and the public at large seized on my book because it described exactly what they were already experimenting with and experiencing. It helped them to understand what had taken place and how they had reached the point they had. But the book had only retrospective interest. In France people dismissed my expositions as the reveries of a solitary walker who prefers the country to the town. No one had any idea that it might change the course of things. I had indeed dreamed of another possible development. I wanted to play a role that no one had assigned to me, that of a watchman. But no one listened to me and the result was inevitable. My main purpose today, then, is not quite the same. I am now looking at the point which we have reached today.

There is a story behind this book. My first intention, stated at the end of *The Technological System*, was to write a book on all aspects of the impact of the computer on society, on technical malfunctioning, and on the way on which the computer might enable us to overcome some of this malfunctioning. I did work on the book from 1978, but I was too late. The world of computers was evolving so fast that I was always two years behind. I could never catch up. I thus abandoned the two hundred pages that I had already written and abandoned the whole project of clarifying the computer jungle and its relations to our world. I obviously never succeeded in mastering the material. It slipped through my fingers as soon as I thought I had grasped it.

After lengthy detours I conceived of another book, which would correspond, in a picture at once broader and more superficial, to an attempt to understand the reception of technique in modern society. The first of three parts would be: Challenges, Stakes, Wagers. Technical progress provides the challenges, the stakes are on the table, and intellectuals and politicians make the wagers. The second part would deal with the changes of society under the impact of technique. The third would have as its theme: The Issues, The Probable, and The Possible. I amassed for seven years an enormous amount of material that I had great trouble in assimilating and arranging, but I worked out a detailed plan and set to work. The first part was almost completely written when a book by Chesneaux, *De la modernité* (1984), was published. It was a terrible blow to me, for it corresponded almost exactly to my second part. It was a good book, and I would have had to repeat it. My own work would have been better documented and would not have put things in quite the same way, but the essentials were already there. I did not know quite what to do, and then, at the end of 1984, there came out the massive tome of the *Encyclopédia Universalis* entitled *Les Enjeux*, which contained a thousand pages and one hundred and twenty articles which dealt expressly with the stakes, the challenges to which they related, and implicitly the wagers. In other words, it contained my first part, and with ten times better documentation than I could offer. I thus had to ask myself whether it was worthwhile persevering in my investigation of technique. Yet I still had the impression that I had something different to say.

Three books drove me by way of reaction into a different field. Thus far I had been trying to work out a theory of the technical society and system. Three books now came out which greatly advanced this theoretical interpretation from a philosophical angle: Dominique Janicaud, *La Puissance du rationnel* (1985), Gilbert Hottois, *Le Signe et la Technique* (1985), and Jacques Neirynck, *Le Huitième Jour de la Création. Introduction à l'entropologie* (1986). I have to say that instead of discouraging me these three works showed me that there is as never before a current of thought which can effectively grasp technique and its implications, and measure its impact and risks, without falling into pessimism or making concessions. I was no longer alone; others were dealing with the issue. Should I myself withdraw? No, after a certain time I began to have a vague sense of another possible approach. Perhaps there might be new orientations in the world of technique, more diffuse and less certain than those that I had thus far analyzed. I was drawn in pursuit of these shadows.

I have entitled this book *The Technological Bluff*. Most who see the title will react sternly. Technique is an area in which bluff is not permitted. Things are plain in technique—either one can or one cannot. When it is said that people will walk on the moon, shortly afterward they do. When it is said that an artificial heart can be inserted, it is done and it works. Where is the bluff? But the problem is one of language. American usage has implanted in our minds the idea that the word *technology* refers to actual processes. This is the way the media use the term. But in a strict sense technology is discourse on technique. It involves the study of a technique, a philosophy or sociology of technique, instruction in a technique. To talk of computer technologies or space technologies when the reference is to technical means of communication or to the building and use of rockets, orbital stations, etc., is strictly ridiculous. I realize that my protest is useless in face of a usage established by collective ignorance, but I have to justify my title. I am not referring to technical bluff. I am not trying to show that techniques do not deliver what they promise, that technicians are bluffers. That would make no sense. I am talking about technological bluff, about the gigantic bluff in which discourse on techniques envelops us, making us believe anything and, far worse, changing our whole attitude to techniques: the bluff of politicians, the bluff of the media, the bluff of technicians when they talk about techniques instead of working at them, the bluff of publicity, the bluff of economic models.

Discovery of this bluff led me into strange areas. The bluff consists essentially of rearranging everything in terms of technical progress, which with prodigious diversification offers us in every direction such varied possibilities that we can imagine nothing else. Discourse on technique is not justification of techniques (which is not necessary) but a demonstration of the prodigious power, diversity, success, universal application, and impeccability of techniques. And when I say bluff, it is because so many successes and exploits are ascribed to techniques (without regard for the cost or utility or risk), because technique is regarded in advance as the only solution to collective problems (unemployment, Third World misery, pollution, war) or individual problems (health, family life, even the meaning of life), and because at the same time it is seen as the only chance for progress and development in every society.[4] There is bluff here because the effective possibilities are multiplied a hundredfold in such discussions and the negative aspects are radically concealed. But the bluff is not without great effect. Thus it transforms a technique of *implicit and unavowed* last resort into a technique of *explicit and avowed* last resort. It also causes us to live in a world of diversion and illusion which goes far beyond that of ten years ago. It finally sucks us into this world by banishing all our ancient reservations and fears.

Pessac, October 8, 1986   JACQUES ELLUL

---

4. A good example of this bluff may be found in the special number of *Match*, Oct. 2, 1987, on "Technique de 2005."

PRELIMINARY THESIS:
THE GREAT INNOVATION

1. Multiple Progress

There is general agreement that the last few years have brought an explosion of techniques, their perfecting in many fields, and the possibility of unheard-of progress: the atom, computer science, the laser, space technology, and genetic engineering.[1] I need not say much about these five areas. Their general aspect is well known, though their technical reality is very mysterious and indeed stupefying, surpassing anything that might have been imagined ten years ago. How far we have come from the elementary computers and rockets of 1950! There has been a ferment of invention in every field providing increasingly powerful and versatile types of equipment. This is wholly in accordance with the law that I formulated in 1950, namely, that techniques advance by a geometric progression.[2] My mistake in 1977 was to think that once a certain stage of efficiency and perfection was reached in a particular area, things would tend to stabilize. I believed that the power and speed of computers at that time would be sufficient, that there would be no need for further advance. I was wrong. Scientists have

---

1. I use the term *genetic engineering* in a general sense to cover such things as in vitro fertilization, cloning, and surrogate motherhood. Naturally, I am not dealing with the moral questions entailed; cf. P. Kourilsky, "Le génie génétique," *La Recherche* (April 1980). We should take note of the warning of J. M. Testart (Sept. 1980) that genetic engineering carries with it great risks, and we need to fix limits for it and declare a moratorium.

2. See *Technological Society,* e.g., pp. 89, 91. [Note that when Ellul mentions 1950 he is referring to when he first wrote *Technological Society;* likewise, 1977 refers to *Technological System.*—TRANS.]

1

found that to make the enormous calculations needed for new research they have to have computers a thousand times more powerful. The same applies in all five fields of innovation listed above.

A few years ago it seemed as if one of these covered the whole technical realm and could transform both society and the technical world: the computer had succeeded the atom in preeminence.[3] The computer seemed to demand a new study of its technique. But the computer is now rivaled in importance by the laser, outer space, and genetic engineering, which can also change our world. We cannot speak solely of a "computer shock." This is only one part of a whole raft of changes. We have also to meet the challenge of space, and we feel our impotence in face of such an explosion at so many different points. As I said earlier, I could not expound properly even the fantastic change brought about by advances in computer science. But how can an analyst grasp the transformation effected by every branch of these techniques? Critics might object to my choice of five basic fields, citing other examples of technical progress: cardiac surgery, agronomic research, discoveries in chemistry, developments in television, etc. The great difference is that the innovations in these areas, unlike the multidirectional innovations in the five noted, are measurable, apply only to specific domains, and cannot be put to more varied use. In contrast, the five mentioned above involve incalculable innovations in innumerable and unlimited domains. True, the five differ among themselves. Three—computer science, space, and genetic engineering—are complex fields of incorporated techniques, while the laser is a single technical tool. Yet all are versatile in an apparently unlimited way.

Some years ago there was an exuberant discovery of the many uses of the computer, and the same is true today of the laser. It seems that there might now be a change in the way in which we describe the evolution of society. In classical times, economic and industrial development might be described in terms of the evolution of energy.[4] Animal energy gave way to coal and coal to oil. When energy was then derived

3. There is an obvious hesitation to characterize modern society. One discovers in extremely learned studies some remarkably idiotic statements. Thus the transition from industrial society to our own has been described in terms of "a logic of development from an industrial society based on mass production to a postindustrial society based on science and technology" (see "Technopolis," a special number of *Autrement* [1985], p. 87). This is absurd. Industrial society was also based on science and techniques, and postindustrial society has mass production as its objective.

4. On energy problems see the special number of *Science, Technology and Society* 5/1 (1985), which contains good studies of the evolution of various forms of energy, and especially the future of electricity.

from nuclear fission (as I predicted in 1950, though I was then treated as incompetent in view of the immense problems that had to be solved), we were still within the normal scheme. One source of energy was succeeding another and making possible a new advance in production and transportation. But unfortunately this atomic age was very brief, for the technical novelty which came to domination in the 1960s had nothing to do with energy; it was the computer. Hence the established criterion for measuring the stages of industrial and technical progress, that is, energy, now proved to be inadequate. A new model of society was emerging. The computerized society, the society of networks, was not a prolongation of previous societies. It became less and less appropriate to talk of the postindustrial society.

Up to 1977 I could still think that the central feature of this society was the computer and that we should analyze the computer society in the same way as I had analyzed the technical society. The change was in effect startling, especially with the development of the minicomputer and the microcomputer. In previous models the dream had been of bigger, of more, and more profitable, material goods. But now in the computer field the advance was to the smallest. Efficiency was not tied to size but to reduction of size. The aim, too, was not to produce material goods but to produce, treat, transmit, interpret, and store information. Even in a world dominated by the automobile, research was now oriented to an ideally simple machine which could unceasingly create more and more complex systems. Saving energy now replaced the growing consumption of energy. The new world of computers, office automation, telematics, robots, etc., is a very efficient world which is very productive but which consumes very little energy.

It is not enough to think rather calmly that we are simply at a new stage of substitutes for people. For a long period human physical force was being replaced progressively by machines and enhanced energy. People now had not simply to provide energy; they had to master machines. They had to think, to devise better machines. But with the computer, with the transition from the production of material goods to the production of information, a new stage of human activity has come. Does this mean that human beings have finally been made redundant? Let us leave this question, which some take very seriously but others trace to the fears of ignorant people who are afraid of anything new. We will take it up later. The major fact that we need to note in the complete turnabout involved in this transition is that in the industrial age of energy, whether coal or oil, more workers were constantly needed. This was part of the trend toward the increasingly bigger.

We recall that the theory of Marx rests on this fact. The working class will grow ineluctably and thus become the most numerous and the most powerful class, though also the dominated class. That domination cannot last indefinitely. In these circumstances the dominated class will necessarily become the dominant class. But in the age of automation and computerization the trend is reversed—there is less need of manual work. In spite of attempted proofs to the contrary, the new machines are laborsaving. Investment in capital increases, and investment in labor decreases. The number of workers declines. In compensation another social category arises which corresponds to the multiplication of services. This has been the major issue with accelerated economic growth. It has become evident that we can support a multiplication of services in every field: care, administration, management, instruction, health, social work. Thousands of jobs are created each year in sectors that are of social usefulness but of no economic value. With mass production, increasing salaries, and mass consumption, along with additional wealth on a national scale, it has been possible to develop these services, including those whose sole purpose is to amuse children.

Computers have played a part in this development. They have invaded every field, especially that of services. With computerization, office automation, and telematics, fewer people are needed in offices. The worst reductions have been in banks and insurance. What people once did, machines now do, and with more inevitability than at first. Where the crisis does not mean overproduction, and labor costs are highest, computers have replaced service people, and once they are bought they cost little to run. What automation did in the industrial world, computers are doing in the service world.

I have read the intensive discussions, with statistics, on whether computers cause unemployment. The statistics do not convince me, but once we get into the arguments there can be no doubt that computers bear a great deal of responsibility for unemployment.[5] Only a few jobs are needed for their building and maintenance. This fact raises the question as to the validity of the French policy of teaching all children the use of computers so as to prepare them for the modern world. It is plain that few of them will find employment in the field. We shall return to this point in studying the bluff of the computer. The multiplication of computers and their application has made our society the computer

5. See R. Rothwell and W. Zegveld, *Technical Change and Employment* (New York: St. Martin, 1979); P. Boisard, "Les 35 heures et l'emploi," *La Recherche* 128 (Dec. 1981).

society. The atom has been dethroned as the essential criterion of our times. This computer society is characterized by two factors that modify the very idea of progress. Progress is henceforth measured by the savings of energy and of labor in producing the same amount of goods. One must never lose sight of these criteria when one hears a discourse on productivity. Earlier, productivity was linked to an increase in labor, but this is no longer true. An enterprise is now more productive and competitive the less it employs human labor.[6] At all events, the new society is no longer dominated by the problem of energy but by that of communication and data processing. This has become so fascinating and allows us to imagine so many applications and to open up so many new fields that people have become bold enough to think that they are real creators. The computer seems able to do everything from poetic and pictorial creation to the management of pollution and even the making of political and military decisions. I could almost accept it when I consider (as above) that the microchip might work in favor of freedom. But then there arises the question of knowing (cf. my *Technological System*) whether, thanks to computer science, one can set up a true feedback in the technical system, that is, whether this system, which evolves and expands without any control, might develop a mechanism of feedback that would control the orientation and rapidity of the functioning and adaptation of the system. There can certainly be no doubt that our society is characterized above all by the computer, and that atomic energy, important though it is, is not what gives meaning to our world.

But a rival for the computer replaced it more swiftly than it did the atom: outer space. Space today involves a number of such new and decisive techniques and opens up so many possibilities and fields to human enterprises that one might say quite definitively that our world is just as much characterized by the opening up and conquest of space as by the computer. What might give the opposite impression is that

6. For this reason two types of argument are ridiculous: first, that to solve unemployment we must make business more competitive and increase production; and second, that the crisis of 1979 was like that of 1929. The accelerated development of new industry solved the 1929 crisis, along with the use of new techniques; the automobile industry gave the economy a new boost. Therefore, according to this incredibly simplistic argument, the development of a new technical apparatus, i.e., the computer, will revive the economy today. The computer industry will enable us to overcome the crisis. It matters little what the product is, or what its features are! Unfortunately, the features in the world of computers are very different from the new techniques in metallurgy in 1930. A new technical development alone will not give a boost to the economy. The two types of techniques have little in common and we cannot expect from them the same results.

the microchip has invaded daily life in the form of desktop computers and Minitel, whereas space is still only a spectacle on television. In political, military, and economic realities, however, and also in the matter of science and information, space is supreme. Our society, then, is just as validly characterized by this opening up of boundless worlds.

Yet a new technique now threatens the two dominant ones, namely, the laser. Ten years ago we hardly knew what this was; it was still at the stage of laboratory experiment. Now it is everywhere: in surgery, heavy industry, music recording, astronomy, military preparations, etc. We have here an instrument of tremendous efficiency, unimaginable flexibility, and a power and precision never previously attained. We cannot as yet conceive of all its immense possibilities. Should we not describe the world that is now burgeoning as that of the laser? In other words, in twenty-five years there has been a total transformation of the data base, the organization, and the activity of our society. It is in fact characterized by four dominant factors that are coherent with one another: computer science, genetic engineering, space, and the laser. These are dominant as regards investment, expected returns, the number of specialists and research workers involved, prestige, etc. And all this seems to be completely new. Yet the more I study this society and its technical system and all the new data, the more I see that it is still the same technical system. It is richer and more complex. It is constituted differently. Yet it is self-coherent. That is, all these great innovations fit in just as well with the general characteristics that I have demonstrated for the technical system as those of technical progress and the mode of technical progression. From the standpoint of basic analysis nothing has changed. If you have a motor, you want to go fast, but whether you do 80 or 120 makes no real difference. Even if you press your engine to do 180 it makes no real difference. Things are exactly the same here. The computerized society, the space society, the laser society—in the last resort they are all just technical or technicized societies. The magnificent innovations change neither the character nor the basis of the problem. It is not they that are the great innovation of our age. The technical changes enhance the impact, power, and domination of technique in relation to all else, but no more. The same applies to the means supplied by research.

Nevertheless, this does not mean that nothing fundamental has happened. It has happened, but in a different way from that imagined. Obsessed by the computer, people want to see something there that evades prior qualification. What they see is true at the superficial level of society's modes of operation. It is true when we think of the change from the domination of industrial production to that of information, though obviously the latter is no substitute for the former, as in the absurd thesis of J.-J. Servan-Schreiber in his ridiculous book *The World Challenge*. As great changes in energy do not affect the industrial system, the computer does not change the technical system; it simply confirms it, develops it, and makes it more complex. At the surface level of what is seen there may be great changes (progress?) in the organization of the system, but these remarkable gadgets change nothing. In spite of that, we have in fact been witnessing a great innovation from the early 1980s, but of a very different kind.

2. Social Discourse

Before studying this innovation, however, I want to show what social discourse relates to it. Our new situation is denoted by the trilogy of challenges, stakes, and wagers. Whether in articles, in publicity, on the radio, on television, in political discourse, or in some serious works, everyone is speaking in these terms. Computer science, space, and genetic engineering all involve challenges, stakes, and wagers. There is the Japanese challenge and the European stake; the Third World stake (for the two great powers), wager (for its development), and challenge (its possibility of growth); and the stake of independence from the USA. Peace is a stake, but peaceful coexistence is a wager. Growth and plenty are both stakes and wagers. The space shuttle is a wager, immigration is a challenge, unemployment is a challenge, politics is a wager. I might give many other examples, but all these are taken from the information that is fed to us and all come through the same grid. But I do not think the categories are imposed by chance. They unconsciously express the collective interpretation of our situation. But what do the three words mean?

The term *wager* suggests an awareness of risks and a readiness to be bold. In a situation that seems to be fixed and closed, we wager on something uncertain and whose outcome is unpredictable. We take risks when we wager. "The game of truth involves risks, the game of democracy involves risks, the game of revolution involves risks, and to play all these games together involves many risks" (Morin). By talking of wagers we want to show that we are taking risks, that we are authentic and democratic. We also want to show that we are free. To make a wager (even though it be fictitious) is to be free.

Again, when we talk of a *challenge* we want to show that we have the courage to take up the challenge thrown down by a disturbed, evolving world of nightmares and apparent impossibilities. We can face the worst of circumstances. Society may be full of rivalry and conflict, but we can handle the conflict and triumph. We will not be defeated. It is not a matter of power but of judicious judgment and bold decision. "All computing, transmitting, and reproducing of information involves the risk of error. The living organization of each of these operations is subject to the risk of error, which is the risk of degradation, disorganization, and finally death. Thus the essential and basic need of every living organization is to combat error."[7] We might apply this idea exactly to the social body. It shows us that if we fail to meet the challenge of circumstances we are rejecting the possibility of error.

Finally, the word *stake* is ambiguous. It means first that everything is to some extent a game. It does not have ultimate seriousness. "Monopoly" was a game in the image of reality, but reality has now become a game in the image of "Monopoly" (or a naval battle). When we think of the contingent character of new computer games we realize that the games that we play do not depend on our skill or knowledge or virtue, but neither are they a matter of pure chance. These games are not roulette.

No matter how serious our situation may be, it is only a game. But the problem is to know definitively what we are putting in play. Our outlay depends on the seriousness or gravity of the game. The important thing is to know the dimension of what we are risking in this game that we do not control. It may involve our whole future, and if the stake is well chosen we can win all or lose all. It is then a game of all or nothing, and the main thing is to see clearly not the rules of the game but what we decide to risk. Ladrière can even talk of the wagers of rationality.

With the three terms, which meet us on every hand with great insistence, we are thus trying to signify our freedom, audacity, and clear-sightedness. This sums it all up. Nothing in our society might not be described in terms of challenge, stake, and wager. Nothing is any longer a mere private adventure; there is no longer any private sphere— we are all in this game. Nothing is indifferent, nothing is outside. The game is so big and so universal that it is communal. There are no individual players; we all play it together. In circumstances great and small, economic and political, we have to win a real battle in the form of a game the stakes of which we cannot appraise exactly. We set the stakes but we are not sure exactly what they are.

Having tried to show what the three terms conceal, I must say in conclusion that they finally relate to technical questions. This is

7. Edgar Morin, *Pour sortir du XXe siècle* (Nathan, 1981).

certainly not obvious. The Third World, Europe, militarization, etc., are all political matters. Inflation, exchange rates, standards of living, and growth are all economic matters. Yet technique has a part in all of them. It is like a key, like a substance underlying all problems and situations. It is ultimately the decisive factor. The new technique is the real challenge. We have only to think of the number of books on the subject: the computer challenge, the genetic challenge, the technological challenge, etc. (cf. Salomon). We see at once that both individually and socially we are not prepared quietly to accept the computer, for example, into our way of life. It is a challenge because it upsets not only our rhythm of management but also our way of thinking. It overturns our bureaucratic and cerebral approaches. On every point it puts thousands of bits of information at our disposal which overwhelm us and accustom us to count in millions and billions as if they were single digits. It changes the dimension and speed of life. Whether we like it or not, it sets up networks in society that have nothing whatever to do with ancient networks or traditional structures. We cannot continue as before. Simply because the computer is there, we cannot ignore it.

When the railroad and the automobile came on the scene, those who wanted could still travel by horseback. But now there is no choice. The computer as such implies networks. A businessman cannot acquire a computer just because he likes progress. The computer brings a whole system with it. The difference is that the technical system has now become strongly integrated. Offices, means of distribution, personnel, and production have all to be adapted to it. If they are not, they run the risk not merely of losing the advantages brought by a fascinating and useful gadget but also of causing unimaginable disorder by introducing computers into an organization or society but not making possible their proper use. I might say the same of nuclear energy or genetic engineering. To bring the latter into play is to challenge not merely social organization but our philosophical concepts, our traditional humanism, our morality.

Everything, then, is challenged. Can we adapt physically, socially, and intellectually to the computer? Can we adapt morally to genetic engineering? These are not rhetorical questions or academic hypotheses. We might see clearly enough how to train children for the computer age, but the other questions are left in the void and are subject to mere reactions. Yet here is the point of real relevance. Let there be no illusion: All the challenges that are called such in social discourse are directly or indirectly the result of technique. The same applies to the stakes. What are these? Among other works there is, as we have noted already, a whole volume of the *Encyclopédia Universalis* which devotes a thousand pages to the stakes in our society. Obviously, these are of all kinds and they give the impression (which is accurate from one standpoint) that we are at a decisive turning point in human history. We need to know what is going to happen in the sphere of communication, or what the simple fact of prolonging human life and the increase in the number of older people will mean for society, or what is the relation between human intelligence and artificial intelligence, or what will become of the arts with all the new processes (e.g., in music), or whether ideologies (religious or political) will disappear when they come into contact with technical reality. We can no longer say that technique is a simple instrument in the service of human thought. But the stake is also knowledge itself. Can we continue being able to know as humanity has known for centuries past?

The stake is also the social and natural environment. Ecologists are obviously behind the times; they cannot meet the challenges. They rightly stress the natural stake, but how about the technical challenge? The stake is the social link itself. Processes of change are modified. The media have now confused what used to be the clearly separate domains of social and private life. The terrible growth of violence and terrorism is not just a political matter but primarily a technical matter, as I have shown in various articles. The stake is information, and disinformation through excess of information. It is the inability of political and administrative structures, of politicians and political doctrines, to take into account the reality of the technical mutation. Will there ever be any politics again but television politics? This is a more profound matter than simply having to make constant decisions about problems that are infinitely beyond us, or being in the presence of things that are impossible to decide. In relation to techniques, where is the frontier today between what can be decided and what cannot? An unheard-of stake. Finally, of course, there are the more banal stakes of the relation between North and South, of the paradoxical and apparently inextricable situation of a world which has become one, both materially through the means of communication and economically through complementary economies (because one must henceforth speak of a world economy, not an international economy), but which is still divided into over- and underdeveloped countries, or over- and underequipped countries.

This is not a matter of justice or of sentiment. It is a matter of technique. It is urgent, for if we do not win the stakes we will die. The stakes, which are multiple in the economic world, may finally be reduced to two that are economically dominant: redressing demographic imbalance and reorganizing the world monetary system. We certainly cannot continue indefinitely to live with aggravated monetary imbalance, especially now that the sums involved are unimaginably high. But finally at issue in each of these stakes is what technique to apply here and now to win this game. In society as it now is our only recourse is to find the techniques that will enable us to evaluate the stakes and win. This is not always perceived if we look at the wagers that are made. When we find this term in discussions, what is meant by it? The wager is that we will be able to control technique. But it takes different forms. Some wager that it can be done by a planned policy. Others wager on democracy, that is, that they can direct technique by democratizing society. We see this in the ideal of controlling energy. Still others wager that law can master technique: global law in the form of new international institutions, international treaties, codes of conduct, rules of responsibility. The globalizing of law will correspond to the globalizing of technique that produces the globalizing of the economy. In the last resort the wager is that technique will issue in the decentralizing and self-organizing of society.

The idea here differs from that mentioned earlier (i.e., the conjunction of microtechniques and a socialism of freedom). Here it is technique itself that leads to a self-organized society. The two wagers are obviously antithetical. In the one case the bet is that technique will be brought under control by politics. In the other the bet is that technique will produce a normalized society.

It is thus wagered that technical growth is the only means to overcome the present crisis and that we must use new techniques to deal with the economic crisis. It is also wagered that we will finally produce a technical culture. This is the decisive wager. In face of the crisis in modern culture, the apparent conflict between culture (in the traditional sense) and technique, and the fact (noted by only a few authors)[8] that in all societies up to the eighteenth-century West, techniques were integrated into a global culture, it is culture that is now dominated or marginalized by technique, so that thoughtful people believe that we must now create a new model of culture that will fully integrate technique and again subordinate it as a simple tool.

The stake of a technical culture is a major one and it is wagered (e.g., in education) that this culture is possible. Bets are placed on the computer. Thanks to it we can supposedly solve all our problems. It is also wagered that there can be a new and adequately balanced world economy on the triple basis of research, development (or growth), and modernization. It is thought that in coming years the contradiction thus

8. I studied the decisive switch in this regard in *Technological Society*.

far perceived between growth and development can be overcome. Qualitative growth can be achieved; Poperen has proposed a "pact for new growth." It is wagered that technique will overcome the innumerable impasses of the Third World. Thanks to new techniques the Third World will finally "take off" (i.e., take the good path of growth, which is that followed by our society!). It is wagered, too, that technology will make it possible to forecast successive stages of evolution, as well as the perverse effects of techniques and the objectives of assured growth. Finally, we are told that to the degree that people are not at a technical level we must invent or create a new people who can use techniques correctly so as to achieve the best result that will not slow progress or have adverse effects. A bet is placed on the possibility of these new people.

These wagers all merit full discussion. I have simply listed them here as the wagers that are commonly made today by those who discuss technique. It is noted that with technical proliferation technique is itself the challenge to all humanity, to our society, and to our economy. It challenges all (to live or die). Technique is also the stake. We must find the most efficient technique. Every form of technique, being the key to all else, is the stake in political, economic, and scientific battles. It is finally the wager as well, for thanks to it one can win the bets that are placed on the future of humanity and society. Everything now depends on technique. We live incontestably in a society that is totally made by it and for it.

In the many works and discussions that center on this trilogy, I am somewhat perplexed by the links that are most often made between the wagers put on the stakes and the evaluating of the stakes in terms of the challenges. For coherence each ought to correspond to the others. But this is the exception! Thus the challenge is to integrate new techniques smoothly into the social order. But the stake is economic, that is, balancing the resources and technical forces between North and South. What, then, is the wager? That law will solve all these problems. I might take each word in turn and show that a feature of technical, scientific, and even highbrow intellectual discussion is that the terms do not adequately correspond to one another, so that the discussion is finally incoherent in spite of an appearance of rigor.

The progress made in these areas is the growing (if tardy) awareness that everything depends on technique (and not economics), but the responses are still halting and so is reflection or discussion on the part of those who bear responsibility in this venture. Throughout this book we shall see, indeed, that the reality is far larger than anything we can imagine or think of mastering. Extraordinary innovations in many fields have forced all those responsible to take note, but this is not enough now. We need to be convinced that what we see clearly or learn through the media is nothing compared to the small changes that are taking place, each important in its own sphere but not spectacular enough to engage the public's attention.

We must constantly remember that when we learn with astonishment that a new space vehicle is used or a new surgical operation performed, these are merely the final results of hundreds of small, patiently accumulated improvements. The multiplying of innovations in many different areas makes it possible for us to speak of a third industrial revolution (though this is not a happy term, since the whole is more than industrial). There is such diversity, such effervescence, that it is hard to describe our society. Is it the media society, developing from the outset data processing in the broader sense? Is it the space society? Is it the society of nuclear power? Or, to revert to older categories, is it the consumer society (Scardigli), or the education society (Beillerot)? In a more comprehensive fashion, let us for the moment speak of the society of progress.

The idea of progress seems self-evident, both as assured direction and as effective progression. Economic growth governs economic development, which in turn governs social and individual development. Quantitative increase in all areas entails of itself qualitative amelioration. This is obvious. Yet at the same time one sees that this technical proliferation involves a kind of neo-mechanism with a panoply of (technical!) sciences of organization, of information, of communication, of complexity, and of the social in general viewed in mechanical terms. For in every way the sciences (as we shall see) have become social. "There seems to be an irresistible attraction of scientists to what is social (biologists, physicists, etc.). Clearly, they are no longer content to elaborate concepts and theories in their own fields; they also try to generalize in the social domain (Varela) or to think in a social perspective (Prigogine)."[9]

But in many new sciences and technologies, whose four main areas I have cited, we have to note the development of a series of second-degree techniques, that is, of techniques supporting technique. For example, studies of the conditions of change in business, of the transfer of technologies, of technological gaps, of the research and development trend, of technology assessment; studies of impact, studies of the psychological effects of technical applications on personnel.

9. See J.-P. Dupuy in *L'Auto-organisation*, Colloque de Cerisy (Paris: Seuil, 1983).

An enormous field of research is thus constituted in relation to the effects of techniques, especially in the four main spheres of innovation. We thus have a sense of tremendous change in the world. This is true. But it is difficult to assess this change. It is not enough to list the technical innovations, for the change is at a different level from the direct level of the innovation itself. It is not enough to cite statistics. These have their value, but we have stated already that the change is qualitative as well as quantitative. We have too weak a view of it if we cite only the numbers and fail to bear in mind the uncertain character of statistics in every field. To measure the change, should we then proceed subjectively, referring to memories, to the experience of older people who have lived through three technical revolutions? This is useful, but it is too fragmentary and haphazard. The measure of change is in every way too vague. It is too limited on the one hand, too sensitive on the other. It is not scientific or comprehensive.[10]

In fact, we cannot measure change and progress. They are unmeasurable in the sense that they cannot be compared with anything else. They are phenomena that we can neither contest nor grasp. There is nothing analogous to them. I said above that obviously everything changes, and yet fundamentally nothing has changed. What counts is the level on which we speak. I have used elsewhere the model of the ocean. A furious storm stirs up its surface, great waves arise and crush everything. But a hundred yards down the water is calm, the storm has no effect. Ocean currents remain the same in spite of the storm. There has been tremendous technical change, but no global technical change, no change in the technical system. No particular technique challenges or opposes the global movement of technique. Even the most dazzling innovations are within the system. There is a vast difference between high-speed trains and local trains, but the railroad system—the rails, the network, and the signaling—is the same for both. There is no change in organization, in timetable, in the material and organizational infrastructure.

We can certainly speak of a "computer shock."[11] Production methods have all been challenged. The computer itself has become an object of production of major importance. It takes the place of other sectors of the production of goods. It modifies our manners and upsets our ways of thinking, reasoning, and understanding. It gets rid of whole sectors of knowledge and creates a new environment and a mode of

10. See Michel Henry, *La Barbarie* (Grasset, 1987).
11. Contrary to Bachman and Ehrenberg, who deny that there is any future shock, *Le Monde*, March 4, 1985.

relating without precedent. It is a shock in that all these upsets come so quickly that we have no time to get used to them. As soon as one begins to get used to some computer equipment, it has already changed.

We could also speak of a third industrial revolution, so long as we retain the word *industrial*. New goods are produced which in the marketplace seem to be competing with those produced for the last hundred years and surpassing them in importance. After the textile revolution and the metallurgical revolution has come the information revolution, that is, the production of goods which themselves produce and manage information. These are radically new products. We might add that this revolution tends to save energy. But it is better to speak of a third economic revolution; it is not strictly a technical revolution, since there is no change either at the level of or in the character of technique. Everything new fits perfectly into the technical system. It follows the normal line of technical development (since it is characterized by incessant growth) and takes on the same features. There may be technical progress and new fields of application, but these do not throw doubt on my earlier analyses of the technical phenomenon. Naturally, I will not repeat those here; I will presume that the main lines of what I found are well known. I refer, for example, to the fact that technique is our environment, the new "nature" in which we live, the dominant factor,[12] the system. I need not elaborate on its features: autonomy, unity, universality, totality, automatic growth, automation, causal progression, the absence of finality.[13] Nor need I discuss afresh the distinction between the technical system and the technical society. I analyzed all these matters in 1950 and 1977, and what I said is still valid and has not been contradicted by any modern technical development.[14]

12. Twenty years ago it was scandalous to say that technique was the dominant factor in society, but now *Le Monde* can have a heading to this effect relative to a Franco-Japanese discussion held on July 1-4, 1985.

13. On the universality of technique cf. E. Zaleski and H. Wienert, *Technology Transfer Between East and West* (Washington, DC: OECD, 1980).

14. See *Technological Society*, e.g., pp. 78ff.; *Technological System*, e.g., pp. 34ff.; *What I Believe*, tr. Geoffrey W. Bromiley (Grand Rapids: Eerdmans, 1988), e.g., pp. 133ff. My views have aroused lively opposition, especially regarding the autonomy of technique; cf. Dupuy, Castoriadis, Lussato, Kemp, Gorz, etc. Basically they are used, e.g., by Roqueplo, *Penser la technique* (Paris: Seuil, 1983), though he does not say so.

3. The Great Innovation

We have tried to show that the innumerable innovations do not of themselves change the existing technical system. Nevertheless, there has been a vast transformation during the last few years, but of a very different kind. It was well known that people adapt badly to modern techniques. In spite of progress in industrial mechanization due to ergonomics, there have been many maladaptations which have produced either disorders—formerly physiological and in modern times psychological—or disruptions in the order and efficiency of techniques. The troubles vary greatly, but the classical problem is that people do not adapt to machines nor machines to people. The ideal goal is a marrying of people and machines. It might be the people who were evaluated negatively: by their retrograde spirit they were hampering the harmonious development of the technical world. Or the blame might be put on the machines: technical growth was crushing spontaneity, imagination, values, the irrational element—in other words, all that makes us human.[15]

Since the opposition between people and machines as it found illustration in Charlie Chaplin's *Modern Times*, the contradiction has changed much, becoming situated in the nervous system (e.g., creating psychoses, loss of sleep), with new techniques being implemented. But it is no less an antithesis, and remedies are being sought without, of course, trying to stop technical progress. On another plane there is opposition between the technical society and the technical system. I have shown that we must not confuse the two. The system has its locus in society. It controls almost all social orientations and structures, but it does not incorporate everything. That is, society remains outside the system; its institutions are not rigorously technical; society carries within it a whole ensemble of ideologies, of survivals of the past, and of myths. As Castoriadis has finely shown in his analysis, it is constituted by "social fantasy." Manners and customs are on the fringe of technique. Society, with the human relations that Crozier sees underlying all else, is not just a simple expression of technique, even though technique is slowly invading domains that it does not yet control. I have also shown that this situation reacts upon the technical system itself: it is developing in a world that does not favor it a priori. It has to overcome resistance. Most of the malfunctioning in the technical system results from the maladaptation of the social body to technique, which otherwise would function without breaks or adverse effects.

15. M. Henry, *La Barbarie*, stops at this point.

In both cases the problem is that of rootage in the past. We go back to a remote or recent past from which (even unconsciously) we take most of our features. Each society is the result of slow evolutions, of progressive creations. It draws its substance from an accumulation of past experiences. It cannot cut its roots without falling apart. Technique, however, is a thing of the present and looks to the future. It gradually effaces its own past. Yesterday's machines are now valueless. The automobiles of 1950 are laughable. Genealogies are not needed to reach the present stage; only efficiency and power count. Today's technique certainly has its origin in yesterday's, but yesterday's, if it is still useful, is incorporated into today's. I think that it is this radical antithesis which brings conflict between people and society on the one side and the technical system on the other.

In 1970 this conflict was the central question of our civilization. One attempt to answer it was made by negating the problem; this attempt was presented in two forms. There were irrational remnants that could be quickly set aside, or technique was not as disruptive as supposed and it could be restricted to enter quietly into the social framework and to fit there. Another attempted answer, once conflict was accepted, was the postulating of utopia, either a social utopia or a human utopia. In literature the former was depicted optimistically in *Brave New World* and pessimistically in *1984*. On the popular level we find the social utopia in films and science fiction. In every case, by a leap past a delicate intermediary period, there is entry into a marvelously equipped and balanced society with machines of unimaginable power that solve all problems either positively or negatively. Good or bad, utopia is a response to the actual situation, though no one can tell us how we arrive at it. I have found it in very serious writers who think that our present-day difficulties are negligible and who look to the future, which will solve every problem.

The human utopia has also been fashionable. It envisions the psychological changes that are necessary for entry into the next century. But since these are uncertain, it also envisions interventions like the implanting of tiny electrodes in the human brain, an operation which is without risk and which can eliminate outdated reactions and emotional adaptations and achieve the behavior that is in full keeping with the environment.

This, then, was the core of all the problems of our Western world, whether political, economic, sociological, or psychological. The sights were obviously set on the main difficulties, which it seemed impossible to overcome.

But it is precisely at this point that, some years ago, what I call the great innovation occurred. This is infinitely more important than all the technological discoveries mentioned above. The transformation was as follows: People stopped looking for direct means to resolve conflicts. They have stopped trying to adapt politics or economics to technique by force. At the same time, they have stopped trying to produce mutants, that is, people perfectly consistent (flawless) with a technical universe. They have stopped trying to clash head on with obstacles and refusals. They have stopped trying to rectify technical malfunctioning by direct action. There has been a transformation the results of which we are not yet able to measure but which I would call an outflanking or encirclement. Obviously, no one has calculated such a strategy, no one has deliberately brought about what has happened or is now happening. To all appearances nothing has changed. Techniques continue to forge ahead. We profit by them. We have the same ideological reactions to them as from the beginning of the technical era.

No one has taken charge of the system to bring about a corresponding social and human order. Things are done, "by force of circumstances," because the proliferation of techniques, mediated by the media, by communications, by the universalization of images, by changed human discourse, has outflanked prior obstacles and integrated them progressively into the process. It has encircled points of resistance, which then tend to dissolve. It has done all this without any hostile reaction or refusal, partly because what is proposed infinitely transcends all capacity for opposition (often because no one comprehends what is at issue), and partly because it has an obvious cogency that is not found on the part of what might oppose it. For what would it be opposing? This is no longer clear, for insinuation or encirclement does not involve any program of necessary adaptation to new techniques. Everything takes place as in a show, offered freely to a happy crowd that has no problems.

This encirclement by what is obvious takes place in many ways and through many voices, but it is possible only through the prodigious development of modern techniques, which, while being so powerful, give us at the same time the sense that they are more close, more familiar, more individualizing, more personal. Here is the true technical innovation, for it is by this basic support of the whole social body and of each individual that the system can develop without encumbrance. This is incomparably more important than the laser, in vitro fertilization, or intergalactic probes.

There is no longer any need for myths or great projects. The transformation takes place in the everyday world. Its very ordinariness ensures its success. To present an image of a changed humanity is to provoke an inevitable reaction. Ordinariness gives reassurance. The genius of technique (not of technicians) is to produce the most reassuring and innocent ordinariness. This is what we are studying under the title of the technological bluff. We must try to make the situation clear. It is not that we or society are better adapted to technical growth, but only that we are, let's say, neutralized in a way that there can no longer be any open or secret conflict.

This encirclement or outflanking of people and society rests on profound bases (e.g., a change in rationality) and the suppression of moral judgment, with the creation of a new ideology of science. But it is effected by the enticement of the individual into permanent socio-technical discourse. This enticement is brought about by deliberate action on the part of those who want absolutely to make the change and by a spontaneous movement on the part of others. The first case has again two possibilities. There is the work of those who as theorists of technique or as superior technicians think that our supreme good is to adapt as well as we can to this ideal of perfection. There is also the intervention of politicians and economists who think that given the crisis, unemployment, etc., extreme technical development is the only solution, so that willingly or unwillingly we must adapt to it. The second case, that of spontaneous movement, has two paths as well. The first is that of those who want to succeed in society and who know that henceforth only those who are technically adept can do so. The second, to which we must devote more attention, is that of the group of what I would call the fascinated, that is, those who are so fascinated by the kaleidoscope of techniques invading their universe that they do not know and cannot want anything other than to adapt fully to them.

These are the different paths that we shall try to explore, pointing out the great distance between technical reality and our present situation on the one hand and the seductive discourse of techniques, which constitutes the technological bluff, on the other. These different paths, these different actions and reactions, will bring us face-to-face with the great technical innovation, the integration of people and society into the technical world. Naturally, this has not yet been achieved. There is still a gap between society and the technical system, between individuals and the technique which surrounds them. But this gap is constantly narrowing and a new model of humanity is emerging in the West. In spite of the pious talk of politicians, the result is an increasing gap between Western humanity and that of other societies.

Excursus on Simon, *The Ultimate Resource*

I want to discuss Julian Simon's book *The Ultimate Resource,* which deals with demographic growth, natural resources, and the standard of living.[16]

The fine title of the French edition, *L'Homme, notre dernière chance* (Humanity, Our Last Chance), might suggest that this is a humanist work amid the perils of the age. It is not. Its focus is on the growth in population. It shows that this might double, triple, and quadruple, but concludes that this is a good thing and raises no problems. I will not go into that matter and may thus ignore the last few chapters, which defend this point of view, attack its opponents, and proclaim the author's own values. The preceding twenty chapters, however, merit attention because I have seldom seen a book which is so absurd in the realms of economics (the author is an economist) and technology.

In the chapters which follow I plan to study the "average" technological discourse in trying to discern the bluff in it. Simon's work offers us an extreme example which it is useful to analyze. To begin with, Simon pretends to be rigorously scientific, and he accuses his opponents, especially in the report of the Club of Rome and the English report on *The Limits to Growth,* of trickery and scientific error. We must examine this accusation. He himself thinks that being scientific means presenting statistics, graphs, and percentages; his work is full of them. But one of the most interesting aspects is precisely the fact that statistics and graphs inserted in false reasoning are of no help. This observation seems to me to be significant: Simply providing accurate data is not enough. To be truly scientific one must also have sound hypotheses and correct reasoning.

The author's theses are quite remarkable because they run up against all that analysts of the modern world more or less admit. This nonconformity would not displease me if it were not so disconcertingly naive. The author is a liberal economist,[17] but of a kind of liberalism that one no longer sees—an absolute liberalism. For him, in all circumstances, the market is the place of equal and perfect competition. The best will always prevail. For him there is nothing to stop the free

16. *The Ultimate Resource* (Princeton: Princeton University Press, 1981); the French edition was published in 1985 as *L'Homme, notre dernière chance.*

17. He is a professor of political economy and business administration at the University of Maryland. His work was very successful in the USA and well reviewed in France, especially in *Le Monde.*

circulation of workers (who will automatically go where the pay is highest) or capital. He sets out his fundamental theses at the start. There is no food problem in the world. The situation improved constantly from 1950 to 1980. New land came into cultivation and agriculture will expand as needed. Natural resources are unlimited and become increasingly available. The future of energy is just as bright; there are no limits to the development of sources. Pollution does not exist. Air and water are purer than in 1850 and will become even purer. There is no reason to want to arrest population growth, for density of population has no pathological consequences and poses no obstacles. These are the principal theses that the author "demonstrates" in the first twenty chapters.

The arguments rest on two foundations. First, experts cannot arrive at any certain results. Their estimates of potential energy reserves in oil, copper, steel, coal, etc., have all turned out to be wrong. The only incontestable criterion is economic. It is the market. We may thus dismiss technical data relating to pollution and the exhaustion of nonrenewable resources. This is interesting in an author who, as we shall see, justifies technical potential. The second foundation is that there are no limits. In every field the idea of a limit is false. For example, in mathematics this term is ambiguous. One might say that the distance between two fixed points is limited, but there might still be an unlimited number of points between them. Again, no one can say in any field what the limit is. What is the limit of pollution? of copper reserves, etc.? No one can say. Hence, there is no limit! "There is no necessity either in logic or in historical trends to suggest that the supply of any given resource is 'finite'" (p. 50). For lack of a precise definition, we may say that an object is not finite.

Let us take oil. The potential of one well can be measured; it is thus limited. But we cannot measure the number of wells in the world, and therefore we cannot know or measure potential production in an absolute sense. Hence the term *limit* makes no sense. Even if one could arrive at an estimate, one would have to add that better techniques might be able to reach new levels more easily, or make it possible to turn coal into oil, or enable us to derive oil from other sources. Nuclear energy is also inexhaustible. Even if sources for nuclear energy ran out on earth, "sources of energy exist on other planets."

The optimism of this economist rests, then, on an absolute belief in unlimited progress. Whenever a difficulty arises, "technical progress will deal with it." We have here an absolute form of the technological bluff. Let us consider some further strange examples of this so-called scientific thinking. How can one measure the scarcity of a product?

The author rejects totally the ability of experts. For him price is the only strict criterion. When something is in short supply, it is costly. When it is marketed well, it is in good supply. Copper is an illustration. It fell in price between 1800 and 1980. More of it is sold today. The price has fallen continuously, and therefore there is no reason why this should not continue, and why copper should not become more abundant. By extrapolation we thus have a proof that copper reserves are unlimited!

The author constantly extrapolates from his graphs. He seems not to be aware that for the last thirty years forecasters and futurologists have not proceeded by linear extrapolation. He uses the same procedure for the food supply. The price of grain has gone down steadily over the last century; this means that grain is in abundant supply and this will continue indefinitely. "It is a fact, then, that the world food supply has been improving" (p. 59). There is no fear of famine or inadequate supplies. " 'While there have been *some* deaths due to famine in the third quarter of the 20th century, it is *highly unlikely* that the famine-caused deaths equal a tenth of the period 75 years earlier. . . . The percentage of the world's population who find themselves subject to actual famine conditions is probably lower now than at any time in the past' " (pp. 61-62, emphasis added). The example of London in 1880 shows that pollution is not as bad as it used to be, as though London were the only place on earth that suffers from pollution!

Let us return to technical ideology. There need be no fear about the food supply because new technical inventions will at least double production. Thus giant mirrors reflecting sunlight on the dark side of the earth might speed up growth. And if a limit is ever reached according to the law of diminishing returns, the whole galaxy is at our disposal. There lie the true pastures in the sky. We can begin mining the moon by 1990. Satellites for solar energy can supply energy needs from the year 2000. Satellites can also be used for agricultural and industrial purposes from the 1980s (p. 89). We will really have to hurry up! Again, we are wrong to fear pollution. But we must begin by setting aside the advice of experts. The fact that their estimates differ proves their incompetence. Happily, the author has other criteria. He starts with statistics showing that people live longer in the West. This proves that the environment is healthier than it was. Simon seems to think that all illnesses come from the environment. When listing those that have disappeared, he ignores medical advances and new medications, e.g., antibiotics. It is hardly believable. "Life expectancy is the best index of the state of health-related pollution. And by this measure, pollution has been declining steadily and sharply for decades" (p. 131). There has been pollution, but modern techniques have improved the quality of the environment. In the USA the quality of air and water is much better. "The proportion of water-quality observations that had 'good' drinking water rose from just over 40 percent in 1961 to about 60 percent in 1974" (p. 133). " 'There is no contaminating factor in the environment . . . that defies a technical solution' " (p. 139). The same applies to the extension of areas that can be cultivated. "The notion of a fixed supply of farmland is as misleading as is the notion of a fixed supply of copper or energy. That is, people create land—agricultural land—by investing their sweat, blood, money, and ingenuity in it" (p. 225). The proof is that the amount of agricultural land never stops increasing. No thought is given to the depopulation of the country and overpopulation of the cities. The author's ignorance is astounding.

Let us take a final example of these pseudoscientific absurdities. As we have seen, adequate techniques can supposedly solve every problem. But for new technical inventions there is need of inventors, scientists, and technicians. One inventor for 10,000 people will mean ten for 100,000 and a hundred for 1,000,000. Thus we must increase the population so as to have more scientists, technicians, also artists, philosophers, etc. This is ridiculous. It rests on the thesis that every discovery or invention will inevitably be positive and good. There can be no hesitation with regard to techniques. Such inventions as dynamite, nonrecoiling guns, rockets, Molotov cocktails, etc., are just as valid as any others. Simon does not even consider any other possibility. I might cite other enormities of the same kind. I have spent time on the work only because it has had such success and seems to me to be a good illustration of the technolatry that is supposed to be scientific and to be based upon facts. It gives us a good start. But we shall have to deal for the most part with more refined and subtle discussions.[18]

4. The Aristocrats

In the West we live in the certainty that ours is a democratic regime. We see an antithesis between dictatorship and democracy. We enjoy various liberties and take part in elections. To be sure, Socialists think that there is no equality and attack middle-class privileges, the unequal distribution of wealth, and the influence of capital. But both these dated evaluations are superficial and largely inaccurate. We are now grasping two equally important ideas. We live in a society where the important

18. I might add at this point the title of a more recent book on technological growth, Erich E. Geissler's *Welche Farbe hat die Zukunft?* (Bonn: Bouvier, 1987).

thing for individuals is not what is inherited from parents but their own knowledge. Knowledge gives us our place in society. This knowledge comes into play at all levels. At the top of firms, one can no longer imagine a son who inherits a business from his father wasting his time at play. He has to know. In the working-class world, too, the old ideas of apprenticeship and practical experience, which were the older form of training, are outdated. There is a need to study, to acquire, if not a diploma, at least theoretical competence, if one is to fill a post in modern industry.

This point, which Touraine brought to light twenty years ago, is now taken for granted. We are gradually moving on to a meritocracy. We shall return to this idea. But a different orientation in the same discussion stresses the existence of a technocracy. For a long time I resisted this term on the ground that technicians have no wish to have a direct influence on the course of things. I used to say that this is not a technocracy because technicians do not play a role in our political parties, and we cannot therefore put technocracy on the same level as democracy or autocracy. But I also doubted the existence of technocrats in the strict sense. I recognized the growing number of technocrats, that is, men and women who thought they could direct the nation according to their technical competence. Naturally, this was not immediate direction. Politicians still had the role that I analyzed in *The Political Illusion*, that of mediating between the social body and the higher technicians. But there has since been the following development. These technicians have come to see that nothing can be done without them. They dictate in full the decisions that politicians and administrators must take. They have multiplied enormously along with the multiplication of every kind of technique. They have increasingly permeated every field of political action. The principal activity of states is henceforth to promote techniques and to engage in vast technical operations. It seems that the whole life of society is now tied to technical development. The technician is the key figure in everything.

As they have become aware of the situation, however, technicians are no longer talking as technicians who in the presence of given problems provide technical solutions, but as technocrats who say: "Here is *the* solution. There is no other. You will have to adopt it." They now add authority to competence. This is what makes them technocrats. Nothing exists apart from what they know and say. The final aspect of the transformation is that on the basis of their competence they make a general sketch of what ought to be, of what form the society of tomorrow should take.

It is most interesting to compare the sketches of competent

nontechnicians (F. de Closets, A. Ducrocq, J.-J. Servan-Schreiber) with those of technocrats (e.g., Bressand and Distler, or the symposium "Technopolis"). The latter sketches are in the imperative. They do not hesitate to go beyond technical logic and say that this is the society that we have to construct, that we cannot escape this model. Yet these technocrats from different fields do not constitute a technocracy in the narrow sense. The phenomenon is less visible and more profound. They constitute a new ruling class, and we are actually living under an aristocratic regime. Technocrats are the *aristoi,* the best people. "The best" clearly varies according to the dominant social criterion in a society, its main objective. In a military society the *aristoi* are the best warriors; in a democratic society they are perhaps those who have the greatest political wisdom or those who have the greatest oratorical skill and can thus gain the people's adherence.

The *aristoi* today are those who have the greatest technical competence, those who are most apt at increasing applications and results of these techniques. But they are not really interested in strictly political leadership. Many of their texts repeat, directly or indirectly: What good is the state? Is it necessary? When one can exercise this power to organize society and make it work without the mediation of the state, then the state can wither away and be replaced by a social organ which is not political but is based on a certain knowledge. (An obvious external sign of the decadence of classical democracy is the impotence of politicians in their use of words; their speeches no longer say anything; they cannot put them over.)

Our aristocrats, then, are those who in their own fields can put to work the most complex and sophisticated techniques that also promote development. For them the slogan that knowledge is power is true. It is not true for those who know Greek or Roman law—this knowledge gives no power (except to pass examinations). In all technique, however, knowledge means power. We should never forget that its only objective is to enhance power. Those who have technical knowledge in any field have power. Those who do not have this technical knowledge have no power, even though they are eminent ministers or generals. For the latter are directly, strictly dependent on those around them that know the appropriate techniques, without whom they would be completely helpless. They are unable to reach any decisions without computers and technical experts. We have here the realization of a meritocracy,[19] but in a much stricter sense. For if

19. Michael Young's *La Méritocracie en mai 2033* (Futuribles, 1969) is the first study of the remarkable phenomenon of a society based on merit that turns into

we cannot tolerate mistakes by politicians, those of technicians are always excused. We shall have to return to this problem of the general irresponsibility of technicians. We must stress it at this point, however, for it is a feature of aristocracy. Aristocrats can never be held responsible, for who is to judge them? The only cases, and these are rare, are those of surgeons, and we know the general concern that arises when a great surgeon is accused. But even in cases of serious error or extensive damage, we never see nuclear or rocket engineers or administrators prosecuted and condemned. A feature of aristocracy is that aristocrats are above the law. The same applies to technocrats today.[20]

For technocrats the law has no value or interest. It is an invention of the past which might have been useful in the 16th century but certainly not today, for it paralyzes progress.[21] We shall have to come back to this problem of law. In any case law is of no account face-to-face with the technical imperatives issued by technocrats (e.g., the many treaties against nuclear proliferation are rarely enforced). Technocrats, like aristocrats, are characterized by the fact that they have exclusive practices.

A feature of aristocrats is that they alone know certain practices or exercises and have the right to engage in them. In technical spheres, beyond a certain point, only specialists can practice. This is not just a matter of knowledge but a monopoly, a barrier between technocrats and the public, who are merely allowed to play around with certain lesser techniques. Anyone can tap on a computer, but only superior technicians can program the complex systems on which economic and financial orientations and confidential communications depend. The essential part of technical use is beyond the reach of average citizens. An esoteric language corresponds to the exclusive practices. All aristocrats speak a language of their own that is not that of the common people. It is not a different language in the sense that German differs from French, but a coded language that only initiates understand. This was true with the medieval knights as well as among other aristocracies, for example, that of the 18th century. The Freemasons carried it to absurd lengths in trying to set up an artificial aristocracy. Technocrats, too, have their own language. They do not have to spell things out to each other. There is a

---

another form of dictatorship. In an ironic vein it has a much clearer and more farsighted vision than either *Brave New World* or *1984*.

20. An important aspect of commissions of inquiry into nuclear accidents or air accidents or landslides or dams bursting is that neither technique nor higher technicians are blamed but "human error" on the part of management or machinists. Technique is always impeccable.

21. Cf. A. Bressand and C. Distler, *Le Prochain Monde* (Paris: Seuil, 1986).

language that is common to all technical categories, and also a language peculiar to each specialized area. This is one of the important aspects of the power that ordinary people do not share. Of course, the secret language, which among other aristocrats might have a certain religious or philosophical character, makes the aristocratic message esoteric. The language that technicians speak among themselves is not just algebraic; it is digital. This transposition from multiform information, transmitted by an analogical method, to uniform information, in the elementary form of "bits," this omnipresence of numerical logic, means that the language no longer has the same consistency. Nor is it merely the presence of computers that makes technical language obscure for nontechnicians. The complication of society, the multiplication of networks, makes necessary a certain type of intelligence to which the language corresponds and in virtue of which one part of a nexus can easily be put into touch with another.

Knowledge, practice, and language separate technicians from ordinary people. But a fourth feature also defines this aristocracy: They have a multitude of functions, all of which are indispensable to the group (just as the nobility had military, judicial, governmental, economic, and monetary functions). Their technical ability is of general application and enables them to exercise a totality of powers. They are at the center of every organism of management and decision. Armaments, space exploration, medical treatment, communications, information, industry, administrative rationalization—all that is power depends on them.

Naturally, those who are behind the times argue that everything finally depends on capital, or money, that the sole aim is to make a profit, and that the capitalist is in command. This is a touchingly simplistic view. There certainly has to be a massive mobilization of capital, as we shall see later. I have never denied the importance of multinational business. Yet this is no longer the catalyst. The proof is, for example, the springing up of the technopolis all over the world, and the more and more frequent possibility for the true high-tech technician to create with very little capital a business which, if the technique is adequate, will quickly become very large (e.g., Apple). This aristocracy creates its own environment, the technopolis.

Americans invented the technopolis.[22] Its first beginnings were

22. Cf. "Technopolis," special number of *Autrement* (Nov. 1985); Everett M. Rodgers and Judith K. Larsen, *Silicon Valley Fever: The Growth of High-Technology Culture* (New York: Basic, 1984); B. Montelh, "L'éclosion des technopôles," special number of *Le Monde*, Dec. 1986.

in California in the 1930s, and the Silicon Valley, started between 1950 and 1960, is now known everywhere. The ideal of a technopolis is complex. Scientific research is put in touch with industries that can use it but also finance it. An organic union must be set up between them, on which a third factor will be grafted: the university, which will be invigorated by the proximity of industry, whose students will be assured of jobs, and which will be intrinsically established by research. This is the perfect technical world: Technicians control research, they correspond with industrial technicians (dealing with the economy, analyzing needs, etc.), and the major function of the university henceforth is to provide technicians. Studies no longer have any justification unless they serve a useful purpose.

The technopolis inclines to become a motive center for society and the economy. What happens comes to expression in the linguistic transition from the technopolis, the city of technique, to the technopole, the pole of techniques. There are many examples of the technoplis in the USA, in Japan, and in Europe, where it is hoped that they will be a source of new vitality. We find them in Britain (Cambridge and Heriot-Watt), in Sweden, and in France (Rennes-Atalante, Metz, Paris-Sud, and especially Sophia-Antipolis). Technological parks are multiplying. The idea is always the same, a cross-fertilization between the university and industry, but with a center outside the town, in open country or a forest. The twofold point of the location is to make a break with the rest of society (which is directed from this point) and to concentrate all that is useful and to exclude all that is not (i.e., cultural, spiritual, and aesthetic factors, apart from those that are needed to produce pleasant and harmonious buildings and traffic routes). Other things do not contribute to the inflexibly serious aim of cross-fertilization.

When a center is set up, it exerts an appeal. Risk capital flows toward it. New enterprises are set up according to an interesting process. Researchers, industrialists, students, and financiers follow one another. The technopole is a rallying point. News bulletins are issued and new businesses nurtured. But one might ask whether the multiplying of such centers (twenty in France alone from 1969 to 1986 and others projected) will not negate their efficiency, which is connected with their character as models. Will risk capital always be available? Will there always be the same flood of technical innovation? And what is happening to Silicon Valley? A breath of panic swept the technopole when the market for home computers dropped in 1985. For the first time, the field of electronics had a deficit. After having been the glorious example, the model, this collapse prompted fear and restraint. Is a purely technological environment really viable? Is not the infatuation bound to fade in view of the impossibility of indefinite growth? Nor can we believe that cities of this type are really good social models, as it was first thought they might be. In reality they reproduce all the vices of a technocapitalist society, with a characteristic gulf between high-tech technicians with fabulous salaries (highest in the USA) and poorly paid employees and workers who live far from the center of activity, who have to commute great distances to work, and who reap no benefits.

The technopolis is a magnified image in the lens of our society. Economic, financial, political, and administrative powers all pass through the hands of groups of technocrats. If one such group goes on strike, a small number of strikers can bring general paralysis. Fortunately, this seldom happens, for technicians have no reason to strike, since they are in control, situated at the hub of communications and activities.

The reciprocal character of the ideology of this aristocracy, which also shows that it is a real aristocracy similar to its predecessors, might be described as its complete disdain for all else and its extraordinary ignorance of the world outside and of other settings. When these technocrats talk about democracy, ecology, culture, the Third World, or politics, they are touchingly simplistic and annoyingly ignorant. Once when I was voicing my objections to nuclear arms, one of them confidently replied that his technicians would answer all my questions. They respect the idea of democracy but in the sense that telematics will make it possible for each person to voice an opinion and share in decision making. We will show later the astonishing cultural ignorance of these aristocrats (perhaps a feature of all aristocrats) and their errors of judgment regarding human nature.

It is astounding constantly to find among their affirmations the belief that everyone will have access to data banks, to all useful information, to any necessary files. But who is this "everyone"? The reference is merely to other technicians. Immigrant workers, poor farmers, and the young unemployed are not going to consult data banks. In France three million people at the most might do so. But these aristocrats ignore the real situation. Once a thing is possible, they think it is open to everyone regardless of the actual situation. This disdain assuredly makes them different from the rest. At times they openly display their contempt for opponents (cf. Illich or Schumacher). They are also totally indifferent to morality. What does it matter if their discoveries serve a destructive purpose (e.g., improving armaments)? Some of them find it good, and a sign of freedom, that devices (e.g.,

Minitel) can be used to maintain pornographic and obscene verbal relations. To see common people degraded is a wonderful pastime for aristocrats.

Yet this ignorance and disdain are also a source of weakness. It is ridiculous to read from the pens of some of them that tomorrow's society will take the form of networks, communications, progress, free circulation, and democracy, when they fail to mention the growth of armaments, Third World stagnation, and the innumerable problems caused by these techniques themselves (of which ecologists have noted only a few).[23] They put all such things in parentheses. The technocrats have a strange blindness to the complex reality of the world and to the lessons of common sense (e.g., that no system can grow indefinitely in a closed and finite universe, a truth that they treat sarcastically). Their great knowledge and narrow specialization prevent them from understanding questions outside their field. Yet they write authoritatively about tomorrow's world. It will be like this or like that, and therefore we must prepare our children for this particular world. (We shall return to these affirmations in more detail.) They are thus plunged into electronics and computers without a thought that perhaps in the future being able to till a bit of ground or light a wood fire or do proper grooming might be more useful than being able to tap on a keyboard. Such is their casual ignorance of most of what constitutes our world. Here is another general feature of aristocrats who are capable of dancing on a volcano.

The compensation of this is the creation of a specific ideology, the ideology of applied science. (They have some reservations regarding basic or pure science, which in its modern formulations seems to them as though it might upset their secure stock of knowledge.) The ideology of the indispensability of technique is also that of the ineluctability of its progress. In the eyes of technocrats it is monstrous to say that we should stop building nuclear power stations and return to the energy consumption of 1954. They immediately retort that what opponents want is a return to the Middle Ages. As they see it, there has to be growth. They will not accept any other hypothesis. They find their justification in the fact that increasingly everything depends on the application of techniques. Not only is technique good, not only is it indispensable, but also, as we said above, it alone can also achieve all that human beings have been seeking throughout the centuries: liberty, democracy, justice, happiness (by a high standard of living), reduction of work, etc.

23. A typical work is that of Bressand and Distler, *Le Prochain Monde*.

Another feature of this aristocracy is that they impart this ideology to society as a whole. It stands behind what the media hand out to us, the ideology of the political parties, and the banal ideology of the person on the street. "The dominant ideas of a society are those of the dominant class." By this ideology technocrats hope to model the future, though this is not apparent. Their ideology, which is *their* justification, is based on their profession, it forms the link between the environment, the thinking, the profession, and the rest of society. We have here the classical procedure of justification. All that is said about technique is said in ideological justification.

Beyond question technocrats are in solidarity with one another. They cover one another and protect one another as needed. This solidarity may be seen at every level. When any of them makes a grave error, there are others to justify it. They can tolerate no criticism. When I raised certain questions regarding Chernobyl, I was put in my place by French technocrats. The solidarity is complex. It rests on a common interest and ideology and produces social affiliations. But can one compare this group to earlier social groups?

This aristocracy does not constitute a social class in Marx's precise sense of the term, for its place in the process of production does not characterize it. Some technocrats participate in this, but not all, and a number are working to reduce its importance. Again, in this aristocracy there is no inheritance of position or power. Yet the problem of heredity is significant, for it exists without existing. That is, the children of polytechnicians are particularly well placed to follow in their steps, though they do not have to do so. The children of great technocrats will almost certainly have a place in the aristocracy, even though it is not that of their parents. One technocrat will not dismiss another technocrat's child (apart from the rivalries of sections and projects which divide this aristocracy, along with the struggles for power which exist in all aristocracies). The problem of a heredity that exists without existing is like that of the *nomenklatura*.[24] Yet there are also major differences. In this case no political orthodoxy is demanded. Again, recruitment is not by internal administrative promotion. The decisive factors are degrees, competition (meritocracy), and adaptation

24. This word is now in fashion. A recent book which enjoyed an immediate success bears the title *La Nomenklatura française*. But the term is quite inadequate, at least in the sense that Michael Voslensky gave it in *Nomenklatura: The Soviet Ruling Class*, tr. Eric Mosbacher (New York: Doubleday, 1984). A similar term that came into vogue after 1946, after the model of Hitler, was *apparat*, which was applied no less indiscriminately.

seminars. Hence one cannot say that technocrats form a *nomenklatura*, even though there are certain points of resemblance.

The closest parallel is Galbraith's technostructure, for technocrats obviously constitute a very solid social structure around which everything is organized but which has its foundation and justification in the exercise of a technique. Nevertheless, the term as Galbraith defines it seems to be too narrow, since technocrats play a key role in all activities, hierarchies, and social, economic, political, and intellectual groups, not just in the world of industry and production. The best word to describe them is aristocracy, so long as we realize that there is no comparison with the futile French aristocracy of the 18th century.

PART I

Uncertainty

As I attempt in Parts II and III below to bring out the character of the bluff of technical discourse, I have to do more, I think, than simply point to the actual reality of inventions, innovations, and applications, or integrate new techniques into my older schemas. What I have to stress is a feature which, the more I know the technical world, the more it imposes itself upon me, but which so far I have never studied. This is the feature of uncertainty. We are moving into a world which is increasingly the product of technique. But we also live in a world which is increasingly uncertain about techniques (not their origin or mechanism, but their effects). I can analyze this uncertainty from four angles: the ambivalence of technical progress; the unpredictability of development; the double feedback which is constituted by the originating factors of technical progress; and the internal contradictions inherent in the system. I will not dwell on the negative effects that are most often studied (pollution, the exhaustion of nonrenewable materials, the imbalance of ethnic groups, etc.), for it is not my object to study techniques themselves. My concern is with the fact that we are moving ahead very rapidly and are unable to say exactly what the goal is or through what stages we shall pass. Uncertainty is the lot of all technicians and scientists, who are occasionally aware of it. But it is also the situation into which the populace is thrust without being in the least aware of it.

CHAPTER I

Ambivalence

From a very elementary standpoint the ambivalence of techniques has often been stressed (as in my own study in 1950).[1] Technique can have both good and bad effects. My friend Duverger, reviewing my first book on technique, wrote that it was like the language of Aesop: the best and the worst of things. It is often quietly added that everything depends on the use one makes of technique. With a knife one can peel an apple or kill one's neighbor. I also tried to show that this comparison is absurd and that technique carries with it its own effects quite apart from how it is used. (Use also enters in, but leads us into a moral question that has nothing to do with the analysis of technique.) If we want to know what the issue is when we speak about technique, we must begin by eliminating the futile argument about its use.

The ambivalence of technical progress is unfortunately more complex than might be suggested by the simplistic view just mentioned. To believe that everything depends on the use is to think that technique is neutral. In 1950 I demonstrated the contrary and thereby caused a scandal. But today most writers are convinced in fact that technique is not neutral. No matter how it is used, it has of itself a number of positive and negative consequences. This is not just a matter of intention. For a time, use may orient this or that technique in a purely positive direction, but this technique has in itself potentialities that are inevitably exploited. The simplest and best-known example is gunpowder. The Chinese used it only for fireworks, but it had the poten-

---

1. I am taking up and developing here my earlier study of ambivalence: "Réflexions sur l'ambivalence du progrès technique," *La Revue administrative* 18 (1965): 380-91.

tialities with which we are familiar and which could not be neglected for long.

Optimists regard technique as essentially good and "globally positive."[2] This is not just a matter of temperament. Most scientists and technicians view technique favorably. Pessimists look at the negative effects, but again this is often a philosophical or political posture. In effect the problem is both bigger and more complex. We are here in a domain that is not really scientific, though it is given scientific features. If we consider only the concepts and standards of well-being—the easiest to grasp and measure—we see how futile it is to try to see and evaluate fully the effects of technique. It is relatively easy to measure the total quantity of goods and services provided. There are also several partial indicators of well-being, but what about effective use, or the possibility of access (including information, education, and the choice allowed by the social structure)? Should we also measure the quality of actual use or simply look at the level and distribution of incomes? But such things are only preliminary to the real study, which we quickly see to be beyond us.[3]

Analyses which appear very rigorous, which are built on statistics, and which make no reference to these problems are the most dangerous. For they, too, are ideological, but they pretend to be purely scientific and have an appearance of strictness that one does not find in more rhetorical but more honest studies. In this area, in which the whole person is involved today, it is impossible to be purely scientific and completely disinterested. We all realize that everything finally depends on the technical venture. How, then, can we keep cool heads and not take sides? The stake is too great and we are too directly implicated in this movement. The transformation is both global (affecting all humanity and all aspects of society and civilization) and personal (modifying our ideas and life-styles and conduct). We have to ask what will become of us in this upheaval. No response that is purely logical is possible.[4] We do not know all the facts. We cannot achieve a

2. See Lech Zacher, "Illusions of Technological Optimism," *Science, Technology and Society* 2/2 (1982).

3. Cf. the remarkable and precise studies of I. C. Merriam, "Concepts et Mesures du bien-être," *Analyse et Prévision* (1969), and J. Desce, "Consommation individuelle et Consommation collective: essai de mesure," ibid.

4. In spite of its declared optimism, I know of no more radical book than Jouvenel's *L'Arcadie* when, in assessing the technical society, it draws attention to the failure to enhance the quality of goods, the quality of the environment, and the rapport with labor, which ought to have been the normal result of the increase of wealth.

truly synthetic perspective as we must, since all the pieces of the technical system hang together and we can say what will become of us only from a global apprehension and not from a sum of fragmented predictions.

We proceed on the basis of extravagant hopes, only too easily sacrificing what have thus far been regarded as truths about humanity (certain values, or an individuality that is now progressively weakened in favor of collectivism). Or else we give way to despair (e.g., the absurdity of the world, dehumanization, the China syndrome) without taking into account the opportunities that we still have. The game is still afoot. It is in this context, which is bound to be a passionate one, that I wish to draw attention to one of the most important features of technical progress, its ambivalence.[5]

My meaning is that technical development is neither good, bad, nor neutral. It is a complex mixture of positive and negative elements. Some are good and some are bad, if we want to use moral terms. It is impossible to dissociate them and thus to achieve a purely good technique. Also, good results do not depend at all on the use which we make of technical equipment. In effect, even in such use we ourselves are modified in turn. In this totality of the technical phenomenon, we do not remain intact. We are not just indirectly oriented by this equipment itself, but thanks to the means of psychological adaptation we are also adapted with a view to better utilization of the technique. We thus cease to be independent. We are not subjects in the midst of objects concerning which we may freely decide how to act. We are closely implicated in this technical universe. We are conditioned by it. We cannot put human beings on the one side and equipment on the other. We have to set human beings, too, in the technical universe. The use we make of equipment is not decided by spiritual, ethical, autonomous beings, but by people within this universe. Thus this usage is as much the result of human choice as it is of technical determination. The technical universe also makes determinations that are not dependent on us and that dictate a certain use.

Regarding this good or bad use we have also to consider that we

---

5. At this point I must take issue with the simplistic view of de Closets regarding the "ambiguity" of technique. For him technique has better and worse aspects according to human freedom. We have seen already how limited is our power in this area, how severely our freedom is limited. The great remedy that de Closets proposes is organization. Techniques do harm for lack of organization; technical specialists know little about this. What de Closets fails to see is that organization itself belongs to the technical system; it does not stand in counterpoise with it.

are necessarily speaking about individuals who use such technical objects and who can thus make certain choices as to what they use and how. But the technical civilization is made up of an inseparable nexus of technical factors, and the good use of one or other of them will change nothing. What will count is the general behavior of humanity as a whole. We shall not insist on this; it does not seem that we are as yet close to arriving at this situation. To solve the problem of good use, people finally have to have clear goals which are adapted to reduce technique to the state of a pure and simple means. In our actual situation the goals are either well formulated in an antiquated fashion, and thus ill-adapted to the situation, or else they are totally vague.

We have stressed already that the effects of a technique are never solely positive or negative. Ambivalence is thus a basic feature of technical progress. But it is not intrinsic; we can see it only as we analyze the effects. Hence we have to study not technical progress as such but its results. It is by examples, then, that we shall bring to light the nature of the ambiguity.

We are set in an ambiguous universe in which each technical advance accentuates the complexity of the mixture of positive and negative elements. The more progress there is in this field, the more inextricable is the relation between the good and the bad, the more choice becomes impossible, and the less we can escape the ambivalent effects of the system. Marcuse said nothing new about the relation between humanity and the system in the light of my analysis in *The Technological Society*, but he formulated very well the ambivalence of technical progress, and his formulation might serve as an introduction to the present chapter. Operational sociology and psychology, he said, have helped to make human conditions more pleasant and are factors in intellectual and material progress, but they also bear witness to the ambivalent rationality of progress. It is beneficent in virtue of its repressive force and repressive in its benefits.[6]

Before proceeding to an analysis of the ambivalence, I must differentiate it from two other concepts. The first is that of ambiguity, which I have already stressed. Whereas ambivalence implies that an object has two precise and opposing orientations, ambiguity is more fluid. It carries the notion of what is confused, indeterminate, vague, and equivocal. But technique is neither fluid nor confused—it is precise

---

6. Cf. William Kuhns, *The Post-Industrial Prophets* (New York: Weybright and Talley, 1971), who describes technique as radically ambivalent in its effects. At first, however, only the negative or the positive effects are seen, the former in the case of railroads and steamships, the latter in the case of automobiles.

and not at all equivocal. We shall have to speak later, of course, about the uncertainty in technique itself. The second notion is that of the perverse effect, which is much in vogue today. The idea here is that there are normal effects and bad effects, which are surprising and perverse. This word has a moral connotation which we must reject in this context and implicitly recall the idea of a use which would be perverse. We need to understand that the good and the bad, the positive and negative effects, lie intrinsically in the very constitution of the technical universe and in all technique. J. Chesneaux had good reason to say that technique (his word is "modernity") is an original combination of two globalities: that which Sartre denounces and that of which Saint-Simon dreams.[7] On the one hand we have the serializing of beings, conditions, objects, and machines like the computer—the reduction to a single pattern of trivialized life. On the other hand the planet is linked together and we have the global interdependence of economies, networks of communications, sociopolitical structures, and the despotism of the global market. Here is an exact characterization of the technical system and technical progress in their ambivalence.

But it is not enough to emphasize the ambivalence. We must analyze it, and we shall do so with the help of four propositions.

- *First, all technical progress has its price.*

- *Second, at each stage it raises more and greater problems than it solves.*

- *Third, its harmful effects are inseparable from its beneficial effects.*

- *Fourth, it has a great number of unforeseen effects.*

We do not have here haphazard data but a first and basic feature of technical progress. According to a formula that A. Merlin used about the crisis in America, especially in the computer market (1985-86), we try to analyze but we do not understand.[8] As he also stated, the most convinced optimists need time to find encouraging signs. No path is without its stars. But the feature of technical progress is precisely that it is without stars. Its growth is causal, not final.[9] Technical progress does not know where it is going. This is why it is unpredictable, and why it produces in society a general unpredictability.

---

7. Cf. J. Chesneaux, *De la modernité: les treize effets pervers de la modernité.*
8. Albert Merlin, "Les entreprises qui n'ont pas d'étoile," *Le Monde*, June 24, 1986.
9. See J. Ellul, *Technological System*, pp. 256-82.

1. All Technical Progress Has Its Price

There is no absolute technical progress. With each advance we can see a certain setback. It is not just that it needs considerable intellectual effort or the injection of capital. Capital is invested with the hope of return. But this is not guaranteed. Often decisions are made to launch technical enterprises that are not economically profitable (e.g., satellites). These two costs of progress are obvious, and we may set them aside. The problem is much more complex. We have to consider first what is destroyed in many fields by such technical progress.

S.-C. Kolm observes humorously: "The United States would seem to have double the national revenue of France, but would much of this gap remain if one took into account social values and deducted a measure of the ugliness of American cities." If we take human life seriously, we certainly have to take aesthetics into account and the quality of the environment. But everywhere technique creates ugliness. This is the price we have to pay. In the cost we have also to reckon what economists now call externals (a problem to which we must return). These are the costs that are not directly linked to the creation or use of technique: pollution, health problems, protective measures, nuisances of all kinds. Growth carries with it costs that change all calculations according to whether we include them.[10] There has been agricultural destruction in order to promote industrial development (e.g., in the Tennessee Valley in the USA or the Jurançon vineyard in France).

A second aspect that we have to consider regarding the price that has to be paid has to do with the replacement of one product by another. In general no note is taken of products that have disappeared. Thus in estimating textile consumption, account is taken only of textiles that are actually used (wool, cotton, artificial products) and none of those that are no longer used (linen, hemp), which were more common than is often thought and much more durable. This is not to deny that consumption is up but to point out that the growth is far less

10. Cf. G. Dessus, "Les Coûts et Rendements sociaux," *Analyse et Prévision* (1966). On this subject F. Partant, *Que la crise s'aggrave* (Solin, 1978), has the original thought that "technical progress has illusory results but is supposedly necessary for social progress. We have to pay for the latter. We can achieve it only by increasing wealth. It will come only if directly or indirectly it contributes to economic growth. It is less an acquisition financed by the value created by labor than a source of productive activities. But our mode of increasing wealth constantly gives rise to new domestic social problems that are even more dramatic on the international scale. Hence social progress has to be an effort to combat the negative consequences of this mode of increasing wealth." We might apply this thought to what we call technical progress.

important than imagined. We have to include products that are eliminated. Mechanized farming may produce more food, but we have to consider that a calorie of food now demands the consumption of a calorie of fuel. A team of oxen or horses would use less.

In this game of substitutions, which are the price to pay for progress, we have also to consider Third World countries. Many of their products (minerals, sugar, cotton, wood) face total elimination. The production of artificial materials might mean total ruin for some countries. Since 1963 they have several times asked the United Nations Organization to see to it that the developed countries stop manufacturing competing products. It is highly unlikely, however, that technical progress can be halted. Already replacements have affected countries like India (indigo), Chile (nitrate), and Indonesia (hemp). We shall have to return to the effects of technicization on the Third World. For the moment our emphasis is that technical progress almost always has negative effects on people. Roads and truck transport in the Sahara have replaced the caravans that were a main source of wealth for the Tuaregs. The increasingly efficient technicization of seal hunting has deprived the Eskimos of something that was basic to their society.[11] At the same time medical care has increased their numbers. Denmark has thus had to feed and clothe them at great expense with Western products that are transported and distributed free (since they have no cash to give in return). This cost of maintaining a people that used to have a stable society has to be regarded as the passive side of the activity of seal hunting.

But that is all far away. We must look at the situation of the rural population in our own Western countries. I will not take up the problem of rural depopulation.[12] If one says that technical progress demands that people leave the countryside, that it gives more leisure time, and that it greatly increases agricultural production (even without the sun), the cost is the influx into cities and increased unemployment, a concentration of rural property, a rural capitalism that shatters rural society, and the use of nature as a place of recreation for those who live in cities. But most geographers, agronomists, and sociologists tell us that the countryside without inhabitants is either uninterestingly incoherent or else it takes the form of a public park. Nor has it any protection against industrial enterprises (e.g., high-tension lines, ski lifts). Only

11. See Gessain, *Anmassalik* (1969).
12. See Le Lannou, *Le Déménagement du territoire*; B. Charbonneau, *Le Jardin de Babylone* (Paris: Gallimard, 1969); J. Piel, *Les Paysans et le Paysage* (1966); O. C. Herfendahl and A. V. Knees, *Quality of the Environment* (Baltimore: Johns Hopkins University Press, 1965).

an adequate rural population cultivating the soil can create the kind of
nature that attracts city dwellers. The destruction of the rural population avenges itself not only with economic disasters (surplus production, inferior products, the replacement of agriculture by agro-
industry), not only with the despoiling of nature as a place where we
can rediscover ourselves, but also with natural catastrophes (e.g.,
devastating forest fires which are due to the existence of flammable
undergrowth that was not present when people lived in the forests and
that cannot be replaced by useful plants through an occasional clear-
ance by bulldozers).

The destruction of rural workers, with all the negative consequences that it entails, leads us to the price that is paid by all workers.
In an older work that is still relevant, G. Friedmann has shown that
automated and computerized manufacturing, suppressing older forms
of work, modifies the attitudes and habits of workers and leads to the
destruction of values that the working-class world used to regard as
essential.[13] The new style of production carries with it modifications
of our whole being: our instincts, perceptions, vision, sense of time,
instinctive conduct, interpretative perception. Our very notions of
fatigue and prediction have changed in both form and meaning. In the
area of the transformation of labor we have a typical example of the
price that has to be paid. One of the grand claims of technique is that
modern machines greatly reduce muscular effort. This is obviously a
gain in the case of exploitation, of exhausting work that is beyond the
threshold of fatigue. But is economy in all physical effort a real gain?
What seems to prove the contrary is the need to find a replacement by
expenditure of effort in sport. This is not a serious matter. What is
serious is that economy in physical effort exacts a price in the form of
all kinds of physiological, psychological, and even sociological ill
effects. Each of these taken alone is less serious than the exhaustion of
underground miners around 1880, but it is of a different kind. The
problem is that of nervous fatigue due to extreme tension in new
factories. We find ourselves in a world that demands more rapid
reflexes, sustained attention continuously, and adaptation to new situations and challenges. Nervous tension offsets muscular relaxation.[14]

13. G. Friedmann, *Sept Études sur l'homme et la technique* (1966).
14. It is now established that nervous tension brings on premature aging. In
a very exact study Viennay has shown that physical labor did not prevent people
from working to age 65, but miners are now old at 45, workers on assembly lines at
40, and women in some industries at 30. Some automated factories have to find
fresh personnel every two years (*Colloque européen sur l'automatisation* [Grenoble,
1977]) Jouvenel *(Arcadie)* brings out the difference between organized, technicized

Two remarkable facts: A 1980 study (by the Confédération française et démocratique du travail) shows that insomnia is the basic problem of modern workers, and epidemiological studies show that heart and nervous problems have now spread to the working class. Yet we have to point out that nervous fatigue is not due solely to the change in work. It is due to the modern life-style in general, the constant need to do everything faster, increasing life rhythms (fast food!), the multiplying of superficial human contacts, the tension of more and more crowded timetables. It is exhausting to live in a world in which everything is timed to the minute, in which there is never any time for rest. Another reason for nervous tension is that in our life today we no longer keep pace with the seasonal rhythms. Artificial light has made it possible to live as much at night as during the day. It has broken one of life's most basic rhythms.[15]

Exhaustion inevitably follows. Our summary of reasons for nervous tension is no mere matter of hypothesis—this is one of the tragic realities of our time. We have here a danger which is linked to technical progress and the answer to which it is hard to see, for it is the result of all the structures of a society that is geared for technical progress. The remedies that might be found are only temporary palliatives. Thus tranquilizers enable people to endure nervous tension while continuing to live as usual. This simply increases the imbalance and produces in the long run an even more serious crisis. We have here no more than one ill effect replacing another. Clearly, this tension due to more activity at night is aggravated by a shift system (three eight-hour shifts) which forces people to work a third of the time at night and which is demanded by the way factories are set up and the maintaining of communications. Machines do not stop, and to achieve maximum profitability they must be run as much as possible. People, then, have to be organized to work as the machines do.[16]

The breaking of seasonal rhythms is a familiar problem. All organisms are at the height of their vigor and powers in spring and summer. In autumn and winter (until hibernation), all organisms lose their vital force. They are reduced, becoming less resistant and more fragile. Rural work follows the seasonal rhythm; there is less to do in winter. But industry, dictated by technique, which allows and now

---

work and free work. Leisure is no compensation, for the person who is free after work is not in the same situation as the one who takes pleasure in work.

15. See n. 17 below on biological rhythms.

16. But this has now been questioned, along with the whole of Taylorism. Cf. B. Coriat, *L'Atelier et le Chronomètre* (C. Bourgois, 1979), which draws attention to the crisis in the scientific organization of work.

demands vacations, reverses the rhythm. Vacations and rest come during the "beautiful season," and there is more work in winter. This disturbs life's rhythm. Some functions are not discharged as well as they once were.[17]

This leads to a larger observation that I believe to be of vital importance as regards the price to be paid. It is well known, and we are all glad of it, that modern life expectancy is greater than in traditional societies. Tables prove that previously the average expectancy was only about thirty years,[18] and that it has now risen to around seventy years. What progress! But a double price has been exacted. There is first the demographic problem. A greater number of old people have to be cared for and supported, and supposedly more children are needed to compensate. That is, the doubling and tripling of children will produce twice as many workers in twenty years to support the aged. But this is the height of folly; it betrays an astonishing lack of foresight. For in sixty years there will be two or three times as many old people. Need we go on? In fifty years there would be almost a tenfold increase in the population of a country. Completely absurd! Second, there is the physiological problem. We now keep alive innumerable infants who "normally" would have died in the first months. Their health will always be fragile, or they will be invalids.[19] Studies have also shown that the human life which is now prolonged is much diminished and very precarious. Our health is much more fragile. We have less resistance to grief (cf. the studies of Professor Leriche), to fatigue, and to privation. We have less resistance to lack of nourishment, variations in climate, and internal and external stresses. We are more susceptible to infections (cf. the studies of Dr. Carton). Our senses are less sharp, especially sight and hearing. Our nerves are much more fragile (we suffer more from insomnia and distress). We have to take more precautions and are more easily laid up by little things. We have more opportunities in life and live longer, but we live diminished lives and do not have the same vital force. We have to compensate for new deficiencies by artificial procedures that in turn produce other new deficiencies.

17. More and more studies have come out on the importance of biological rhythms. There is a daily rhythm that cannot be broken; medications work differently according to the time they are taken (La Recherche 132 [1982]). Cf. A. Reinberg, L'Homme malade du temps (Stock, 1979).

18. I am of course skeptical regarding the figures for the 16th or 17th century, which can hardly be viewed as scientific.

19. There are actually some 15 million handicapped people in the USA who would have died at birth.

As a final example of the price to be paid I might refer to the functional constraints in our society in face of the triumphant song of freedom that we sing concerning technique. As M. Massenet has strikingly shown, "We have to pay in functional constraints for apparent freedom at the level of objectives. The more society is collectively voluntary, the less it is individually so. The more society controls its objectives, the less it controls its processes."[20] Technical progress demands collective planning and organization. But the technical process, the means, limits the possible choices. Moreover, there are choices only at the level of global society, which can think that it is conscious and free to fix its objectives, but which is really governed by the means that give it the illusion of this consciousness and freedom. The more it plans, the more it imposes functional constraints and reduces the voluntary aspect of individual decisions. The mechanisms of determination increase in this way, producing at the same time the appearance of freedom and its restriction, but situated on different levels.

Looking at another example of the growth of social controls, we recall that one of the serious problems engendered by technique is overpopulation. But how are we to control this growth? It is precisely in terms of the "price to pay" that Edward Goldsmith and others have analyzed this question.[21] To control population growth we have to suppress individual freedom. The cost is not too heavy, he thinks, if we consider the long-term results. But where will the suppression of individual freedom finally lead us? To the control of individuals, and technique certainly makes possible a total control. Surveillance of private life, of births, the regulating of social behavior and finally of all human conduct. To save the species we must sacrifice the individual! But all individuals? And the essential characteristic of the species? Goldsmith is hardly consoling when he tries to justify this total control by saying that it is also present among the aborigines of Africa and Australia, ignoring the differences between the two types of social control. Is it not precisely the only real progress of the human race that it has freed individuals from social controls?

The price to be paid seems to be enormous. The problem of paying it is always a tragic one. I have in mind a saying of the Brazilian

20. Massenet, "Du changement technique a l'éclatement social," *Analyse et Prévision* 11 (1971). Cf. also the thesis of C. Castoriadis that there is a power which is impotence as regards the essentials of contemporary technoscience, "Voie sans issue?" in A. Jacquard, ed., *Les Scientifiques parlent* (Paris: Hachette, 1987).

21. E. Goldsmith, et al., *Blueprint for Survival* (Boston: Houghton Mifflin, 1972).

Minister of Finance (M. D. Neto) in 1975 to the effect that only an authoritarian regime could ensure economic rectification and technical growth. Technocrats alone could establish an industrial Brazil. Economic expansion would be the response to social problems. The price to pay was an absolute regime with all the "flaws" that this entails: a police state, imprisonments, torture. This was Stalin's dilemma. Rapid technical growth involves this cost.

To conclude our look at the first aspect of technical ambivalence, I would say that technique obviously brings with it considerable, indisputable benefits. But it destroys other good things that may be just as important. We can never say that there has been real progress (with no counterbalancing factors), nor deny it, let alone quantify it. Naturally, we must not try to apply our formula too strictly. I am not saying that technical progress means the payment of one value exactly by another, that there is precisely the same amount of destruction as creation. There is obvious growth or progress in the material sphere. There is more energy, more consumption, more "culture." I am not saying that each thing has its own price tag (especially since the price is hard to calculate). What seems to be certain is that technical progress is much less at its level of consumption than it is said to be at present.[22]

There are several reasons why it is hard to grasp this first feature. First, we have to consider the global situation. In most cases the price is not of the same type as the gain. We have to look at the phenomenon in its entirety in order to grasp the compensations which are made. But this is seldom done; people look only at facts of the same category. This is poor methodology. With technical progress we face a transformation of civilization. Since a civilization is not made up of juxtaposed elements but of integrated elements, we have also to take into account all the reactions that may be traced to a single technical advance. This is why studying the phenomenon is such a delicate task. Yet it is at the global level that we can maintain that all progress has a price. The difficulty is that of estimating what is gained in relation to what is lost, since the two things are of different types and cannot be compared. We must not let ourselves be caught in the trap of either the necessity or the possibility of exact comparisons in this field.

A final difficulty in grasping the reality of this compensation is

---

22. For an interesting demonstration of this relativity see the University of Bordeaux thesis of J. Gellibert, "Le Choix de la biomasse comme énergie . . ." (1986), which by estimating the power output of watermills and windmills in the 16th century shows that between 250,000 and 500,000 tons of petroleum would be needed to produce the same amount of energy. The relativity of technical progress!

that it is deceitful. It is always a matter of vague phenomena, which are significant only by their bulk and their general nature. They present only rarely an explosive or tragic aspect, but eventually give a certain negative style to human life by the accumulation of details that all move in the same direction.

2. Technical Progress Raises More and Greater Problems than It Solves

In October 1965 Admiral Rickover, the father of the nuclear submarine, made a shattering statement regarding the "technological monster." Speaking from experience, he expressed the disquiet which was felt by many scholars and higher technicians and which had nothing whatever to do with a romantic antitechnical position. He stressed two aspects: on the one hand the irreparable character of the damage done by uncontrolled technical growth, and on the other the immensity of the problems raised by technique as it becomes more and more complex.

At much the same time Jouvenel *(Arcadie)* made a similar point, showing in detail the relation between the problems that were solved and the complex problems that were caused (e.g., in matters of speed, overcrowding, nuisances, etc.). We might refer also to Kahn and Wiener, who made a useful distinction between an intrinsically dangerous technique (e.g., which contributes to environmental degradation) and a technique which creates problems through dangerous individual choices.[23] But we have to keep in mind that human choices may bear a confused aspect. They thus concluded that technology raises such serious problems as nuclear proliferation, attacks on private life, excessive power over individuals, and centralization. Changes are too big or too rapid to be assimilated smoothly. There are new possibilities of harm to nature. The pace of change increases exponentially. This leads to internal tensions, and things are not perhaps in the best hands.

For a global presentation we might refer to a defender of technique, Elgozy, who asked in 1970 whether the computer might not create more problems than it solves. Might it not be better at the first opportunity to replace the computer with a less presumptuous system of data processing? But to go back is unthinkable. We have to obey the technical rule of the primacy of means. We must accept the growth of

23. Herman Kahn and A. J. Wiener, *The Year 2000* (New York: Macmillan, 1967); idem, "Pouvoirs forestiers et choix humains," *Analyse et Prévision* (1969).

problems.[24] To grasp the situation we must obviously start at a plain fact that we must now discuss more closely, namely, the fact that every technical advance does solve some difficulties and some problems. Each time we clearly see a problem or danger and decide to respond to it, one can say that it is virtually resolved or met. In our world, if we decide to meet a difficulty and devote sufficient technical resources (i.e., people and capital) to doing so, we can eliminate it.[25] Each technical advance is designed to solve a certain number of problems. More precisely, in face of specific and limited difficulties or dangers, we have to find an adequate technical response. This is bound up with the movement of technique itself, but it is also in keeping with the profound and widespread conviction in developed countries that everything is ultimately a technical problem.

The movement is as follows. Every problem—social, political, human, or economic—must be analyzed in such a way that it becomes a technical problem (or nexus of problems). Technique is then a perfectly adequate means to solve it. We might take as an example the crisis due to the rise in the price of oil. The result of the debate touched off by the expression "energy crisis" was the creation in many countries of new nuclear power stations (about 3,000 over the whole world in the last twenty years). This is significant from two angles. First, ideological and political differences were shown not to count: China, like the USSR, the USA, and France, thought only of a solution in terms of nuclear energy. Second, this solution was accepted because it was purely technological. Other solutions were supplementary because they were less advanced technologically.

The debate itself is very hard to pin down. There were in effect three debates. The debate among pure nuclear technicians related solely to the elimination of hypothetical dangers. The debate between nuclear technicians and technicians in other energy fields (solar, geo-thermal) related solely to means, an increase in energy consumption being accepted as the objective. The final debate was between nuclear technicians and so-called eccentric groups (ecologists, humanists, etc.) or the public, who were accused of ignorance or unfounded fears. Politicians were unquestionably trapped in the first debate and their

24. Cf. F. Partant, *Que la crise s'aggrave*, who points out that our mode of enrichment constantly creates new social problems; cf. also Giarini and Loubergé, *La Civilisation technicienne à la dérive*, who argue that new methods of production can have the opposite results from those intended, e.g., the reducing of whole areas to deserts by overexploitation of the soil.

25. Cf. D. Hafemeister, "Emerging Technologies for Verification of Arms Control Treaties," *Science, Technology and Society* 5/4 (1985).

discussion resulted in commitment to the most immediately effective solution. We tend today to see human phenomena solely from a technical angle, and it is true that technique can solve most of the problems that we confront. Having conceded this, I will now take some examples to show that the problems which technique raises are much bigger and harder at every stage than those that it solves. Readers will no doubt find this surprising.

The proletariat offers the first example. In the 18th century capital was available, new technical inventions were there to exploit, and new products could be manufactured for which an absolute need was not yet felt (except in the case of better armaments, faster ships, etc.). Hand in hand with this came the creation of new industries (textiles first, then metallurgy). Workers were needed for the factories. The first migration to the cities followed. Farm workers became artisans. Marx has fully shown that the changing of workers into the proletariat was not just the doing of capitalists who wanted to increase their profits (the only aspect that is generally remembered!). Above all, it was the result of mechanization and of division of labor, which are two technical advances. One might say that these two advances are the foundation of all the rest. Marx has brought to light the strict relation between the technical phenomenon and the creation of the proletariat. Capitalists were simply the intermediaries setting in motion the forces of production. The proof of this is that the analysis applies equally well in noncapitalist countries. The technicization in the USSR demanded the creation of a proletariat at least as unhappy as that of England in 1850. The same is taking place in Third World countries that are taking the path of industrialization and technicization.

The technical society has to respond to a certain number of specific but nonurgent problems, to needs of varying importance, the goal being the creation of material happiness. But in the process a new problem arises, that of an exploited class that is more unhappy, being uprooted and plunged into an inhuman situation. It seems impossible to break this relation, for reasons that are well known and are independent of the economic or political structure. The problem is that human beings find it hard to adjust to machines, that workers have to be concentrated, that there is a very rapid transformation of social structures, groupings, and ways of thinking, that external disciplines have to be adopted, and that the consumer products are not adequate. It is recognized today that the situation of the proletariat is not just one of insufficient consumer goods, of unfairness in the distribution of the products of their labor. That is part of the question, but to some extent it can be rectified by an appropriate social regime. The rest of the

equation, the problem of the change in consumer goods as the traditional ones are replaced by new ones which are not satisfying because
they do not meet real needs, is a most important factor in the situation
of the proletariat. Even if technically advanced countries have been
able to some extent progressively to eliminate some of the disadvantages caused by the relation between human beings and machines, this
cannot be done at once in Third World countries, for example, by
automation, which presupposes an infrastructure that does not exist
there. It is simply not possible to save time and bypass various stages.

The Soviet experience, with the extreme misery suffered by the
whole population from 1917 to 1940, suggests that the faster we want
to go in technicization, the more intense is the global misery (which
creates the true proletariat in Marx's sense). Speed simply creates an
even more intolerable proletarian situation. Mechanization and technique have thus brought great gains and responded to many human
needs. But it is incontestable that they also gave rise to the main
problem for Western society throughout the 19th century. Nor could
things have been different, as recent experiences have shown and as
Marx himself thought. I do not think it is an exaggeration to say that
the problem that has been raised is much greater than those that have
been solved. It was too great, however, to be related directly to technical
progress.

The insoluble problem of a proletariat is not just a nineteenth-
century problem or a problem of immigrant workers today. We recall
that miracle of progress, Silicon Valley. The success and growth of this
new conglomerate produced a new and completely crazy proletariat in
1950, with salaries the lowest in the USA, overpopulation in a distant
suburb, extreme industrial pollution, a high rate of divorce and infan-
ticide, high-density population, etc. This is the result of high tech at
its extreme.

A second example of the law that problems grow with the growth
of techniques may be found in the challenge to the natural world in
which we have to live.[26] The ecological problem with its many aspects:
wholesale pollution, nuisances, the production of new chemical elements that do not exist in nature, the final exhaustion (at a time that
cannot be calculated with exactitude) of natural resources, the great
threat to our water, the destruction of the countryside, the wasting of
tillable soil—these facts are hard to dispute today, in spite of some

---

26. Of the many works see especially F. Ramade, *Écologie des ressources
naturelles* (Masson, 1980); against ecology cf. Missica and Wolton, *L'Illusion écologique* (Paris: Seuil, 1979).

opposition. They are all the result of frenzied growth, of the unrestricted application of techniques.

There is a tendency to divide up the danger, for example, water pollution, or the ultimate exhaustion of copper. This is a technocratic mistake. We must look at the ecological question in its entirety, with all the interactions and implications, without reductionism. We then see that the problem raised is a thousand times more vast and complex than any of those raised in the 19th and early 20th centuries which techniques have been able to solve. It is much harder to purify the Mediterranean than to make an airplane fly. The danger is so great that people prefer to ignore it. After a period of awareness between 1955 and 1970, the public has lost interest and governments do their best to deny the danger. Those who have really studied the situation regard the danger as such that immediate measures are necessary on a global scale if we are to restore ecological balance to our environment, since we are dealing with an ecology that is now socio-agro-industrial.[27]

To illustrate the great gap between the enormous problems created by technique and the many isolated benefits that it has conferred, I will choose only one other example: overpopulation (although I might easily adduce others, such as the effects of automation, new illnesses, urbanization, transport, etc.). I certainly cannot share the optimism of Sauvy and other demographers and economists who argue that the earth can easily support a hundred billion people. We remember that there is no such thing as overpopulation in the absolute. There is overpopulation only in relation to the possibilities of subsistence. In the long run Malthus seems to have been right, not in an absolute sense, but in a concrete sense. In two or three pages I cannot bring out the extraordinary complexity of the matter. I will keep to three aspects. First, demographic growth is the result of techniques: the prolonging of life, the keeping alive of infants who would "normally" be dead, vaccines, the eliminating of epidemics, and hygiene. Technical progress has brought with it the astonishing proliferation of the human species during this century. The difficulties that have been created are due to so-called good or positive techniques, not negative or aggressive techniques but ones that are designed to serve and protect us. It is this that has produced the impasse, showing how hard it is to distinguish between good techniques and bad.

This absolute demographic growth raises the question of sup-

27. Cf. G. Lambert, *Le CO^2 aujourd'hui: les controverses scientifiques;* R. Delmas, "Le gaz carbonique naturel et artificiel," *La Recherche* 114 (Sept. 1980).

port.[28] Again we have arguments among the specialists and experts. Some think there is still much arable land that can be used (up to twice the amount already in production). Others regard it as madness to try to bring most of this land into cultivation, since it would involve disastrous deforestation. We should also note that during the thirty-five years that would be needed for this doubling of cultivated land, the population, according to every calculation, would also double. In absolute figures there would thus be twice as many undernourished people as today. Again, some argue that there are inexhaustible food resources in the oceans (algae, plankton), but others point out that the level of radioactivity is increasing rapidly in the oceans, and this radioactivity particularly effects algae and plankton, thus rendering them useless as food. As Professor Furnestier has also pointed out, we cannot develop aquaculture when our oyster beds are threatened with asphyxiation by the overflow from campsites. Account has also to be taken of underwater drilling for oil and the pollution of our coasts. In Japan industrial effluents at nontoxic levels effect plankton and through them fish and people. Increased cultivating of algae and plankton would obviously involve solving hundreds of problems of incredible difficulty. It is generally conceded that we will have to triple food production in twenty-five years, but no one knows how.[29]

The problem is, in fact, harder than we suppose, for we have to distribute as well as produce food. The transport needed to take it to famine areas is generally lacking. We have also to fight the policies of

28. The growth is aggravated by continuing growth in the Third World and stagnation or even regression in developed countries. People in the West live with the firm idea that we have obtained a spectacular prolonging of life, and that this is one of the incontestable triumphs of of the quality of our life and of technique. But this does not give a totally accurate picture. The reduction of mortality in the industrialized West goes hand in hand with stagnation and even regression in the population, and now there is even a fall in average life expectancy due to more male deaths, the development of harmful sociocultural behaviors (alcohol, drugs, accidents), the aging of the population, and less efficient sanitation systems (even though we have the opposite impression). The demographic gains are very fragile and the amelioration of health depends increasingly on unstable social and cultural factors. We have little reason to boast. Cf. J.-C. Chesnais, "La durée de la vie dans les pays industrialisés," La Recherche 147 (Sept. 1983). R. Lewontin (in Jacquard, ed., Les Scientifiques parlent) adds that population does not depend on fertility but on such resources as food and space. Furthermore, an increase in the number of young people can lead to a decline in the adult population. These are vital points that are regularly overlooked.

29. It is true that proteins can be manufactured by microorganisms and that some fast-food steaks contain up to 30 percent of such proteins. Such is our future! See Bourgeois-Pichart in Jacquard, ed., Les Scientifiques parlent.

monopolists and national leaders who sidetrack aid funds to their own pockets. It has been often emphasized, too, that direct food aid to Third World countries is ultimately disastrous in most cases, since the products compete with local products and bring ruin to Third World farmers. It is hoped that new chemical products and technical advances will bring solutions. Research goes on and artificial food that is high in proteins and vitamins can be produced. But experience shows that undernourished people do not readily accept these substitutes. The main difficulty lies in the field of traditions, attitudes, and social, political, and religious beliefs, which give rise to strong resistance. We must not think that people who are the victims of famine will eat anything. Western people might, since they no longer have any beliefs or traditions or sense of the sacred. But not others. We have thus to destroy the whole social structure, for food is one of the structures of society (see, of course, Lévi-Strauss).

This will mean destroying groups and personalities. Again people will be fed materially at the cost of inner psychological and social destruction. This time, too, the price to pay is very high; it is qualitatively immense. It cannot be compared with the benefit of chemical nutrition. Here we have a general feature of the price: It cannot be compared with the positive technical gain. Nor is every discovery good. We recall the extraordinary finding of Dr. Gudmand Hoeger that most of the people of the Third World are allergic to milk. Millions of people do not have in their digestive systems the enzymes that enable them to assimilate it. Milk causes them serious gastrointestinal troubles. If this finding is confirmed, we can see what its effects will be, for example, on the use of powdered milk in aid, or on the policies of agricultural development. It is in keeping with the native resistance to powdered milk (as though this food were bewitched) that has often been noted (in Africa especially). The technical calculating of calories and economic orientation can obviously have unforeseeable results.

The World Health Organization has attempted to reduce the birth rate by setting up research centers to stimulate research and to influence public opinion and legislation. But psychological, cultural, sociological, and interrelational problems have increasingly been encountered in teaching methods of contraception, sterilization, etc. In two excellent articles Dr. Escoffier-Lambiotte stressed that a profound change in traditions and scruples is entailed.[30] There has to be an

30. "Le droit de procréer," *Le Monde*, Jan. 1972.

ethical revolution or else a dictatorship will be needed to suppress the right to procreate, with the attendant abuse of compulsory sterilization.

For his part A. Sauvy drew attention long ago to the psychological and sociological disasters that a sudden stoppage of demographic growth might bring.[31] We must not believe that the only danger is the bomb. There is the second danger of growth. But checking growth by drastic measures will entail no less serious consequences, except that they will be in other areas and less obvious than famine and material want. This is not to speak of the spiritual impact on populations which are driven by strong and vital religious convictions toward procreation, which regard an infant as a gift from God (Judaism, Roman Catholicism, Islam), or as a condition of the transmigration of souls (Hinduism). We cannot dismiss such convictions with a wave of the hand, stating that they are ridiculous and must be eliminated. The price that has to be paid in this sphere is surely enormous. We have here a typical case of the insoluble dilemmas that technique thrusts upon us when it is applied to some problems.

These, then, are three examples of the gigantic and unforeseeable problems that technique raises. We can say that every technical advance carries with it problems of the type that we have just sketched. These problems are the fruit of the industrial era (with its obvious progress in metallurgy, transport, medicine, etc.) and of the early technical era up to about 1970. What vast problems will arise with the new stage of the technical system, with genetic engineering, computers, the laser, and space? It is just as impossible to answer this question as it was to foresee the creation of the proletariat in 1800. Nevertheless, in the course of this work we will try to give some hint of the problems that might arise in the future. We are certain of one thing: they will be more difficult, more extensive, and more complex than their predecessors.

3. The Harmful Effects of Technical Progress Are Inseparable from Its Beneficial Effects

We have seen already that it is almost impossible to distinguish good and useful techniques from those which are bad or useless.[32] It is futile

---

31. Alfred Sauvy, *Zero Growth?* tr. A. Maguire (New York: Praeger, 1976).

32. On this subject cf. the works of B. de Jouvenel, especially *Arcadie*, and B. Charbonneau's *Le Jardin de Babylone* (1969). One of the few writers to see how positive techniques become negative is Donald Schon, *Technology and Change* (New York: Dell, 1967). In his last chapter he emphasizes that we are beginning to realize that we cannot regard the bad effects of technique merely as bad application, but

also to speak of a "good use" of techniques. Simplistic evaluations of this kind rest on very summary, general, and abstract views of techniques without any real study of the technical phenomenon and technical progress as such. In fact, things become much more complicated the moment we cease looking at an abstraction and philosophizing, and instead consider the functioning and development of a specific technique. Classification, then, is by no means easy, for one technique entails a multitude of diverse effects.

It is not at all a simple matter—in spite of appearances—to distinguish peaceful techniques from military techniques. For some time past I have tried to show that the atomic bomb was not at all the product of wicked warmongers but a normal result of atomic research, a necessary stage, and that the enormous effects for the human race go far beyond the bomb, there being many peaceful applications of splitting the atom. I will not go over that material again. But one might take any level of technique, the lowest or the highest, and one will find that nothing is univocal. The techniques of exploiting riches are good for us? No doubt. But what if they involve the exhaustion of riches, their unchecked exploitation? The techniques of production are unquestionably good. But the production of what? When techniques make possible the production of all kinds of things, if we give people their freedom, it will be used to produce things that are absurd, empty, and useless. A remarkable truth thus comes to light: Producing is regarded as good in itself.[33] No matter what is produced! The only function of technique is to increase production. And since our only function is to work, and to participate in the development of production is our means of livelihood, we go on to produce absurd, empty, and useless items, but items that are infinitely serious because our life and work are devoted to them, and from them we earn our living.

Let it not be said that this is not an effect of technique or that things might be different. With a totalitarian government and an authoritarian control of production, such objects are not produced (rather, tanks and missiles with nuclear warheads). But dictatorship does not sit easily with technique. Production techniques work properly in nondictatorial regimes. Let it not be objected, then, that we ourselves are to blame, for in the long run we must see people as they are. One of the great weaknesses of those who separate the good results

that technological changes themselves sabotage the good effects that are expected from progress to the benefit of the great society.

33. I will simply refer to Galbraith's *The Affluent Society*, 2nd ed. (Boston: Houghton Mifflin, 1970).

of technique from the bad is that they constantly think of people as wise, reasonable, in control of their desires and instincts, serious, and moral. Thus far experience has not shown that the growth of technical powers has made us more virtuous. As of now, to say that we must make a good use of techniques is to say nothing at all.

I want to show, however, how the very core of technical mechanisms produces without distinction both good and bad results quite apart from human intervention. Here again I will give examples. Let us take first a very complex instance that I will sketch very simply. I refer to a cycle of bad effects linked to positive effects which has been notably brought to light in an excellent article by Nathan Keyfitz.[34] We might sum it up as follows. Europeans settled in countries with scattered populations and developed single-crop plantations or exploited the raw materials. They required more laborers, diminished the mortality rate, and increased the population. In time they found substitutes for the raw materials, and they were driven out of their colonies, which for the most part they no longer needed. But the population growth, once started, did not stop. It is one of the essential factors in underdevelopment. Thus the good effects (hygiene, better techniques, discoveries, decolonization) are inseparably tied to bad effects (loss of outlets, loss of a subsistence economy, dangerous overpopulation).

We will cite other examples. One of the constants of technique is the growth of rhythms and complexities. Every economic, administrative, managerial, and urban operation becomes more and more complex as a result of the multiplication of techniques. Every field demands knowledge of more and more techniques. This extraordinary extension of techniques brings with it increasingly advanced specialization. It is virtually impossible for a single person to know several techniques, several methods. Processes are increasingly refined, complex, and subtle. We have to restrict ourselves to one if we are going to master it. In this situation we have to know perfectly the technique that we use, because this gives greater efficiency and greater speed, but a single mistake can be catastrophic. The faster the machine, the more serious the accident. The more subtle the machine, the less forgiving the error. What is obvious at this mechanical level is equally true in every technical field. Technicians become increasingly narrower specialists. Yet the system cannot function unless the operations parcelled out to specialized technicians are related to one another, are literally connected. Within the different operations of an automated chain, each

34. Nathan Keyfitz, "Développement, démographie, technologie, isolement des pays tropicaux," *Analyse et Prévision* (1967).

operation successively controls and determines many others. Similarly, in a technicized society all the work of a technical specialist has to be coordinated with that of others if it is to be effective and to mean anything. Working together, specialization and coordination allow an acceleration of rhythms but also cause congestion.

This phenomenon of congestion, to which I have already alluded and upon which P. Massé has insisted for many years, shows itself to be one of the unavoidable but disastrous effects of technical improvement.[35] We find it everywhere, not just on the roads. It is the surcharge on academic programs, the ransom that must be paid for the growth of knowledge. I admire those who reform our curricula. But do we have to stop teaching the rudiments to children? If we reduce these rudiments to useful, technical knowledge (which does not develop the intelligence but is essential for entry into this society), we have demented programs that simply serve to crush children's personalities and sensibilities. This is the congestion of sounds, images, and written material. Paper (which is far from being dethroned) congests all activity, which loses its significance by being smothered in papers received and papers given out. There is a mania for regulation, by which we think we can control the proliferation. We draw up rules, make organizational charts, set up groups, and are convinced that in this way we can see clearly what we are doing. The result is a multiplicity of regulations which are both finicky (for how can one get a handle on the proliferation without going into details) and contradictory (for there is no longer any possibility of synthesis). This regulation finally becomes totally detached from reality. By its density and complexity it becomes itself a real hindrance and a source of other hindrances. In such situations the number of decisions one must make multiplies endlessly. It is an illusion to hope for a machine that will make these decisions. Human beings themselves, politicians or managers, must do so. But the decisions are increasingly inadequate and confused, and managers are increasingly crushed under their weight.

These phenomena of congestion are simply an inevitable consequence of positive technical progress. Even though each element might be useful, the totality is inhuman. As Massé says, it crushes the individual and dislocates social life. The result is twofold containment. The greater the congestion, the more difficulties of communication confine us in restricted localities. The era of planetary exploration translates itself into increasing immobility at the level of daily life. Again, the mass of knowledge acquired restricts us all to special fields

35. P. Massé, "L'homme encombré," *Prospective* (1969).

with their own secret codes. We must insist on the fact that these results are inextricably linked to one another.

The same applies to the problem of transport—the means to escape, to freedom, to knowledge of the world. In the insoluble problem of heavy traffic, noise, and loss of time in commuting, the intermingling of positive and negative effects is evident. It is less so, but the more tragic, when one considers the effect of the growth of rhythms and complexity in work. There is undoubtedly more efficiency and increased production, etc. But it is this that impressively adds to what we have to call human scrap heaps.[36] In our technical society we meet with increasing numbers of men and women who cannot adapt to specialization, who cannot follow the general rhythm of modern life. This is not merely true in capitalist countries, as may be seen from the testimony of Rudenko, Soviet Minister of Labor, in 1961. Nor does it apply merely to the aged. An increasing number of young people are called "maladjusted." We are in the presence of a whole population of semi-incapable people. They are so, not intrinsically, but relative to the technical society. There are exhausted men and women, nerves stretched tautly, able to work part-time (the question of part-time work arises not merely for married women), unable to concentrate or to do precision work for long. There are also the slightly unbalanced, who can do slow and simple work (which no longer exists in our world), and the aged (we recall that with the present rhythm of work and the constant flow of new techniques, people are old at fifty, and that people have to be continually retrained to learn the new techniques in their own fields).

There are not so many of these human scrap heaps in a traditional society because conditions of work that are not technical permit the employment of all kinds of people; anybody can be used. Our own society, then, more and more strictly separates the qualified from the unqualified. Freely supporting a host of the unqualified is no doubt possible in a very productive society but it is reprehensible from a human standpoint. We do not refer to exceptional cases but to the general difficulty, which may be temporary (but for how long?), of living in a continuously mobile society that has no frame of reference, in accelerated changes in forms, in work, in speed, in employment, and in knowledge. The state of permanent disposability that is imposed on us is shocking to us all. There are too many new ideas, novel situations, unconnected techniques, towns without roots (Massé).

The risk is that there will be created what Mendras calls a

36. Cf. Willard Wiatz, *Report to the Secretary on the State of Labor* (USA, 1965), in which he concludes that there is a growing "human scrap heap."

countersociety made up of those who cannot follow the rhythm. We should also include in this category what Keyfitz calls the unexploitable, that is, those whom it is not worth employing. The situation of those who are exploited in the capitalist world is ultimately less serious than that of the unexploitable, of those who are useless, of those who are not worth employing, whom no one is interested in employing even for the minimum wage, who have nothing to do even in Socialist lands. This is not the leisure society, which demands an income on which to live even while not working. In the movement from production to productivity, technical progress makes whole categories of people unexploitable before it reaches a level of production that would make it possible to support them for nothing. The situation is even more startling when one looks at it globally, taking into account the relation between the advanced technology of the West and the demographic growth of the Third World.

The more technique advances, the greater the dangers. This is true in practically every area. The bolder the technique and the greater the achievements, the more unheard-of the danger. A decision has been made to put a road across Amazonia which will link the Atlantic to the frontier of Peru. The forest will have to be cut for 3,000 kilometers, a gigantic technical feat which endangers all the Indian populations of the area. N. Neto's description is both plain and terrible. The Indians will lose by mere contact with civilization. We need to take into account what this road will mean in terms of a human disaster.

A spectacular example is the pollution caused by the fight against pollution. The methods used to purify the air in the USA are very effective against smoke pollution, but unfortunately they eliminate certain solids, so that in combination with air and water the emissions form acids which the solids previously prevented, the result being that acid rain falls on houses and crops. Rainwater has sometimes been no less acid than citrus juice. Studies have shown that the situation in Norway is catastrophic in this field. Fish and forests are affected and corrosion is reported. But the clouds that are charged with acid come from places at which attempts have been made to solve the problem of smoke pollution.

In this period of the diffusion of means of communication I might also study, as an instance of the inseparability of positive and negative effects, the inextricable mixture of information and propaganda. But I will not dwell on this matter here, since I have looked into it elsewhere.[37] I will conclude with three general remarks. According

---

37. See J. Ellul, *Propaganda*, tr. Konrad Kellen (New York: Knopf, 1968).

to the fine study of M. Micaleff, the idea that technique serves human needs to increase individual well-being is illusory.[38] Economic growth has become detached from improved well-being. All remunerative activity is viewed as added value that generates well-being. In many cases, however, the activity represents deducted value. Thus investment in the anti-pollution industry does not improve well-being. It is a supplementary cost of production which reduces well-being. At times the deduction exceeds the addition. E. Morin also theorizes that the principle of entropy tends to degrade the original meaning of an action, to sidetrack it, and finally to dissolve it in the play of interactions. In a haphazard fashion every action enters into a complex and multiple play of interaction which the action does not control and in many cases does not even suspect.[39] This applies in an exceptionally exact way to technique and to actions that use technical means. Morin was probably thinking of technical actions, since others today are of no importance.

We may conclude this account of the link between the positive and negative effects of technique with the profound thought of Jouvenel: "We are spoiling our environment not merely as individuals, when we act like ignorant brutes, but also as agents serving a useful social function when we do things in a way that is rational in terms of the objective but ill considered and damaging from the overall standpoint." The problem is to know whether, given technical action, it is possible to act in such a way that we take every factor into account, or whether the "overall standpoint" would not paralyze technical action.

4. Technical Progress Has a Great Number of Unforeseen Effects

Unpredictability is one of the general features of technical progress. We find it at the commencement, at the stage of invention and innovation, then in the course of application, and finally at the end, at the stage of effects. I will study general unpredictability in the next chapter;[40] here

---

38. Micaleff in Jean Touscoz, et al., *Transferts de technologie* (Paris: PUF, 1978).

39. E. Morin, *Pour sortir du XXe siècle* (Nathan, 1981).

40. I will leave aside the unpredictability due to the autonomy assumed by technical instruments, as in the story of the robot revolt. I will simply emphasize that the greatest uncertainty obtains in this regard. For one thing machines are simply machines and cannot be autonomous. No machines can think or create. Nevertheless, there is a tendency toward autonomy and a creative capacity, e.g., in the aesthetic realm. Furthermore, "as computers become more self-programing they

I am looking only at the effects and not at the total movement. This section should thus be read in the perspective of the chapter that follows.

The technical phenomenon never has the simplicity of a diagram. All technical progress has three kinds of effects: the desired, the foreseen, and the unforeseen.[41] When scientists do research in a technical sector, they are often looking for a precise and sufficiently clear and obvious result. The kind of problem might be that of how to drill 3,000 yards to get at a pool of oil. Various techniques can be used or new ones invented. We have here desired effects. When there is a discovery, scientists see how it can be applied, work out the methods by which to apply it, and expect and obtain certain results. The technique is fairly sure and yields the expected results. There might be uncertainties and setbacks, but one can be assured that technical progress will eliminate the zone of uncertainty in each field.

Each technical operation has a second series of effects. These are not sought but they may be foreseen. Thus a great surgeon once said that surgical intervention replaces one infirmity by another. To be sure, the new infirmity is less serious than the one it replaces, or it is perhaps local instead of being a general threat. Similarly, very effective medications might have serious side effects. Accidents are numerous, leading to some 10 percent of all the hospitalizations in France. The use of such medications is justifiable only if the risks associated with them are less than the benefits they confer. A calculation has thus to be made of foreseeable risks and benefits. These are effects that we would rather not have, that are negative, but that are also inevitable, that are known and kept within bounds. In every technical operation, we have to be as clear-sighted as the surgeon and recognize effects that are not sought but that can be foreseen. The fact that many people are not so clear-sighted we saw under our first section. If they are not, they cannot properly evaluate what they are doing or strike a balance between the positive effects and the negative.[42]

---

will increasingly tend to perform activities that amount to 'learning' from experience and training. Thus they will eventually evolve subtle methods processes that may defy the understanding of the human designer" (Kahn and Wiener, *The Year 2000*, p. 90).

41. Schon, *Technology and Change*, offers a good analysis of risks (the probability that an action will produce an undesired effect) and uncertainties (situations in which we cannot analyze risks or there are unlimited possibilities and applications). He stresses that technical innovation is the biggest factor of uncertainty, since it unceasingly creates situations in which one cannot say what the real issues are.

42. There is still a good deal of contingency in this regard. In the economic

But then, third, there are totally unforeseeable effects.[43] Yet we must still distinguish between effects that are unforeseeable but expected and effects that are unforeseeable and unexpected. In the former case we can see the possibility but not give an exact prediction. Thus, in housing, we can see that unit housing might have profound psychological and sociological effects. People who live in large unit housing will undergo changes, but how and in what respect we cannot foresee with any accuracy. There will be some change in behavior and relations and amusements, but what these changes will be no one can predict; one guess is as good as another. Ironically, the conclusions of Francastel on this subject are diametrically opposed to those of Le Corbusier. All that we can be certain about is that there will be changes. Leisure provides another example. If it is true (and this is not absolutely certain in spite of the prophecies of a number of believers in technique who never explore anything in depth) that we are moving on to an era—a civilization?—of leisure, we can be sure that this will produce great changes in people, but no true prediction is possible. This is a very hypothetical domain. Our concrete knowledge of psycho-sociology is still uncertain, and we cannot go on to make a prediction. We can only extrapolate, and since the data are limited and relatively uncertain, we are only guessing.

There are also unexpected effects that might have been foreseen. Among dramatic examples I might adduce Venice, the *Torrey Canyon*, and those that have followed. I am deliberately taking older instances. The troubles in Venice might have been foreseen. The unexpectedness was due to a lack of information and a refusal to foresee what is negative. The slogan that the worst is the least certain is the best pillow for sloth. It could easily have been foreseen that changes in navigation would cause damage. In a city exposed to salt winds, the combination of salt and carbon deposits has, it seems, a pulverizing effect on marble.[44] The sculptures, pictures, and frescoes have suffered more

---

sphere external effects have been the most studied. A barrage at the mouth of a river can cause changes in the movement of the tides and affect ports. These effects have now begun to be taken into account; cf. D. Pearce and S. G. Sturmey, "Les effets externes et l'antagonisme entre bien-être individuel et bien-être collectif," *Analyse et Prévision* (1967).

43. See F. F. Darling, *Wilderness and Plenty* (Boston: Houghton Mifflin, 1970), who shows especially that each liberating technique always brings new constraints. See also A. Mendel, *Les Manipulations génétiques* (Paris: Seuil, 1980), who raises basic and pertinent questions regarding genetic engineering.

44. This process accentuates the damage caused by automobile and industrial fumes as these come down on stone monuments in the form of rain and have a corrosive effect; cf. Y. Rebeyrol in *Le Monde*, Jan. 1, 1970; A. Jaubert, "Venise sauvée des eaux?" *Le Recherche* 122 (May 1981).

damage in the last twenty years than in the preceding five centuries. We can well imagine the effects of the constant waves in the canals washing against walls than never previously knew anything but the slow motion of waves in the lagoon. Again, it was easy enough to foresee the disastrous effects of an oil spill of the magnitude of the *Torrey Canyon*. These effects were unexpected only because it was not known when the accident would happen. What is known is that oil spills are likely.

Then there are results that are totally unforeseeable and unexpected. I recall an example that I have quoted elsewhere: the cultivation of cotton and corn.[45] This represented incontestable progress for many new areas. The cutting down of forests seemed to be a good and profitable venture from every standpoint. Here was technical progress. It could not be foreseen that the cotton and corn would destroy the soil, not only robbing it of its richness but attacking its very structure. The roots destroyed the organic link of humus. Areas that were brought under cotton and corn were reduced to dust after thirty or forty years. The wind had only to blow and nothing was left but rock. This happened in the USA around 1930, but the phenomenon is global. We find it not only in the USA but also in Brazil and Russia. One of the battles between Khrushchev and certain agricultural specialists in the USSR related to this problem. Experience served so little in this domain, even though the danger was known, that there was no hesitation in starting the cultivation of corn. Khrushchev himself was passionately aware of the possibility of impoverishing the soil, and many Soviet agronomists were hostile to him for this reason. The battle lasted for three years (1960-1963), ending with the famous declaration of Khrushchev to the Central Committee on December 10, 1963, that we have not sworn eternal fidelity to corn and hence must reduce the areas on which it is cultivated. He stopped the experiment in time, but by then the effects were not unforeseeable; they were known very well. Incidentally, this brings to light one aspect of the development of techniques: the obsession with efficiency is so great that increasingly serious risks are taken in the hope that they will be escaped. At first the ill effects of corn were not known. The facts were unforeseeable; they could not be known until twenty-five or thirty years later. Pessimistic evaluations (e.g., by William Vogt) lead to the conclusion that the arable land in the USA has been severely damaged (if not destroyed) by about 20 percent.

45. Ellul, *Technological Society.* The same for all the manipulations of the genetic genes: one never knows the exact consequences; cf. "Les manipulations génétiques: des risques encore mal évalués," *La Recherche* 107 (Jan. 1980).

In the same field agronomists are now alerted by the fact that the embodying of massive amounts of chemical fertilizer in the soil, mainly nitrogen, begins by greatly enhancing the yield but then destroys later crops by developing by-products and polluting neighboring lakes and underground water. Above all, it is realized that new strains demand enormous quantities of water and fertilizer. In some countries in which they have been introduced water is scarce and the people are too poor to buy sufficient fertilizer. The poisonous effects are so great that the talk now is of a "satanic trilogy": improved strains, fertilizers, and pesticides.

There are literally innumerable examples of unexpected secondary effects. The whole world has been talking about a miracle in the case of the green revolution, but after three years it was seen that the new strains of rice carry resistant parasites that are disastrous for traditional strains and that they produce rice which is nutritionally inferior. There is always a qualitative price. Bringing large new areas under cultivation even under expert control always has disastrous results. Nepal has made a big effort in this direction. Excellent! But it was necessary to cut down Himalayan forests, and catastrophic floods and devastation have been the result in Pakistan and Bangladesh. The massive extension of irrigated lands has also entailed grave ecological distortion. The Aswân Dam has brought epidemics to peasants in the Nile valley, and by retaining the soil it has reduced fertility. A. Toffler, with his usual illogicality, said that some people were arguing that the dam, "far from helping Egyptian agriculture, might someday lead to salinization of the land on both banks of the Nile. . . . But such a process would not occur overnight. Presumably, therefore, it can be monitored and prevented."[46] He wrote this in 1970, but already by 1973 it was possible to see the extensive and disturbing effects of the dam on the whole Nile valley. Phenomena which disrupted the whole of life and culture had now become irreversible. Modern fishing techniques (deep trawling) have destroyed the beds of algae where the fish seek their food and spawn. Using techniques always pays off in the short term and then brings disaster. I am not passing ethical judgment, but I identify very closely with the assessment of Mme. Ferhat Delassert that "technological solutions work for good or ill according to whether they preserve or disturb the basic ecological balance."[47] This is indeed an excellent principle by which to judge. The examples that I have just given show that it is not always easy to

46. Alvin Toffler, *Future Shock* (New York: Bantam, 1970), p. 444.
47. Ferhat Delassert, *Analyse et Prévision* (1970).

detect quickly the profound effects of techniques and their unforesee-able consequences.

It is especially in the field of chemistry that we find unforeseen and unexpected results of this kind. We see it first in the use of medicines. No matter how serious or cautious researchers may be, they cannot possibly do all the experiments that are needed to discover every possible effect of a medicine. Psychological effects cannot be discovered from animals. Physical effects may also be unexpected. Experiments cannot go on long enough to show what the long-term effects might be. At issue are the effects on descendants, the effects after long years of use (e.g., of tranquilizers), and the effects some time after the use of a powerful medicine that might modify a physiological function.

Do we have to recall the unexpected side effects of penicillin, or the dreadful scandal of thalidomide? In the latter case, to save the face of science, it was argued that there had been no negligence at the experimental stage. There had been six years of laboratory tests on animals. But it was not possible to imagine every possible result. The case of thalidomide is particularly well known because of the media campaign and the trial, etc.[48] We must not forget, however, that there are more instances than we think. In 1946 another drug (triparanol), which had been tested by serious laboratories, had to be withdrawn from the market because of very serious side effects.[49]

Nor is it just a question of medications. In many other areas the development of chemistry has had unforeseen effects that are very dangerous. Even with careful controls, medications can be harmful, but in the case of other chemical products the controls are much less careful. The typical example is that of DDT. From 1941 to 1951 it was claimed that DDT had no harmful effects on warm-blooded creatures. Alarm was raised in 1951 when it was discovered accidentally that it caused rickets, and since 1958 all kinds of other harmful effects have come to light. Now its use is so extensive that an official analysis by the U.S. Department of Health in 1965 showed that average Americans have twelve parts per million of DDT in their tissues, while a report to the Council of Europe tends to show that forty-four parts per million can be fatal. Since 1968 there has been a very violent reaction to DDT

48. Cf. J. Paulus and J. Rozet, *Le Procès de la thalidomide* (1963). There were 4,500 instances of infant deaths or deformity. At the trial expert witnesses disagreed, some saying that the drug was not at all harmful in normal conditions. We find the same defense of DDT (cf. Y. Rebeyrol, *Le Monde*, Aug. 1973).

49. Outside the field of chemistry, sonograms, while posing no danger to adults, might be very harmful to the fetus; see *La Recherche* 106 (Dec. 1979).

in the USA, and its manufacture and use have been strictly controlled since 1969.

The more we advance, the more we bring to light the unpredictable and disturbing effects of many chemical products that have been in use for many years. Almost every day an attentive reader will find the denouncing of very dangerous products that have been in common use. Thus biphenyl has a harmful effect according to the Academy of Medicine (February 1970), and phenacetin might be said to bring death by ignorance to those who take certain remedies that contain it.[50] We might also refer to chemical products that are not taken. Thus it was discovered in 1962 that certain plastics are not stable and can finally have effects that are dangerous to the human organisms. Detergents, too, are by no means harmless. Their abuse has harmful effects on watercourses, whether through the effluents of the factories that manufacture them or through the contamination of city water supplies. Massive amounts of detergents in rivers destroy all life and according to some experts in 1963 may even threaten the continuity of the water cycle (evaporation-precipitation). It is now admitted that even in very small doses (a tiny fraction of a milligram per liter of water) they are fatal to fish. The best analysis of the problem may be found in a German study of 1965 which led to a law regulating the manufacture of detergents and which showed how complex the problem is, the need being to manufacture a product that is profitable, effective, and yet harmless to bacteria and fish. Thus far all satisfactory products are either too costly or ineffective. A detergent that is harmless to fish is harmful to bacteria, which have an indispensable role. Furthermore, detergents are still present. Indeed, they are accumulating and their toxic effects are still to be feared.[51]

As regards the toxicity of detergents (which remains in spite of every attempt at cleansing), the French Committee on Detergents published a report in 1963 which stated that there is little acute toxicity and that chronic toxicity is not any cause for disquiet. But new and more powerful detergents have not been tested from this standpoint. It is hard to relate to human beings the results obtained from animals,

50. Escoffier and Lambiotte, *Le Monde*, Feb. 1970.
51. Unpredictable side effects of this kind have produced some notable theses. Thus I. Cheret in *L'Eau* (1968) adopts a position that is gaining ground in the USA, namely, that when a new product is invented there ought to be verification before it is put on sale to make sure that it can be introduced into the natural cycle without doing harm, being not only nontoxic but also ecologically sound. This policy would be disastrous for plastics and detergents. It would mean, however, that unpredictable effects are dealt with by reintegration into the natural cycle.

nor is it easy to calculate long-range effects. We have to recognize the honesty of such conclusions, but specialists on toxicology have raised questions. Direct toxicity is certainly rare, but granted the cancer-forming properties of some detergents, they agree on the important point that detergents can cross the intestinal barrier that agents normally cannot pass. They are weighing the seriousness of this fact. Fournier and Gervais show that no chemical product is without danger for any of us, that products created for profit create new dangers, and that there are so many associations that no one can detect them all.[52] They point out how some controls, for example, the use of computers, the setting up of more centers for toxic detection, etc., can mitigate the dangers. But the unforeseen must always be expected. Most cosmetics are more or less dangerous. It has also been discovered (after many years of use) that aerosol sprays, used in painting, lacquer, perfumes, pesticides, etc., are dangerous because of the "neutral" gas that is present in them and that can cause unexpected accidents. The most serious gas, Freon, is not toxic, but it has been found that it releases gases (in refrigeration, air conditioning, etc.) which are attacking the ozone layer that encircles the atmosphere and which have already caused holes in this layer. For us these are very harmful effects since that layer absorbs most of the ultraviolet rays that can disrupt the structures of living matter. Hence the holes in the layer are fatal, but who of us think that we are participating in suicide when we use aerosol sprays?[53]

Controls of new products are now stricter. Y. Rebeyrol can thus assure us that with progress in the chemistry of pesticides and general mistrust of these substances the launching of a new product is now preceded by many years of research and testing. Groups like Pepro (of Rhône-Poulenc) and Plant Production Limited (of Imperial Chemical Industries) study seven or eight thousand new chemical products each year, and only one of these will reach the production stage after seven or eight years of research. The effectiveness, specificity, and acute and chronic toxicity of a new molecule are studied first, then they are tested in all kinds of different conditions and climates, and the effects on the microorganisms of soils and species are verified. Finally, there is research into the residues and derivatives in the cereals, fruits, and legumes that it is desired to protect. Experiments will show what is the

52. Fournier and Gervais, *Le Monde*, Sept. 1972.
53. Cf. P. Aimedieu, "La disparition de l'ozone (essential)," *La Recherche* 186 (1986). Supersonic aircraft contribute to the destruction of the ozone layer as well. A conference on the subject at Vienna in 1985 was fruitless, but a protocol of agreement was signed at Montreal in 1987.

ultimate toxicity of the residues and determine the maximum tolerable
dosage with a security coefficient of 100. All countries submit a new
pesticide to a strict procedure regulating the way in which it is used.

This is all well and good. Yet other technicians are impatient and
find these delays ridiculous. Doctors who know that a medication is
being studied in a certain sector urgently want to use it.

We have also to take into account two factors which are becoming increasingly disturbing: hazardous waste and accidents. The question of waste in general (to which we shall have to return) is an
agonizing one, for the more we produce and the more the public has to
buy, the more there is to throw away, to discard. For the economy to
function well we need to replace cars and television sets every two
years. The waste is thus accumulating. Here, however, I want to speak
about hazardous waste, for example, dioxin, or potassium salts (cf. the
Rhine), or nuclear waste. Disposing of these things is very difficult if
harm is not to be done. We must also look at accidents: Three Mile
Island, Chernobyl, Bhopal, the emission of dioxin at Milan, etc. Such
accidents are not very numerous, it is said.[54] So be it. Yet this is not
wholly true. Every year there is one serious accident per 500 reactors,[55]
and if the risk is so great, can we engage in such operations without
allowing for accidents? The accidents of new technologies are not like
airplane or road accidents. They have long-term effects, perhaps for
many generations. They have unknown secondary effects that are
discovered only after they happen. Of course, the question is whether
we can arrest progress (the course of technique) because of the serious
problems it raises. This is a choice of civilization.[56]

One might say that unpredictable effects will finally come to

54. But most of the accidents are ignored or concealed by the media.
C. Castoriadis (in Jacquard, ed., Les Scientifiques parlent) has revealed that the USSR
is manufacturing bacteriological weapons and that two serious accidents have taken
place (at Novossibirsk and Sverdlovski) with the escape of a virus. Thousands died
in these two cases. Delayed fallout has also to be taken into account, as at Three
Mile Island (cf. La Recherche 113 [1980]).

55. See La Recherche 137 (1982).

56. Cf. Patrick Lagadec, La Civilisation du risque (Paris: Seuil, 1981). A good
example of unexpected effects may be seen in the seemingly innocent field of
electronics. Acid baths are needed in the manufacture of chips to impress the
circuits on the ceramic, and strong detergents are needed to eliminate impurities.
An inquiry at an important factory in 1983 revealed that it was storing reserve
products in containers that did not prevent infiltration into the soil, thus contaminating the water, and causing seven deaths in a neighboring town and many birth
accidents and defects. Accident? Negligence? No doubt, but the more such operations, the greater the probability of such accidents and negligence. Cf. "Technopolis," special number of Autrement, 1985.

light and can then be contained, analyzed, and avoided. But we must qualify such optimism. There are some irreversible effects. There are irreparable accidents in the individual sphere (the victims of harmful products). It is no real answer to say that progress demands victims. In many cases it is impossible to go back, to clarify matters by saying that a fresh start will be made. The effects are there. DDT is fixed in the human organism; radioactivity is slowly increasing. Often, too, the process is irreversible. We cannot stop using pesticides. The insects that have been affected display by way of compensation an increase in fecundity and resistance, so that they would proliferate on a gigantic scale if we ceased to destroy them. If this is true, it shows in what direction we have to proceed.

There are also phenomena which are so big and have such social implications that it is impossible to go back even though their harmful character is perceived. Can we even conceive of stopping the manufacture of detergents, aerosols, or insecticides? We face here an industrial and social complex which is too important to be called into question. We might improve the products, or withdraw some medicines from circulation, but such acts would only lead to more unforeseeable effects. In other words, we have less and less mastery over the techniques that we use. If one product that is secondarily toxic is taken off the market, at the same time a hundred more are put on it the effects of which we do not know, since they will become evident only two or ten years later. Finally, we have also to take into account the element of prestige and technical fervor.

An aspect that is by no means negligible is that of the opposition between immediate needs and long-term effects. Technique tends to reply to immediate needs. I am not saying that all real needs are taken into account, or that technical development is in terms of needs, but when it does take needs into account they are always immediate needs, and it is in relation to these that we assess its effects and that it sees its justification. But this being so, little account is usually taken of long-term effects. We are indifferent to these so long as our immediate needs are met. We are also indifferent to the promise that technique will satisfy our needs in ten or fifty years. It is now recognized that the liberal promise that increased production will finally end the misery of the disadvantaged is a snare and delusion for the poor. "At once" is an important feature in the technical mentality. But this implies that only immediate effects claim attention. Long-term effects? What matter? Let us eat and drink, for tomorrow we die.

But the harmful effects are long-term effects. Even though they are set forth, people find it hard to focus on them. How can arguments like those that I have advanced prevail against the evident need for technical progress? We will constantly come back to this point. The problem of nuclear disintegration offers the best example. There is endless discussion of the dangerous thresholds of radioactivity. Innumerable experiments contradict one another. Nevertheless, there is the real possibility of a dangerous accumulation of radioactivity and of the threshold of tolerance being lower than some say.[57] The Chernobyl accident has given new life to the discussion.[58] We face a decisive and irreversible eventuality. If we were in the least bit rational this (by no means hypothetical) eventuality would cause us to stop all nuclear research and application. But even though the world perish, we will not arrest technical progress.

We can actually formulate the principle that the greater the technical progress, the larger the number of unpredictable effects. To complete our demonstration we would have to give a detailed inventory of the situation. This is not possible. But the importance of the examples given seems to me to be enough to make possible a generalization, and their quality authorizes it. Ours is not a method of rough approximations. Instead of drawing up statistics or collecting unimportant facts, we have here significant facts that carry considerable weight. It seems to me that the analysis of the ambivalence of technical progress from this standpoint enables me to evaluate very accurately the reality of our society and our human life in a technicized world without engaging in value judgments or obeying concealed presuppositions.

Meadows had good reason to say that in the presence of every technical invention, before adopting it and putting it to general use, we ought to ask the following questions. What are the parallel effects, both socially and materially, of the general use of the invention? What are the social changes that are necessary to put it into use properly (the price to pay)? If it is really effective and enables us to remove an obstacle to material growth, what will be the new limit that the system will come up against in expansion? Should we prefer the constraints inherent in this limit to those that the invention is designed to overcome?

---

57. M. Errera (*La Recherche* 168 [Aug. 1985]) has shown that the effects of even low-dosage nuclear radiation are by no means negligible, and we have to take into account its progressive increase. Furthermore, for many pathologies there is no threshold of tolerance. Some authors have also noted mental retardation among descendants of the survivors of Hiroshima to two generations. When scientists are not sure, they ought to admit that there is no dose of radiation that is without its effect.

58. Cf. Ellul, "Incertitudes," *Sud-Ouest Dimanche*, May 1986.

These are vital questions that sum up the problem. Unfortunately, in fact, we cannot answer them. Even if we could, rational decision would never be possible. I will add these questions to the principle that I believe to be fundamental, namely, that if a venture carries with it a considerable potential risk, even if this is not normally foreseeable or short-term, the course of wisdom is not to undertake it. This principle would suppose full mastery of the situation and disbelief in progress. For this reason there is no chance that it will ever be applied.

Naturally, the facts that I adduce are simply examples of the profounder reality that we are trying to trace, namely, the permanent phenomenon of the unforeseeability of the effects of each new technique and their constantly renewed seriousness. In other words, the illustrations are not designed, as in most works that deal with such things as pollution or pesticides, to stress the seriousness of the particular situation. They are designed to stress the constant factor of ambivalence. In effect, each time we become aware of one of these dangers we fairly quickly find an answer. I am not unduly afraid of the current situation of pollution or of poisoning by pesticides. It will be aggravated for a time but solutions will be found. As regards pesticides, I realize that there is advanced research on replacing them by parasites that prey on the species we wish to eliminate. A priori these should have no toxic effects on other living creatures. But listing the disasters shows us at once that each time the problem raised is more difficult and the remedy more costly. The constant feature is that we never know what we are starting. We cannot even imagine it. This fact sets at once a very strict limit to our ability to predict. If we look at studies of the evolution of techniques, we note that normally they can see extensions or, on the basis of current scientific research, applications. But what threatens totally to upset the calculations is the appearance of problems or dangers that result from the techniques themselves, that cannot be imagined in advance, and that demand a great deal of time, of research, of money, which ought to be devoted to foreseen technical applications. This ambivalence of technique constitutes the true limit of the possibility of prediction.

I must now draw some conclusions from these fragmentary data.

1. There is no progress that is ever definitive, no progress that is only progress, no progress without a shadow. All progress runs the risk of declining. There is a double play of progress and regress. The 19th century ignored the shadow of industrial development and we today basically ignore the shadow of technical progress. But this progress entails and produces specific regress. Technocratic thinking finds a

place only for what is vital, both anthropologically and socially. Its only logic is the simplistic logic of artificial machines. Technocratic competence is that of experts whose general blindness envelops specialized lucidity. Socially and politically, technocratic action can only be mutilated and mutilating.

2. Ambivalence can take the form of stunning reverses. We shall have occasion to list some of these. In this context let us simply say that technique has always functioned in a mode of rationality but that at the present point of development it is falling into irrationality and at times delirium. It has always had utility in view and followed the criteria of utility, but it has now reached a climax in generalized inutility. It has always sought value, but it now functions in a way that contributes nothing of value (mere services and data processing). It has always tended to be constructive, but the potential for destruction is now its main development. Irrational reactions in individuals (music, sports, social maladaptation) have always counterbalanced it, but now irrationality lies in technique itself, in its processes and results, so that it includes the irrationality of the reactions themselves.

Ambivalence confronts us with one of the chief questions put by technique. We refuse to see what real technical progress is. We refuse to see its real consequences and the way in which it calls into question all that we are. We refuse to pay the price that technique exacts. When one draws attention to this price, we call it pessimism. We also refuse the possible technical remedies to the problems that technique causes, for we want to regard these problems as accidents, and to think that we have opted for the good side of technique. We are always too late, therefore, when we try to respond to specific technical challenges. In the face of technique, and of our inability to confront it, I would say then, with Livy, that we can endure neither our evils nor their cures, and with Tacitus, that the weakness of human nature means that cures always lag behind evils.

I know how the defenders of progress usually reply to the problem of ambivalence and unpredictability. We are called, they say, to more responsibility, more choice, more freedom. We have to prove ourselves worthy of what we create (as is said increasingly about the pill). I wish we could, but who is to guarantee that most people will quickly rise to this high level of awareness and responsibility? We face a whole host of dangers caused by technique, and the most serious of them are as yet only hypothetical. But is not the hypothesis enough that someone might put a given technique to evil use? Technological instruments become increasingly powerful and therefore pose greater and greater dangers. If at the extreme there are instruments of absolute power (not merely the H-bomb but computer systems that might control whole populations, or chemical intervention), who is to guarantee that an absolute government will never use them, that it will never, for example, make use of electrodes in the brain (Rorvik's ideal)?[59] As we know less and less about the results of our innovations and are increasingly unable to find the necessary remedies, little more is needed to bring us face to face with an absolute risk. This is why all technical development that increases infinitely a risk that is hypothetical but absolute seems to me to be totally reprehensible. This is the first time that I have said this.

5. Our Lack of Awareness

An embarrassing question remains that I shall treat only briefly. If technique has such negative effects and raises such dangers and threats, why do we have so little awareness of it? Why do most people not sense it or see it? Why is there this headlong rush into technical progress? Why do only a few specialists know it? Soveso, Bhopal, Chernobyl—but we are assured that these were simply rare and local accidents. They are no challenge. There are, in fact, many converging reasons that combine to prevent awareness even apart from the great machine of advertising and propaganda that plays such a big part in molding public opinion.

As always, I prefer to examine the mechanism of social organisms rather than the evil or self-interested intervention of specific groups. Furthermore, we shall be looking at advertising later. The first factor in suppressing awareness is very simple. The positive results of a technical enterprise are immediate. They are felt at once, as in the case of electricity or television. The negative effects, however, are long-term and are felt only with experience. I have sufficient confidence in the honesty of researchers and technicians to be convinced that if they discovered the dangers in time they would not market a product. But the negative effects come to light only many years after the product is on the market and there can be no going back. Automobiles cause terrible slaughter (12,000 deaths a year in France), but this does not halt our love affair with them. The fact that the negative effects

59. We shall take up later the problem of decision, but I might quote here the question of Castoriadis: "Who is to decide on in vitro fertilization? Who is to decide that the way is open for genetic engineering? Who is to decide on anti-pollution practices that produce acid rain?" (in Jacquard, ed., *Les Scientifiques parlent*).

come much later is decisive. Second, we have to take into account what has been called the paradox of Harvey Brooks,[60] namely, that whereas only a small fraction of the population usually has to bear the costs and carry the risks of a new technique, the advantages are widespread. Often, too, the disadvantages for the local group are barely perceptible (to public opinion) and the global advantages for the population carry much more weight.

A chief engineer at one of the most polluting factories in France once told me with a frankness which I appreciated that the area around the factory was obviously dangerous in spite of precautions, for it was a traditional place of pasturage and farmers from a village close by continued to lead their cows there. Occasionally some of these cows died, but it was much less costly to compensate the owners than to install a more complex system of depollution. One might add that often specialists themselves do not see the dangers or drawbacks. The public certainly does not see air pollution and is not aware of the pollution of water tables. If there are no immediate ill effects, it is hard to impose what seem to be useless measures (e.g., catalytic converters). The public will not accept the changes that are needed. Intellectuals even less so! As they prepare to "enter the 21st century" (to quote the title of Morin's famous work), what they think are the problems of society are already outdated and their responses are inadequate. The grasp of things is increasingly behind the times even in supposed looks at the future. The problems that arise are thus increasingly difficult because there is public awareness of them only when they have become vast and inextricable.

A third feature is to the same effect. Except in the case of accidents the problems and dangers are very diffuse and there seems to be no clear causal relation between the technique and its effects, for example, between industrial techniques and the creation of the proletariat or medical techniques and the population explosion. The more hazy and hypothetical or contested a problem is, the less it affects the public. People prefer not to see it or to hear about it. Those who speak about the dangers are labeled ignorant or pessimistic. The advantages are sure and plain to see; the disadvantages are diffuse and uncertain, all the more so since there are experimentations. The pill is glorified in the name of women's liberation. Women can have children as they choose. Those who oppose it are sexists or moralists or enemies of progress. The risks of cancer ought to be considered but are ignored.

60. H. Brooks, "Science, technologie et société dans les années 80," OCDE Report, *La politique de la science et de la technologie dans les années 80*.

Even serious studies discount also the cardiac risks. But the danger has come to light where it was least expected. There has been an explosion of venereal disease, which has again become a scourge, and the fact is plain that it is a secondary consequence of the pill. Yet it is a diffuse effect, and the causality is not apparent to the public.

A final factor is that the advantages are concrete but the disadvantages are usually abstract. Motorcyclists take pleasure in their engines and the pleasure is doubled if they make the maximum noise. But there is now increasing alertness to the great danger that noise presents to our society, affecting hearing, the heart, and the nerves. Noise is regarded as one of our greatest scourges, but in spite of its confirmed effects, which are precise and concrete, the danger seems to be an abstract one to the public. The same applies to the dangers of television. These are obvious examples. In many cases, however, the danger is not apparent at all (and this is one of the inevitable setbacks to serious ecologists)—it comes to light only after long arguments, by means of a specific method that must be used to present to the public problems that they do not understand, in studies that demand a certain competence. This is why the public can have no awareness of the negative effects of technique.[61]

Yet this is not the end of the matter, for even if awareness developed it would come up against three decisive obstacles. First we have the existence of the so-called military-industrial complex, which really ought to be called the technico-military-statist complex. The original term applies only to a capitalist organization and even there it is too narrow. Not industry, but the technical system is to blame, along with the state, which is the engine and primary user of techniques and which organizes the military. The larger term embraces the socialist world as well. This complex will prevent awareness from developing and having an impact. Its power is unlimited, and scientists and groups of militants are helpless against it. All the opposition to nuclear power stations has been of little avail even though this subject did attract public interest. The interests behind technical operations are so important that opposition is useless and is regarded as retrograde.

Second, we must add that the operations involve enormous amounts of private and public capital. They must not stop merely because the public is uneasy. The use of a technical means must not

61. In rebuttal one might point to the relative success of the Greens in Germany. But this success does not represent any true awareness of technique. It is due to the fact that the Greens have become a political party of the left with some fragmentary policies regarding technique.

be abandoned because it has harmful effects. The situation is the same as in the 19th century when the exploiting of mines went on even though miners suffered from pulmonary ailments. Investments have to pay dividends (to the state or the firm). The harmful effects are less important than this imperious demand. At best they will be assessed financially and compensation will be paid. But the work will go on.

This leads us to the third point, namely, that damages and dangers are assessed only in money. Methods cannot be changed, manufacturing stopped, branches of production abandoned, except very rarely. Compensation is given, for example, to the inhabitants of polluted coasts or the sick at Bhopal. This forms part of the general expenses and it might make the economic situation a little more difficult, but the facile conclusion is that we do not stop progress. This is why technique is intrinsically and unalterably ambivalent. Its negative effects are never suppressed nor are they a reason for suppressing operations that have positive effects. A balance ought to be struck at the outset between the advantages and the disadvantages, *all* the disadvantages, those that cannot be assessed in terms of money and also those in the realm of psychology and social grouping. But this is unthinkable. According to Salomon, we need new rules such that unfavorable effects are always less than they would be if competition alone were at work, and we need to define these rules at an early stage in the process before interests are in place, situations are set up, and the dynamism of competition renders their obligatory application impossible.[62] This is why all the dissertations on autonomy (individual and institutional), decentralization, personalization, the growth of liberty, the opening up to small groups, and democratization thanks to new technologies—and these dissertations have multiplied infinitely over the past few years—are absolutely futile and inconsistent. For they ignore the feature which is intrinsic to the very being of technique: its irrepressible ambivalence.

---

62. J.-J. Salomon, *Prométhée empêtré* (Paris: Pergamon, 1981).

CHAPTER II

Unpredictability

1. Introduction to Unpredictability

We live in a society in which looking ahead has become an absolute necessity. In earlier societies farmers and seafarers needed to forecast the weather a few days ahead, merchants had plenty of time to assess market needs, workers had no need to look ahead, and although politicians needed to foresee the reactions of their partners in a limited, well-structured arena, they too had plenty of time to do so. Since our invasion by technical objects, however, along with the intensity and complexity of relations, the rapidity of reactions, and the growth in population, looking ahead has become essential in all matters and all the time. We have to look ahead in an automobile. We have to provide safeguards and assess risks. Every business has to arm itself against competition; it has to prepare for the future. All enterprises invest and borrow. The future, then, has to be as they forecast. If not, they can lose their investments and be unable to repay their loans. Without foresight an economic debacle is certain. The state, too, must foresee growth and prepare for it or else its budget will become enormous. If production does not increase, the state's drain on the economy will be excessive, which will help to bring on recession. There has also to be global forecasting. "Society accepts the present only as preparation for the future. We all accept our economic and social status only because there is hope of improving it. Everything is viewed in relation to the future, hence growth is an economic necessity. . . . Our economic and technical construction is in reality only a movement. Our wealth exists only if it continues to grow."[1]

1. See F. Partant, *Que la crise s'aggrave* (Solin, 1978).

The urgent and universal need to look ahead applies in all systems. A liberal system of government? This, too, rests on calculations of strategy and probability. The stock market is continuously a school of forecasting. Buying and selling shares take place on the assumption that they will rise or fall. We are also aware of the economic effects of predictions which by their very existence bring about the effects that they regard as probable. We know the effects of predicting devaluation, or how entrepreneurs are encouraged or discouraged to launch a particular business according to their anticipation of possible profits. In effect, no social reality is independent of the anticipations, predictions, and general indications that are offered in relation to it. A prediction may become "true" simply by reason of the actions or reactions that it induces.

Morton has studied this remarkable fact of the "unanticipated consequences of intentional social actions." What we have here is "self-fulfilling prophecy."[2] To understand the prediction of one thing, we need to understand many others. A technical innovation has no chance of being accepted by a manufacturer and marketed unless its chances of being accepted and bought by the public have been assessed first. Agriculture no longer depends on the weather but on a vast international market which has to be predicted. This need to forecast is vital in a liberal order. It is on this condition that the invisible hand orders everything. It is also on this condition that individual disorder becomes collective order. It is here that we find Friedrich Hayek's intermediary category in which the social machine is self-organized and can engender forms that no one can control but that come into play to the degree that individual and group predictions of probable evolution prove to be right.[3] This implies an ideal of total transparency.

However that may be, in our society, which is neither liberal nor interventionist, there is an increasing need for forecasting. The military must predict the potential of an enemy. There has to be forecasting of economic growth and of technical possibilities. As technique becomes more powerful, forecasting becomes not only more necessary but also more accelerated. The more powerful the technique, the more serious the effects of an error in forecasting. (At 35 miles per hour an automo-

2. Cf. the colloquium of Cerisy, *L'Auto-organisation* (Paris: Seuil, 1983).
3. Hayek divides phenomena into three classes: the purely natural that are independent of human action; the purely artificial that are humanly produced (prediction being relatively easy in this field); and the intermediary, which comprise unintentional configurations and regularities in human society and which social theory has the task of explaining, since they are the result of human action but not of human design.

bile crash might not be too bad, but at 150 miles per hour it would be fatal.) This applies on every level of life and for every activity, whether individual, collective, national, or global. The situation is much worse under more or less totalitarian, statist regimes with a ubiquitous administration, in which the state plays the role of providence and the economy is planned. Planning is not the same as forecasting. It fixes long-term objectives that are to be reached sector by sector. State planning has to be transformed into individual planning to achieve a plan that takes account of objectively calculated needs according to a well-known double procedure. But it has been progressively shown that this system does not work and that attention must be paid to what is viable. Hence forecasts have again to be made. Thus in French projects, where plans have always been flexible, it has been seen that everything depends on a better forecasting of technical progress.

France used to think that economic growth could be controlled, but as Chesneaux has said, the plan finally works only for the state itself.[4] The state is an indispensable aid to economic growth. The plan decides how much the state will give and to whom. Yet it has less and less control over the growth. Thanks to television it has impressed upon the public the image of a consumer society but it has not been able to stimulate growth. Interestingly, the Socialists, with their desire to direct the economy, have failed. Big business has a monopoly of big technical projects. J.-P. Chevènement could state that the task of the Socialists was to organize the profound technological transformation that France needs.[5] It is obvious that they have not succeeded and that the change that has come about is quite different. What was needed was to forecast the consequences of the new technological beginning, but in fact practically nothing was foreseen: not the revolution in personal habits, not the upsetting of relations between regions and generations, not the risk of higher unemployment. At all costs we need to foresee, but the government was incapable of doing this, and the plan that should have produced a rational economy "opened up an abyss before being a failure" (Chesneaux).

Yet the failure was not because the most scientific or technical methods were not sought for forecasting. Forecasting by linear extrapolation had long since been abandoned. Instead, two great branches were developed. One was futurology. This takes account of all possible scenarios in an effort to foresee which are the most probable. G. Berger

---

4. Jean Chesneaux, *De la modernité* (Maspero, 1983); cf. M. Moraucsik, "Can We Plan Science?" *Science, Technology and Society* 4/4 (1984).

5. J.-P. Chevènement, *Le Monde*, Sept. 1982.

worked out this method toward the end of the 1950s. It is not a science but a rigorous exercise in thought. Present action is based on hypotheses regarding the future. An evaluation of the true long-term stakes is thus demanded. The core of the method is the discovery of the stakes so that each group concerned may devise a strategy to win these stakes. Futurology looks at all possibilities in the actual situation and tries to work out the evolution of the most probable. But we have to say that failure will often be the outcome. E. Morin rightly points out that in the 1960s the method presupposed that the past and present were known, that society was stable, and that on a firm foundation the future would develop dominant trends in technique, the economy, etc. In its feeble optimism this type of prospective thinking believed that the 21st century would reap the ripe fruits of the progress of humanity. In fact it constructed an imaginary future on the basis of an abstract present.[6]

The other branch is that of models. This system analyzes the existing situation, establishes the main parameters, considers what might happen if one of these parameters is varied, and then looks at all possible variations in all the parameters. But this approach gives a very broad canvas of possible developments without saying which has the best chance of coming to pass. It can hardly offer much guidance for action. In "Futuribles," a movement of forecasting launched by B. de Jouvenel, a different process has thus developed, that of proposing objectives instead of simply assessing what is most probable. First we have to consider what might happen if we do not intervene, then we have to consider what result we desire. One thus takes account of the difference between the probable and the desirable, and devises the best strategy to reduce the difference. (I will not go into the techniques that are used.)

The vital point as I see it is that the different methods of forecasting meet with almost constant failure. Need we recall what seems to have been the most serious effort, the most rigorous, the one which uses so many facts and parameters, the famous report of the Club of Rome with the help of MIT? Ten years later its forecasts had all been successively demolished. The authors themselves had to admit that their work offered no guarantees. Need we recall the no less serious errors in the forecasts of Hermann Kahn? Or those of Fourastié, who after the war wrote a little work predicting the civilization of 1960, no part of which has proved true? Or the fact that economists failed to predict the great economic phenomena between 1950 and 1980 (the impact of oil, the collapse of growth in U.S. productivity)? One might

6. E. Morin, *Pour sortir du XXe siècle* (Nathan, 1981).

extend to every economic and technical domain the detailed studies that have been made concerning forecasts about the price of oil. In 1974, 1977, 1980, and 1983, all the predictions of all the experts proved false.[7] The experts were regularly mistaken about the economy and indeed in every technical field, according to Robert Gibrat.[8] This is a cruel fact. One might analyze the sources of the errors, but the fact remains. We have only to reread economic and technical forecasts twenty years later.

I must devote a special note to the evaluation of prediction in A. Bressand and C. Distler, for these authors offer a remarkable analysis.[9] They regard prediction as essential, and most of their work is a prediction of what the world will be like in 2000. They rightly note that there have been periods in forecasting: prediction (and preaching) in the 1930s; then after 1960, with Daniel Bell, an attempt at more scientific forecasting that sets aside mere fantasy and personal judgment. According to Bell, research into economic growth makes forecasting necessary in every society. We cannot foretell the future but we can describe alternative futures that depend on choices, along with the consequences of the choices. This is the epoch of P. Massé, G. Berger, and the Hudson Institute Report (1967). But none of this work really forecasted what actually took place between 1972 and 1980. The interesting point is that the two authors do not regard it as of any significance that the forecasts failed to predict the rise in prices, the oil crisis, or the problem of natural resources. After all, foreseeing the energy crisis in 1967 would not have changed anything! The strategies set to work in the 1960s were strategies of growth, not of managing scarcity. Hence foreseeing the crisis would not have influenced those who made the decisions, especially coming only from individuals or small groups. The authors argue, however, that the works were justifiable and necessary, for they gave prominence to the idea of the strategic hour. The concepts used in forecasting are dependent on the problems, themes, and questions of a given period (which, according to them is a decade or so). Moreover, the 1960 forecasts are closer to what has been happening after 1980 than to the situation in 1970! The stress on technological possibilities rather than on constraints, and the attention paid to the great powers (USA, Japan, USSR, etc.) rather than

7. See Arthur Andersen and Cambridge Energy Assoc., "The Future of Oil Prices: The Perils of Prophecy," Le Monde, Feb. 1985.

8. R. Gibrat, "La prévision à long term est-elle possible?" Le Jaune et le Rouge (1980).

9. A. Bressand and C. Distler, Le Prochain Monde (Paris: Seuil, 1986).

to North-South relations, bring the forecasts of 1960 into line with those of 1980.

Here are two important points. First, in such scientific forecasting errors of fact are supposedly unimportant; trends are what count. Second, the North-South relation is of no significance (according to Bressand and Distler). The forecasts of H. Kahn *(The Year 2000)* are regarded as technically noteworthy (though only in relation to the field of computers and not as regards surgery, space, new materials, etc.). But according to the authors' own criteria Kahn's work is viewed as weak in economic forecasting since it deals only with productivity, tables of growth, etc., and fails to identify such new variables as the cost of money, debt, etc. Yet the two things are inseparable. The only point at which I agree with the authors is that Kahn fails to see that the information explosion is not merely in the cultural domain but fully in the economic as well. Finally, the authors stress that Kahn thinks the value of economic efficiency will be reduced, when in fact our age is more devoted to efficiency than ever. We thus see the errors of one of the most famous futurologists as they are pointed out by new futurologists. But the new ones are also much deceived if they think that the technical system has ever obeyed any other imperative than that of efficiency.

In other words, I can conclude without misgivings that on the one hand forecasting is more than ever necessary in our world but also that economic and technical forecasts in fact are always inaccurate. This leads us into the twofold question of our failure to foresee and to predict.

2. Our Failure to Foresee and Relative Unpredictability

There is a lack of foresight when one might foresee something but does not. There is unpredictability when in spite of every effort future events are obscure and one cannot give the probable course of their development. This happens often. We fail to foresee because there are so many things that we have to foresee. The lack of foresight may sometimes be individual and due to emotion or passion. For example, we now know that the speed of automobiles is the cause of fatal accidents, yet drivers who are sure of themselves and intoxicated with power drive much too fast and cause terrible crashes. Here is a complete lack of foresight. It is all the more serious when we find it in big business (e.g., polluting factories, though perhaps often we should not speak of lack of foresight, since accidents and pollution are perfectly foreseen).

Finally, public agencies have a very serious lack of foresight when they devise inefficient plans, fail to fund them properly, provide inadequate materials, and do not fully consider the methods to be used. Some of the resultant accidents are almost inconceivable: forest fires, the running aground of oil tankers, nuclear power accidents, disastrous floods, etc. How are we to respond to all this? The point is that most of the problems could actually have been foreseen. Experienced sailors told us that tankers of over 350,000 tons present great difficulties in steering, stopping, and changing course, but in spite of that, tankers of 500,000 tons were built. The shipping industry also lacked foresight when they equipped very large ships with automatic pilot systems and put their full confidence in them. Here we run into a problem which applies especially in the computer field. The functioning and results of machines have to be controlled by specialists. One cannot put blind confidence even in fourth-generation computers. I might argue that whenever such confidence is placed in a device or a machine, there is a lack of foresight.

One instance of lack of foresight demands particular notice, however, because it is not local like the others but global. The more industrial equipment is produced by techniques with a scientific character, the more rigid the economic system becomes and the harder it is to correct if it is moving in the wrong direction. In particular, market mechanisms are slowed down by technical inertia (and the economic system thus becomes more vulnerable). This is why there are so many (discreetly concealed) failures in what seem to be correct computer programs. People have often failed to include in the calculation of objectives a recognition of all the actual factors present that have already irreversibly programmed the future, the length of time they will last, and the probability of events that are still unpredictable.[10] The more we advance our equipment, the more the social and economic system is subject to inertia, to inevitable stickiness. It is an irreparable mistake not to foresee this.

More serious, of course, is relative unpredictability, that is, the occurrence of things whose existence and date one cannot reasonably foresee. Everything is functioning normally but we know that an accident is likely to happen. We can then take one of two courses. On the one hand we can be always on the alert and take drastic and costly measures to prevent an accident (which of course may never occur).

10. See Giarini and Loubergé, *La Civilisation technicienne à la dérive* (Dunod, 1979). See also Simon Charbonneau's excellent studies of the management of risks.

Or we can let things be, thinking that accidents are rare and usually not too serious. We can talk of unpredictability in a certain state of knowledge, corresponding to a larger uncertainty. We know that statistically a certain number of traffic accidents will occur in this or that country. But what can we do about it? We speed along, not knowing the when, how, where, etc. One might say that amid this fairly general uncertainty the Western world has tried to control, not uncertainty, but the specific risks to humanity (Giarini).

The most striking example of unpredictability due to uncertainty may be found in the area of nuclear energy.[11] In the French program we find an excess of production of electricity, an extreme rigidity (well studied by Granstedt in *L'Impasse industrielle*), a lack of certainty about the real effects, an inability to arrange for long-term disposal of wastes, a gamble on future operations when a plant reaches the end of its useful life and the core has to be shut down, and a failure to foresee the possibility of triggering some irreparable processes. All this, in France, goes hand in hand with a lack of respect for the laws (e.g., regarding the sale of enriched uranium) or for the many nonproliferation treaties, which are violated as soon as they are signed.[12]

Chernobyl is an example of unpredictability along with uncertainty. This accident brings to light the general uncertainty about nuclear power that seems most typical. It was not just a matter of the USSR being slow to pass on information or its tardy circulation in France. Such things were only secondarily to blame. The essential points are the ignorance of experts and the uncertainty of the public. This is where the unpredictability resides. Here is a basic uncertainty which puts the human race in a far worse situation than any seasonal disasters or famines. Nor was this accident by far the most serious that might be imagined. We can understand the reasonable arguments of scientists. Here is a single accident among hundreds of nuclear power

11. See F. David, "Le retraitement des combustibles nucléaires . . . ," *La Recherche* 111 (May 1980). When supergenerators were initiated, the price of electricity was supposed to drop by 1990 (Giscard seeing an outpouring of energy equivalent to that of Saudi Arabia), though both the left and the right agreed that the main benefits would not appear until 2200. In 1984 Fabius signed an agreement with Italy, Germany, etc., for Super-Phoenix, which *"theoretically* would free production from reliance on uranium supplies" (Louis Puiseaux, formerly the person in charge of forecasts for Électricité de France).

12. Cf. B. Goldschmidt, *The Atomic Complex* (American Nuclear Society, 1982). Excess production was finally recognized (with bad grace) in September 1987, when the president of Électricité de France admitted the need to slow down (industry would not permit stopping) and not to press ahead with a second supergenerator (*Le Monde*, Sept. 20, 1987).

stations in the world that are functioning with no problems. It happened seven years after the accident at Three Mile Island. One accident every seven years is a tolerable risk. And how many victims were there? The Western press announced that two thousand had died, though the USSR at first said two, finally seven. The figures have undergone later adjustment and no one knows for sure. But two thousand would be no more than the number killed every two months in traffic accidents in France. This is not a disaster out of all proportion. Even if seventy-five or eighty thousand people had been killed, this is no more than might be killed in a violent flood or cyclone. We have thus no reason to regard that nuclear power station as the Devil. Nor must we rush into extreme measures or give way to panic as in Germany—a panic exploited if not provoked by the Greens. All necessary measures have been taken to protect against radiation.

If there is no panic, however, there is anxiety in spite of all the explanations and reassurances on radio and television and in the press. I think that this anxiety, which might engender a mad panic, is present, perhaps unconsciously or in hidden form, in every inhabitant of Europe, and its main source is uncertainty.[13]

We do not know what is dangerous. We see clearly that the experts do not know either. This uncertainty is the disconcerting factor. A cloud, pushed by the winds, covers most of France. No one is permitted to divulge its presence. It seems to be as innocent as other clouds. It then withdraws. But what exactly does it carry? The public by and large knows that there are four kinds of radiation and radioactive products. Some are terrible but last only a few hours. Others are less harmful but can last for hundreds of thousands of years. What is in the cloud? When do we think it will go? Is there going to be a long game of hide and seek?

We can also state that this accident has brought to light something that ecologists have known for a long time, namely, that scientists are in disagreement. Most biologists think that some radioactive fallout can cause cancer in those exposed to it within ten years. Others argue that infants exposed to the bombing of Hiroshima in 1945 have had no significant increase in death due to leukemia or other cancers. The same applies to babies born to pregnant women who were exposed to radiation at that time. They have not had more cancers than others, and

13. According to a poll by the Institut française d'opinion publique on October 23, 1986, 93 percent of the French people think that they have not been well informed about the problems caused by nuclear power stations and radioactivity.

they are normal. Unfortunately, laboratory experiments on mice do not back up this argument. And naturally the long-term results of radiation are totally unknown.

There is uncertainty again, and contradiction among experts, about the maximum doses of radiation that are bearable without danger. I have heard discussions among German, French, and Italian experts; none of them agrees with another. The Germans are much stricter, claiming that doses 30 percent weaker than those that the French regard as dangerous are harmful.

A little fact that is not recalled is that thirty years ago the French, too, regarded as dangerous much weaker doses of about the same strength as those suspected by the Germans. But the limit was suddenly raised when the decision was made to go ahead with nuclear power. The information disseminated by experts leaves us perplexed. Ten days after the accident at Chernobyl there was talk of a point 150 times above the mean but an alert would be sounded only for 500 times above!

Another uncertainty was that radioactivity had always been measured in millicuries or rems but the reference now was to B litres (B: Becquerel). What was the relation between the scales? Intellectuals did not know, nor, of course, the public. The intention, it seems, was to reassure at all costs. This is understandable. But in the debate about the threshold that we must not pass it is very clear that the French experts are tolerant. The reason is that France has the greatest number of nuclear power stations. Scientific objectivity has nothing whatever to do with the different measurements. They depend on politico-economic factors.

Another uncertainty lasted for many weeks. Was there a meltdown? a complete loss of control of the chain reaction which would lead to such high temperatures as to force the core deep into the earth with all kinds of pollution to water tables? Some said yes, others no, and most specialists said nothing, since they did not know. In any case, there was uncertainty in the information, which the government came to recognize was slow and watered down. At first we were given the absolute assurance that the cloud had not reached France, and then ten days later we learned that four-fifths of the country was covered by it. We were also uncertain what to do in case of radiation. Argument raged as to whether we ought not to drink milk or to eat salads or river fish.

These were matters for general political decision, but individuals had no idea at all what to do finally to protect themselves. After all, it made no difference. Providence in the form of the state and science would look after us. But is this providence as sure as that which one attributes to God? Three months ago I read a report by the director of a nuclear power station in the USA which came to the confident conclusion that another accident like Three Mile Island is totally impossible. It is also considered that a more serious incident with an effect equal to that of one or several nuclear bombs (I realize that a power station is not to be compared to an atomic or hydrogen bomb) is to be ruled out. But is it totally impossible? No one can say. No matter which way they turn in this field, citizens come up against such differences, gaps, and possibilities that they live in complete uncertainty, the uncertainty of being delivered up bound to a kind of destiny that is beyond their control and that might be triggered at any time. It is this uncertainty which causes panic, and, as I see it, it is likely to continue for many years.

3. Conditions that Make Forecasting Impossible

An optimistic note about the future is often sounded,[14] but I would draw different conclusions on the basis of several conditions that make forecasting difficult. There is the difficulty of getting information. We are overwhelmed by writings and by data from data banks, but it is hard and costly to get the relevant information (I will return to this matter at more length). There is the difficulty of projecting oneself into a different situation, whether in space (what does a reporter understand when he arrives in a distant country?) or in time (how does one project oneself into the year 2000? The well-informed work of Ducrocq is an amusing example of the limits in this regard).[15] There is the difficulty of understanding: Science and techniques develop so fast and become so complex that they are beyond the competence of futurologists who honestly recognize their limitations. There is the difficulty of imagining scenarios, for we are constantly coming up against negative probabilities and we cannot take refuge in a dramatic and inexorable science fiction.

Even what seem to be the clearest developments must be treated with caution. Interestingly enough, great uncertainty obtains even in the field of computer science. Scardigli has said that given their cost, and the overturning of mentality and habits that they presuppose, the new techniques in computer science and telecommunications will

14. See, e.g., *La Prospective à la sortie du désert* (Société internationale des Conseillers de synthèse, 1985).

15. Albert Ducrocq, *1985-2000, le Futur aujourd'hui* (Plon, 1984).

develop much more slowly than their backers think; by the end of the century they will hardly rival the automobile.[16] We will see in effect a kind of state terrorism forcing people to go into computer science.

I will begin with the impossibility of being well informed in the present state of affairs. But how can we make a forecast unless we are well informed about the situation, the present situation? Everyone agrees that accurate information is not possible. This was the great scandal at the time of the book by Morgenstern,[17] who pointed out how haphazard are even the best statistics. But the more we advance, the more we have to admit that the data on which we try to establish exact knowledge of the present are fragile. Morin rightly stresses that the progress of the media in setting up an extraordinary network of information has helped to promote disinformation and ignorance.[18]

I have studied elsewhere disinformation through excess of information,[19] but in this flood we have to take into account the flattening out of all information. It is materially impossible to pick out in the flow what is important and what will change overnight. It is impossible to discern what bears on the future and will fairly certainly affect its development and what is dramatic but has no real significance (e.g., walking on the moon). Morin also shows that the powers that are threatened by the power of information have no recourse but to change it into an instrument of obfuscation. We shall return to this point. For the moment, however, we have in mind only deliberate deception. According to Morin, "Stalinist history in the USSR is authenticated by doctored photographs which have left out the faces of the old Bolsheviks whom Stalin condemned. Nothing is more misleading than documentaries on China, Siberia, or Cuba with a camera as witness [I might refer also to irrefutable television reports on South Africa in 1985 and 1986]. The powers systematically practice pseudo-information. The spread of lying in this field is the answer to the potential spread of truth through the rise of the media. Lying has spread because the media permit the spread of truth."

It is hard to know what is real today because of systematic doctoring everywhere. But even more serious is the impossibility of knowing what is real, due to an incompatibility of criteria.[20] Thus it is

16. V. Scardigli, *La Consommation, culture du quotidien* (Paris: PUF, 1983).

17. O. Morgenstern, *On the Accuracy of Economic Observations*, 2nd ed. (Princeton: Princeton University Press, repr. 1972).

18. E. Morin, *Pour sortir du XXe siècle*.

19. J. Ellul, "L'Information aliénante," *Économie et Humanisme* 192 (March-April 1970): 43-52.

20. See F. Partant, *Que la crise s'aggrave*.

totally absurd to reckon the income of a peasant, especially a Third World peasant, in U.S. dollars. How can we evaluate the standard of living of one living in southwest France in this way when we do not take into account the abundant picking of mushrooms (about 20 pounds in one outing), or the gathering of crayfish, or hunting (real hunting and not organized massacres)? Peasants might not have many skills, but I know firsthand that they can live well with little money in the form of income. And what about Third World peasants? F. Partant has rightly noted that a theoretical study has yet to be done on the concept of wealth in our societies. When we consider the patrimony of peasants which has been built up in a significant way to assure the growth of agricultural production, we are surprised to find that it is no higher in relative value than half a century ago. Different crops are grown, but the value is no higher. Evidently, southwest France may have more corn to sell, but hardly any mushrooms, fish, or wood-pigeons.

It is thus impossible to get accurate information in matters of this kind. And once a crisis comes in a country it is impossible to get any serious information at all. I have shown this in the case of the war in Lebanon, and J. L. Seurin has shown it in the case of New Caledonia.[21] It is necessary and easy to show it in the case of South Africa. All information is given a bias by the warring parties and interests. A good example is the passionate hostility of Le Monde to Israel and South Africa. I might say that all of us (including politicians, specialists, and experts) are poorly informed about the world as a whole. We have to know all the facts (political, economic, social) to be able not merely to know but to evaluate a society: the positive or negative results of modern techniques; the probabilities of technical evolution; the insertion of techniques into the society. How can we predict if the basic, primary facts are either missing or contradictory?

In writing this section I have had in view the remarkable analysis of information in economic practice by Ingmar Granstedt.[22] Some call for the disseminating of information, an end to secrecy, access to all the facts. Others rationalize communication systems and computers. But the problem lies elsewhere. With the endless growth of facts that have to be taken into account in all economic activities, something has burst. The integration of information has moved beyond the limits at which it is reliable. As the contexts engendered by our instruments of

21. See J. Ellul, Un chrétien pour Israël (Monaco: Rocher, 1986); J.-C. Seurin and Couteau Beguarie, L'Antipode de la démocratie (Lieu commun, 1986).
22. Ingmar Granstedt, L'Impasse industrielle (Paris: Seuil, 1980).

power have increased beyond measure, we continue to put the innumerable variables in the hands of ever more exalted officials. But synthesis is beyond us. What the best-performing means of communications dispatch and treat and stock is no longer faithful to the reality of data that are constantly expanding. The integration of information is not possible. To the size and growing complexity of the economico-technical world we inevitably respond with differentiation, with an increased fragmentation of the information necessary to each laboratory, office, or firm. With the massive preponderance of the mode of integrated production, the contexts become enormous. It is no longer possible to track directly all the variables. External "observers" try to do this, their task being to pass on the information obtained. Information networks receive, synthesize, and filter what is learned. The breakdown occurs at the center where the fragmentary data ought to converge. A programmed reconstitution of the contexts is impossible because our mental faculties of assimilation and communication are limited.

The data that information networks make accessible for the guidance of a business are far beyond what can be assimilated. The integration of information has been pushed beyond the breaking point. We can extend it indefinitely but it will no longer function. It is no longer reliable. Ignorance becomes chronic. We see it clearly when we read even the best journals. Crushed by information, those in charge realize that they are constantly underinformed. At the extreme this might mean the exclusion of people. Computers will talk to computers, for they alone can take everything in. But then decision making would also pass into the hands of computers. We have not yet reached that point. But a number of economic decisions are taken without a proper knowledge of the matter. What Mr. Chirac says is much the same as what Mr. Fabius says and he in turn reminds us of Mr. Barre. They are all using the same data. In the process of decision the essential data are not known and the basic variables are not perceived (even though they may be found somewhere in the enormous network). The "integration of information" means that we now know only the data relating to our own area. No human brain or committee can master the whole. In other words, the chronic ignorance of those who make decisions in every field is not due to the absence of data or to lack of access, but to the disproportion between our limited mental capacities and the unlimited complexes that we think we can daily assimilate.

In the preceding remarks I have been thinking especially of the economic and political realm. I have not been referring directly to technique. Technicians, of course, are specialists, and (with increasing difficulty) they can master the information in their own fields of microspecialization. But the problem here is not that of a single technique but of technique as a whole. Unpredictability in technical matters depends to a great extent on unpredictability in economics. Strictly one might estimate (like Ducrocq) that in ten years we will have this or that gadget or innovation or means of transport or system of communication, but are they all economically possible? At this point our ignorance is total. We recall that technique has three phases: invention, innovation, and diffusion. The first two depend on effectively applying prior techniques.[23] The first also depends on contingent political directions and decisions. The third depends strictly on the economy. Will there be venture capital? Will there be a market? Since in this regard we are in a totally unpredictable situation, there is necessarily unpredictability in technical development. When the industrial use of nuclear fission coincided with the early stages of the computer, no one could say which would win the market or become the dominant force and shape society. The latter won, but no one could foresee this.

4. Absolute Unpredictability

I will not be referring here to the unpredictability of scientific invention, which is always possible. The discovery of penicillin or the silicon chip was not foreseeable, but we must always leave the field open in techniques. We shall see the importance of this later (allowing that an unforeseeable invention can often be blocked, or that there is no one who tries to apply it, so that it becomes the victim of a blackout). In the preceding section we have dealt with the unpredictability that is due to excess of information. We must now have a look at absolute unpredictability[24] if we are to cover all the parameters. When the report of the Club of Rome came out, in spite of great efforts to collect data and the scientific apparatus, it soon became evident that because of faulty analysis the authors had forgotten what are to my mind certain essential parameters of the technical process. It seems that on the one hand the complication of the calculations became insurmountable if the whole was to be kept in view, and on the other hand that the refinement of analysis has limits, and here we cannot trust computers to do the work.

23. See J. Ellul, *Technological System*.
24. This must obviously be differentiated from accidental unpredictability. I have in view an irremediable unpredictability inherent in the technical system.

What is more, the longer the term of the forecast, the more account has to be taken of a great number of parameters of which some are not yet known and have not yet even become visible. Moreover, since each of the parameters evolves on its own for its own reasons, combination is doubly impossible.

What happens in fact when we try to give a long-term forecast (and history has now speeded up, so that the long-term of a century ago is no longer long-term: a forecast for the year 2000 is now long-term!) is simply that we reduce the number of parameters so as to be able to combine and normalize them, other things being equal. All long-term calculations relate to precise points that need not be combined with others. In 1995 France will have the largest and most modern airport in the world. That is an example. I would say that the shorter the term of forecasting, the more it can be global; the longer the term, the more it has to be restricted. Furthermore, if we want to take into account the negative effects of an innovation, we are tempted, in long-range forecasting, to neglect some of the effects, whether through a belief that technique will annul them, or through the fact that, since they will not become apparent for twenty or thirty years, we do not regard them as important.

Truly to understand absolute unpredictability, we have to look at three decisive factors. E. Morin has excellently analyzed the first of these. It is not enough to have a correct view of the present, for this contains as yet invisible microscopic germs which will develop. Furthermore, it is obvious that innovations, inventions, and creations cannot even be imagined before they appear.[25] We can imagine only the consequences of actual inventions. Innovations not only modify given factors as they enter into new and unforeseeable combinations; they also cause the very principles of evolution to evolve. At first an innovation is always a deviation. If it succeeds it can become a tendency. It can then form a new norm of evolution. This seems to me to be vital in technique.

Technical (or as Morin would say, techno-economic) thinking sees the world in terms of power (we shall come back to this), of rates of growth, of gross national product, of speed, of consumption—that is, inevitably in terms of secondary phenomena. It is radically incapable of thinking about technique itself. It is thus incapable of

25. I have never been able to understand the enthusiasm over the last fifty years for Jules Verne as a pioneer! He was no more a pioneer than Cyrano was for going to the moon or Icarus for flying. All the machines that he imagined are completely absurd, and much less practicable than those of Leonardo da Vinci.

dealing with the root of any problem in the modern world, since all these problems are at root technical. In all the circumstances and conditions of the modern world the root evil is of a technical order (I do not say it is provoked by technique). Technical thinking is incapable of thinking about technique. This is the real problem of the so-called technical culture that we shall have to study at length. Technical thinking thinks only in terms of the progress of techniques. It cannot think about the general phenomenon of technique. If we document the malfunctions or negative effects of technique, it is incapable of making any real response. Technical solutions bring with them the very evils they are supposed to remedy or produce worse ones in another area. Along similar lines, technical thinking is incapable of foreseeing anything new. It can foresee only the extension and perfecting of what already exists. It cannot think in terms of a new paradigm, an unpredicted event, a true invention, a social revolution. It is shut up in its own limited logic. I would add that this inability is present not merely when there is a change of scale or domain. We find it in its own domain as well. Forecasts are fantastic once we leave the laboratory where experiments are made of which this or that result can be expected.[26]

When there is a need to forecast (even at the technical level) what is likely to happen in the next fifteen years, the predictions are contradictory, for in the order of global technique we do not know exactly what we are capable of doing. Uncertainty reigns not merely regarding the dangers of nuclear power but also regarding space, genetic engineering, and computers. The advice given by experts in these areas is contradictory. There is no real knowledge of what is possible, what is probable, and what is merely desired. The impossibility of knowing exactly what we can do is very plain in the political game that utilizes technical potential. The great debate about star wars is significant in this regard. No one knows precisely what this is, or whether it is technically or economically feasible. I will discuss this further in Part Two below.

Finally, there is a third factor of unpredictability that D. Janicaud has brought to light.[27] We are now in a complex of power, in its absolute novelty and vertiginous unpredictability. The novelty may be noted in the irreversible character of phenomena. Many points of no return have been passed. The unpredictability is due to the incommensurable

---

26. Technical forecasts are grotesque in public journals; cf. *Match*, Oct. 2, 1987.

27. D. Janicaud, *La Puissance du rationnel* (Paris: Gallimard, 1985).

growth in the risks that are taken and the specificity of the new phase of potentialization that has been initiated.

A powerful idea as I see it is that when a system achieves a certain degree of power, forecasting becomes impossible (as we see in the case of nuclear power and its problems). In his study of power Janicaud rightly emphasizes that we cannot control the exercise of this power, since its degree of rationality is problematic. Never before has there been calculation so precise, so incomparably exhaustive. Power itself is incessantly adjusted, computed, and reassessed, but precisely in these circumstances (Janicaud cites Castoriadis) no rational calculation exists that can show that a temporal horizon of five years is more or less rational than one of a hundred years. Decision has to be made on a basis different from the economic or technical basis. The basis is an incalculable postulate of development. Power carried to an extreme and seen in terms of indefinite development is no longer quantitative. It is qualitative, for it becomes a real quality. The gigantic (which we shall find in techno-economic relations) becomes as such incalculable.

To conclude our discussion of unforeseeable aspects, I will also note the uncertainty which results from an important change in the order of economic and sociological thinking. There has been introduced the flux and vision of a fluid world. We have been accustomed to viewing the world as an object made up of fixed objects and quantities. In this world forecasting was possible. But this conception has been overturned by the discovery that our world and society are not made up of objects but of a flow, of currents, changes, and combinations. We no longer have to study money but the flow of money. This is so obvious that we might ask why it was not seen at once. Probably people refused to see it because of the complexity that it entails. Everything is moving and changing and taking on new aspects according to currents that we have to discover and mark. But obviously if we put technique in this varied flow it is almost impossible to follow its shifting course. For technique is not uniform. As we have seen in thinking about the four main modern techniques, it is diverse even as it uniformly pushes on to absolute power.

This analysis of absolute unpredictability leads us ineluctably to lateral but essential reflection on our relation to time in terms of techniques. Lewis Mumford showed already in the 1930s that it was this relation above all that early techniques changed, and that it was from an accurate and widespread measure of time that the whole technical system could be developed (the clock and the watch). Today the basic problem is the same but with much more acute transformations. I am following here the notable study of J. Chesneaux with its new insights on the subject.[28] Chesneaux starts with the diffusion of the computer and data processing in our society. The computer is a machine which compresses the time needed for development, production, and management. It reduces time to smaller and smaller units. It thus gives the present an absolute primacy over the past and the future. It is the main indicator of social evolution, which takes its rhythm from "generations of computers" (a vital observation that we need to explore).

The primacy of the computer makes forecasting not only uncertain but futile. In a modern technical society what we have to do is to integrate time, that is, integrate the past and future into the present, which alone is real. A nuclear power station is a model of the integration of time as well as of centralized social control. It cannot stop at will. It cannot vary its rhythm in producing energy. It produces energy even when energy is not needed. Its clients have to adapt. Factories must work round the clock. At this point we leave the domain of prediction and arrive at a more global view of the relation of technique to time. Technique does not predict; it programs time. Everything must be ordered according to a single temporal axis, that of the organization, functioning, and production of technique. This does not allow for contradiction or for dialectical evolution, and it seems to me to be significant that at the very time of this transformation of all society, philosophy and scientific theory rushed into thinking in "loops" and "vortexes" and "gulfs," that is, the very opposite of the actual reality that had been constructed.

Technical programming is infinitely broader than planning, for in its totality it includes even living elements. General synchronization results (Granstedt). Real time, in which the computer now functions, is a time looped in advance and made instantaneous. One must eliminate dead time, compress deadlines, increase the pace. We have moved beyond our obsession with speed to demand the instantaneous. This is no longer the present, for it is gone even before we are aware of the present. There is no longer any delay or tarrying. The ideal is that of the instantaneousness of the computer, which shortens the time of execution by apportioning out the tasks; or of the digital watch, which shows only numbers and eliminates the element of space denoted by the movement of the hands; or of television, which unlike the cinema constantly presents us with new pictures; or of fast food. "The instantaneous achieves hegemony to the point of almost literally dissolving

28. J. Chesneaux, *De la modernité* (Maspero, 1983), chap. 2.

the natural time accumulated over the centuries. A forest becomes a daily paper, which is almost immediately thrown away."[29]

Naturally, this forms part of the basic unpredictability as the time of observation is compressed. All our indicators are falsified by the brevity of monthly or weekly reporting unemployment rates, prices, trade, public opinion polls on international political decisions or the popularity of a politician. We need to realize that all these numbers and percentages mean absolutely nothing.

From instantaneous use (objects immediately ready for use and then discarded) Chesneaux also draws another conclusion that seems to me to be decisive if we are to understand the change in our society in relation to time. There has taken place a radical inversion between the time of use and the time of elimination. Up to the 19th century objects were made to last. They were kept as long as possible. When they were no longer of use, being degradable, they could be disposed of without problem. Today we have reversed the situation. Each machine or appliance becomes obsolete in a year, as a much more efficient one replaces it. But the vast number of discarded machines produces a buildup of scrap, which takes a long time to eliminate. Increasingly, indeed, we are making products that cannot be eliminated. This inversion seems to me to be very significant. It shows how technical rhythms are unable to fit in with the rhythms that are natural to us, to the world, and to its possibility of a future. We are now accustomed to recalling that there cannot be infinite growth in a finite world. This applies to space, but we have to realize that it applies to time as well. Fanatics for progress think there will always be more. There is talk of exponential progress. Chesneaux shows that when we look at the real curve of the evolution, at the host of technical inventions over the last twenty years, both great and small, the curve is "superexponential." There has been infinite growth in finite time, and this "will pitilessly flatten out the vertical asymptote." I believe that the complex transformation in the relation to time, which is not yet fully felt, will produce psychological disorders which will contribute to the disintegration of society.

5. Foresight

What I have just said leads on to another point. If forecasting is scientifically necessary but no less scientifically impossible as regards both specific events and larger trends, and if this fact is related fun-

29. Ibid.

damentally to the change in our relation to time, we have to abandon any illusions we may have about our grip upon the future. Instead we must appeal to a different quality, that of foresight, which functions precisely when there can be no sure prediction. In the 1930s it was thought that a society could be set up which would not only foresee the future but also prepare for any accidents that might occur (insurance companies in the 19th century spoke plainly of taking thought for the future). We had to be protected and secured at all points. There thus came into being life insurance, social security, unemployment insurance, and pensions. But this system, which relates primarily to accidents in private life, can hardly deal with collective disasters; note the extreme difficulties of compensation for oil spills or for disasters like Bhopal. Insurance companies, indeed, have to have growing reserves to handle the risks of terrorism, the taking of hostages, and air crashes. The risks that we have to deal with are different from those in the 19th century. We should also note that some Christians argued at that time that they ought to express their faith by refusing to take such precautions and insure themselves. That was the noble age of the *Devoir d'imprévoyance* of Isabelle Rivière. But if Christians have constantly to affirm and assume and live out this duty of improvidence in virtue of their faith, they must not impose it upon responsible politicians or all other citizens. It is within this general framework that I would say that since foreseeing is definitely impossible, we now need to manifest foresight.

But what is that? We must begin by stating that in the case of serious accidents, whether natural or artificial, it is never possible to find an adequate response, whether economic or technical. Foresight comes into play when we recognize that we have now created a civilization of risk, according to the fine formula and demonstration of P. Lagadec.[30] Lagadec's theory of "major technological risk" seems to me to be irrefutable, even though understandably it has been poorly received. His analysis of social attitudes in face of this risk is to my mind exemplary. Industry has to produce at all costs, no matter what serious risks it creates. The state wants to protect productive activity and refuses to alarm the populace, stating in each case that the situation is not really serious. The people are ignorant and impotent and finally accept little-known risks as the price they have to pay for the pleasures that the technical society hands out to them. As for experts, they are always on the margin of the world of risk. They study each case and regularly conclude that there is no problem, or that it is a chance

30. P. Lagadec, *La Civilisation du risque* (Paris: Seuil, 1981).

incident. No one will accept the idea that technique has placed us effectively within a ring of volcanoes. At most there is interest only in the risk of war or military armaments, since here we are on familiar ground; we have lived through war, and arms today are infinitely more deadly. As for the rest, we prefer to ignore it, not taking up the very difficult task of securing information when we meet with secrecy on every hand: the secrecy of laboratories, of administrators, of experts, of technicians, of politicians, etc.

We have seen that forecasting is practically impossible, and information is barred to us by secrecy. In the first instance, then, foresight is the virtue of accepting, on the basis of incontestable experiences in the immediate past (at least two in the last twelve years),[31] that we are living in a civilization which is at serious risk as a result of technique, and that the more technique advances the more serious the risk and the greater its probability. It will naturally be objected that with hundreds of nuclear powers stations in action an accident every five years is negligible, especially when there are few victims. But more powerful instruments push us toward a risk that might be final. In another work Lagadec has shown that we cannot go by probability alone in face of these disasters.[32] Scientific rigor relativizes what is usually said about the safety of high-risk installations on the basis of past results. The question of possibility eclipses that of probability. Furthermore, statistical reasoning loses its value when a single serious incident can upset the curve, as is always possible.

Foresight should then develop attitudes and institutions and instruction based on the constant possibility of a serious accident. It can be objected that the worst might not happen. This commonsense slogan is no longer valid. What we ought to say is that the worst has become much more probable. Foresight demands that we now take the step of regarding the worst as probable, not by a calculation of probabilities to which circumstances no longer correspond, but because high risks are accumulating. To give only one example, we all know that nuclear power stations are built to function on average for thirty years. Many of them are now reaching that age. What are we going to do with them? We cannot simply erase them. Are we going to build concrete mounds to isolate them and then build new ones alongside them? It is the function of foresight to consider all essential facts so as

31. In addition to Chernobyl and Three Mile Island, there would be many more if we count forest fires due to technique or technical blunders.
32. P. Lagadec, "Faire face aux risques technologiques," *La Recherche* 105 (Nov. 1979).

to have a response to the worst scenario in each case. We have to weigh what is probably the worst in each situation and then find a solution for it.

This is the foresight which ought to replace forecasting if we have the least sense of responsibility. Along this line, the only serious one, we ought to do the very opposite of what we find in most books on technique. The worst thing that we can do is declare that all is well or act as though there were no risk. I have already criticized the work by Simon, but I might equally well take to task that of A. Ducrocq *(1985-2000, le Futur aujourd'hui),* in which we find hundreds of pages on the marvels of technique and the glorious future that is before us, but only a page or two on star wars, only one page on nuclear weapons (which will happily lead to the disappearance of states!) and nothing at all on pollution, or on the growth of armaments, or on chemical dangers. Nothing! Ducrocq simply tells us that there will be some difficult transitional years as we move toward the ideal society that technique permits, but he says nothing whatever about the enormity of the risks. His books are a public menace, for they lull his readers to sleep and prevent them from achieving the indispensable foresight which is our only chance of survival.

CHAPTER III

Double Feedback

In *The Technological System* I emphasized that if technique in its complexity were finally to be regarded as a true system in the scientific sense (according to Ludwig von Bertalanffy), it would have to have a regulatory mechanism, a mechanism of feedback which would react to the source or origin or cause of malfunctions, deviations, and negative effects and correct them, thus conferring order and balance in growth. Thus far the system had in fact functioned in a completely anarchical and spontaneous way; no one could control or direct it. But it already seemed to me that there was a possibility of creating coherent feedback thanks to computers. In effect computers enable us to record all the snags and malfunctions and perverse effects in every area, to trace them back to their source, and thus to eliminate, correct, or bypass them with some degree of certainty. The decision to do so, of course, is still a political one. Computer feedback on the technical system can only inform; decision rests with us. The computer has the task of showing us where the problem lies and what to do about it, in this way preparing the ground for decision.

To arrive at this stage, however, there needed to be awareness that technique does constitute a widespread system which affects political circles, so that they realize that it is the key problem of our society rather than armaments or production. It was also necessary that computer scientists should orient their researches to this enormous (and not very profitable) work of research and should be able to subjugate technique to the general interest. This was the most urgent task in this field. But speed was necessary in view of the rapidity of development in every direction. There was a need to think about mastering the technical system, and the computer was the means. I used to think that it would be possible to get a hold on technical growth so as to be able to direct it during the 1970s. It did not happen, of course. In spite of the determined proclamations by intellectuals, in spite of the intentions of the socialists to direct the economy, no one saw the need for an operation of this kind. The socialists were immersed in the big problems of the 19th century (nationalization, the battle against large fortunes). As for computer scientists, they do not have any larger vision. They do not see the greater potential of their tools. They thus continued to manipulate their instruments so as to offer various services, from the reserving of theater seats from the home to the guidance of rockets. Instead of mastering the technical system, computers entered into the system, adopted its features, and simply reinforced the power and incoherence of its effects. In fact, I think that the game is lost. With the help of computer power, the technical system has definitively escaped from control by the human will.

Once again the power of events has prevailed over free human decision. But if my utopian dream of feedback control has not come to pass, this does not mean that the technical system is without feedback. Feedback has come into it spontaneously, but in a very different way. The true problem is that over the last few years we have witnessed the formation of a double feedback. This familiar mechanism is that among the effects of an applied force there appears automatically one that reacts on the origin of the force and changes either the force or its orientation. This is a retroactive control, or self-regulation. But feedback of this kind can have two consequences. There can be positive feedback, that is, acting positively on the force in question at its origin, and consequently reinforcing the action and its effects. There can also be negative feedback (and we usually think of this), that is, the force is moderated, the effects are kept at a certain level, and indefinite development or acceleration is checked. This feedback is a regulator. Now when we consider the technical system, we note that spontaneously, by the very pressure of things, two types of feedback have appeared, a negative and a positive. The one tends to check the acceleration of technique in every direction, the other tends to increase it. These two reactions are completely beyond our human will or human control, although naturally we are their agents. Here is a major difference from what I used to think. We are no longer masters of the speed and direction of the system. We are no more than intermediaries. As I study the creation of this double feedback in these chapters on uncertainty, it is obviously impossible to know precisely what is the real effect of the two contradictory actions in combination. Positive feedback arises out of the relation between politics and technique and science and technique. Negative feedback arises out of the relation between the economy and technique. In all three cases the process is much the same. Technique has consequences for politics, science, and the economy, and because of this they, in turn, react upon technique.

1. Positive Feedback

I will first study the relation to politics. We begin with the effects of technique on politics. Clearly, technique provides politicians with extraordinary means to achieve their projects, whether they are of the right or the left. Technique is an extraordinary means of unification. Even in liberal or federated states the trend toward unification is inevitable because it makes things easier. It is much easier to govern a unified whole than a collective in which people speak different languages, work in different ways, use different currencies, and receive autonomous instruction. Such things are an almost insurmountable obstacle to government. Technique, however, necessarily overcomes divisions. The networks of railroads and power lines mean unification. Yet technique brings to the state the twofold possibility of concentration and centralization on the one hand and deconcentration and decentralization on the other. The one superimposes itself on the other. Whether the organization is viewed as centralized or decentralized finally comes to the same thing if unified techniques of management, communication, and information are used. Widespread and unified communications make it possible for the central power to know at once all that is happening within a nation and thus to exercise control even though the administration is deconcentrated and power is decentralized. We do not have here an evil will but a process of simplification, of unification, of facilitation, which seems to be inherent in the human creature (at least in the West).

Technical means also improve the means of control and accelerate the process. And what political power can be exercised without means of control? The media are essential both in order to know all that is taking place within the nation and also to control the public. In this field technique is of great assistance to the state. It seems to make forecasting possible (though I have proved the contrary). It thus seems to give us a handle on the future. It shows those who have political power how they may act in new spheres of the social body. Technique gives substantial and increasing aid to political power. In return this power ascribes to technique exorbitant qualities. For years we have been told that we must develop technique in order to solve the crisis. All economic problems come back to technique, thanks to which productivity will increase, unemployment will supposedly disappear, and the deficit in the balance of trade will be overcome, etc. The state finally finds its legitimacy in science and technique.[1] This relation needs to be explored. The legitimacy of power is no longer religious or democratic. Power affirms itself scientifically. Science validates it because it can do nothing without power. For the public, science is the great goddess which it cannot question and which validates those who serve it.

Science and technique thus act upon politics and politics functions as positive feedback. It throws all its weight into the development of technique. For the most part it does not commit the mistakes regarding technical research that I stressed in 1950. In only a few cases (e.g., the decisions of the U.S. government on NASA) does it try to direct research. Instead, it engages in what is at root a full-scale acceleration of scientific and technical research. The state is now convinced that all development depends upon this research. It provides considerable funds to promote it. Research is indeed very expensive and demanding. Much of it in the USA is funded by powerful companies, but the government offsets the lack of funds in neglected fields.[2] In France most research and development is state-funded and there could be no new development without state aid. The state thus accelerates the movement, counting on economic benefits and the strengthening of its own means of control.

The second positive feedback is that of science. For many years a distinction was made between pure and applied science. In 1950 I showed that this is not quite accurate. From the very first science has been dependent on technical possibility. The interaction is very significant. Science can advance only through technical improvements, whether it be a matter of discoveries in space, molecular structures, or the effects of developments in chemistry or mathematics, etc. By 1970 computers were functioning so rapidly that they could meet all "useful" needs in management, in social and economic information, and in the handling of this information. But mathematicians and physicists were posing problems that were so complex and demanded such great calculations that the computers of the day were not adequate. Another generation of more powerful computers was needed that were in fact of use only to these scientists.

All real scientific progress, in biology, chemistry, physics, astro-

1. See M. Barrère, "Les limites du secret scientifique," *La Recherche* 151 (Jan. 1984).
2. The U.S. government spent $250 million for basic research alone in 1985.

physics, microphysics (and, more relatively, medicine), depends solely on technical equipment. This does not mean that simply having the equipment will make progress possible. Science has to throw in all its political and social weight, its prestige, and the support of scientists for technicians to amplify and accelerate technical progress.[3] It is also thanks to the discoveries of science that ultramodern technique is possible. Some seemingly very remote sciences like linguistics make possible the discovery of how to talk directly to a computer. Analysis makes possible the establishment of systems experts, etc. There is mutual support. It is very significant that the launching of space laboratories is justified on the ground that only in this way are certain experiments possible (e.g., physiological observations to see how we might live in space), or the making of new chemical products that cannot be made on earth but are needed in technical application. There is thus a strict circle. Science accelerates technical progress and technical progress reacts by making possible new discoveries. Much technical progress is of value in making scientific development possible. And since science is the final justification of our Western world, money devoted to these technical exploits seems to be legitimately spent.

2. Negative Feedback

Over against this remarkable acceleration, however, there seems to be another force, that of negative feedback, which tends to put serious limits on technical growth. We refer to the economy and finance. Research is increasingly expensive. Various questions arise. Is there certainty that the relation between research and economic development is sure, clear, and unrestricted? This is still the belief in the world of Western politics, but much less so in the USA.[4] Again, from the economic point of view, are research and technique really profitable? The costs are so great that it is hard to meet them and avoid inflation. Many projects have had to become international, like the Concorde. Even the USA has to make choices (e.g., the NASA budget was severely

---

3. See P. Papon, "Pour une prospective de la science," *Recherche et Technologie: les enjeux de l'avenir* (Seghers, 1983); P. Fasella, "Une stratégie européenne pour la recherche," *La Recherche* 150 (Dec. 1983); G. Price, "The Politics of Planning and the Problems of Science Policy," *Science, Technology and Society* 2/5 (1982); D. Collin Gridge, "Decisions on Technology and Policies," ibid., 3/2 (1983).

4. Cf. J.-J. Salomon and G. Schmeder, *Les Enjeux du changement technologique* (Economica, 1984), which deals almost entirely with this problem in the USA.

cut in 1983). But costs continue to mount as techniques are perfected and scientific demands become heavier. The vital fact is that technical progress is much faster than economic growth. For some years the latter has leveled off almost everywhere except in Japan (we shall have to look at this in more detail). But technical growth is exploding in all directions and needs financing. Can one expect, however, that in three or four years the money spent on technical projects will bring the fresh economic growth that will cover the investment made?

In the main the system works as follows. Technique makes economic growth possible. But it demands such enormous funding by the economy that the economy reacts by putting a brake on it through forcing it to make choices.[5] The following facts make this inevitable.

First, technique takes up projects concerning which no one can say whether they will have any economic value. Space is an example. The building of rockets and satellites undoubtedly has benefits for some firms and workers. It might also attract foreign orders. The rocket Ariane can carry a Brazilian or Nigerian satellite. It thus brings profit to France. But it has no economic value, none that consumers can use. At the very most there are some salaries and profits, which will be used up eventually, but this is money made for nothing.

Second, the growth of techniques means diminution for the secondary sector and temporary gain for the tertiary. But the tertiary sector creates no economic value. Services are useful but they do not produce anything. This is constantly forgotten. Here again technical growth produces no real wealth.

Third, technique has external effects and its costs are thus much more complex. The more it progresses, the more it creates global problems, pollution, potential dangers, the exhausting of nonrenewable resources. If we are to find the real costs we have thus to take into account the precautions that have to be taken and the cost of looking for substitute materials. When a technique carries with it great risks of toxification, protection has to be provided and institutions must be set up to exercise control. These are hidden costs. It is the famous question of the internalization of external factors.

Fourth, techniques produce increasingly powerful and expensive armaments. Someone might intervene and say that this is a simple matter for political decision. But that is false, for one must apply here the same argument applied above to space. Making these weapons provides jobs and makes exports possible. It thus helps the balance of

5. We shall discuss this point in detail in our study of economic absurdity; see chapter XI below.

trade. But in this case the negative effect is global. Most of the countries that buy arms are Third World countries, and the purchases plunge them increasingly into debt. (In 1985 Latin America had an external debt of $300 billion. For all Third World countries it amounted to $620 billion. I find it hard to grasp what these figures mean.) Economically, the whole thing is increasingly impossible. In other words, these techniques mean growing difficulties and impasses in economic life. Concretely and within budgetary limits financiers and economists have to say to technicians that they cannot do all that techniques make possible. They have to choose. At one and the same time it is not possible to make very expensive surgery possible for everyone, or to provide every hospital with every form of modern equipment (lasers, scanners, ultrasound, etc.), and yet to supply all the necessary hospital beds. In relation to technique the economy acts as a brake or buffer; that is, it represents negative feedback.

I deal with this problem in the present chapter because positive and negative feedback increase uncertainty. We do not know what is the true impact or effect. We do not know how far state support is restricted economically. We do not know exactly how feedback functions as a whole. No one can hazard a guess. At best we can only be aware that the phenomenon is global and that the effects are uncertain. In any case, the whole technical system today is subject to this feedback, which both complements the system and also tends to derail it.

CHAPTER IV

Internal Contradictions

A final factor that increases uncertainty regarding the future of techniques and their progress consists of the internal contradictions of the technical system and society. The first contradiction arises out of the fact that on the one hand the conflicts which divide multinational concerns, supranational movements (Islam, Communism), and nations are now extremely violent, a violence both expressed and enhanced by the multiplicity of techniques, and yet that on the other hand the violence of the confrontations masks the nullity of the stakes. This may well sound scandalous, but it is the exact truth. People are opposed to one another at every point but only to do the very same thing. The conflicts have no real point; there is simply a passion to lay hold of nothing. We have reached the extreme limit of progress (that of modernity or ultratechnical nature) when the object plays a role by its mere absence, being the fictitious stake of a mimesis (Girard) of appropriation. At the beginning of this chapter we need to understand that it is "the presence of the absence" of a purpose in all the conflicts that renders them the more violent. Terrorism is typical.[1] Studying the point, which would merit a separate book, leads us to three themes: thresholds of reversal, fragility, and compensations.

---

1. On this vital theme cf. P. Dumouchel and J.-P. Dupuy, *L'Enfer des choses* (Paris: Seuil, 1979). We find there the very illuminating formula that if I do not see that I am blind, I am blind. This formula might be applied to the illusion of technical progress and the violence of the conflicts all around us.

1. Thresholds of Reversal

Ivan Illich was the best if not the first of those to emphasize thresholds, and I myself have studied elsewhere the difference between finitude, thresholds, and self-limitation.[2] The threshold of reversal is an essential concept in our age. The first model came a century and a half ago with the law of diminishing returns. This is very simple. When we increase quantities in any field to gain greater results, a time comes when the process reverses itself and a lesser result is achieved than that sought. This law is a decisive one in studying technical progress. We will take some simple examples. The desire fully to rationalize political and economic organization and the desire to rationalize human behavior will always lead to a point of reversal and an explosion of the irrational. The process of rationalization leads to a specific irrationality. "The rational and total organization of the conditions of life produces of itself the arbitrary and irrational rule of organization."[3] What is irrational is the actual desire to rationalize! The most familiar and vivid example is Soviet planning. But we must take note of similar effects in relation to the spread of computers. The introduction of computers into schools and homes might make things more efficient but this turns inevitably into irrationality. Why, for example, save time if the time thus saved is empty and meaningless? The more accurate our clocks are, the less we know the value of time. It is the clocks, I would say, that prevent us from using the time that we have and reflecting upon it.

We thus arrive at a vital global formulation given by Janicaud: "The power of the rational is subject to the law of reversal from the rational to the irrational to which no one has the key and of which no one can finally say what is the meaning." Technical power is the ideal model of this reversal. It is this which in effect associates rationality and power, and in its unlimited growth, which is regarded as valuable in itself, leads to the basic irrationality of destruction, privation, and insoluble disorder. But given the extreme complexity of the technical system, we are unable to foretell what is the threshold of reversal, the more so as it might be an accident that brings it to light. The transition from the rational order to incoherent disorder which will not create

2. See Ivan Illich, *Tools for Conviviality* (New York: Harper & Row, 1973); J. Ellul, *Les Combats de la liberté* (Paris: Centurion, and Geneva: Labor et Fides, 1984), pp. 70ff.; cf. idem, *The Ethics of Freedom*, tr. G. W. Bromiley (Grand Rapids: Eerdmans, 1976), pp. 344ff.

3. Janicaud, *La Puissance du rationnel* (Paris: Gallimard, 1985); B. Charbonneau, *Le Chaos et le Système* (Anthropos, 1973).

future order is totally unpredictable. We cannot formulate or quantify it. Taylorism, which was supposed to promote the happiness of workers (I am not jesting!), is a first example of total reversal. But our system of maximum programming by technique leads to much more serious crises. We need to meditate on Janicaud's formula that reversal becomes the limit that no coherence or relation can remove. It is a constitutive limitation of the power of the rational. It is clear enough that already in our world, production does not keep pace with the raising of capital devoted to it. But we shall have to return to this point. Let us give a few more examples of thresholds.

Beillerot has fully demonstrated this reversal in the excess of pedagogy.[4] This has the aim of making life easier for children, of promoting their psychological, moral, and intellectual development, of transmitting knowledge with less difficulty. But the perfecting of techniques in some sense means total control of the children, which is logical. The more pedagogy becomes scientific and technical, the more efficient it is, and the more it takes complete possession of the pupils. At each stage in their development the children meet specialists, and the same will soon apply to adults, too, as they pursue their education. Pedagogy supposedly has nothing to do with the acquisition of knowledge. It is a psychological and political action, a regulatory action. It is in the service of belief and duty. Teaching and learning are with a view to belief and service. Beillerot fully shows that the technicization of pedagogy, while certainly making it more efficient and better for the general good, also makes of it a perfect instrument of social control. The more information I receive, the more I think I know, and confusion between the information I am given and my knowledge brings me into a "spiral illusion." The more I receive information, the less I think to believe, whereas experience proves the exact opposite. The transmission of knowledge seems to serve the development of social adhesion, of democratic consensus as opposed to conflict. Universalized (and diffused) pedagogy does not really serve democracy but the techno-economy. More extensive transmission of knowledge reinforces the adhesion of everybody to the same norms and values. The more pedagogy becomes technically perfect and institutional, the more it becomes a political weapon of government. It is the generalized pedagogical conformity that is significant. All these theses needed to be proved, and Beillerot has done it. The converse is true: Ignorant people

---

4. Beillerot, *La Société pédagogique* (Paris: PUF, 1982). See also the remarkable book of Jézéquel, et al., *Le Gâchis audiovisuel* (Éditions Ouvrières, 1987), which shows that in television the more noise there is, the emptier it is.

cannot be brought under social control. "Knowledge" is requisite for adhesion (and hence for manipulation and control).

It is not just knowledge, techniques, and science that are not neutral, but also their transmission. A perfected pedagogy is not neutral. Management by knowledge increases social control. A threshold is thus passed. Pedagogy, which ought to liberate, becomes an instrument of blind and accepted servitude.

I will give two other examples of thresholds of reversal. The one concerns freedom. What outbreaks of joy there were when freedom of radio transmission was granted in France! It did not matter what ideas would be broadcast. Groups that had not been able to speak could now be heard. What an advance in liberty! But experience has proved the contrary. First, capital had to be found and the state gave only a little initial support. The independent local radio station, then, depended on a big newspaper or a financial or industrial corporation. But this was not the most important feature. Three kinds of independent local radio stations quickly developed. A first group wanted to adhere to the original ideal of providing relatively fresh local and national news and serving the cause of culture and education by discussions, commentaries, etc. A second group corresponded to public groups whose choices and opinions are well known (e.g., Roman Catholics, Jews, or Socialists), and which had an audience in advance that was ready to support its radio (though it is not really the radio of those who do not speak but of the party or church speaking through its specialists). The third, very large group of stations would essentially transmit only music (rock) and sporting news. The problem that soon appeared was that money is needed to transmit. The only true source of money is advertising. But advertising needs a sufficient audience, and competition developed between groups one and three. Statistics generally show the proportion between them to be one to ten. The battle of advertising thus ran in favor of the rock stations, not the cultural and research stations. The latter were for the most part either forced off the air or compelled to conform to the prevailing model of music and sport.

This, then, is our divine freedom, the great advance made thanks to "independent local radio stations." We find a similar disaster in the case of private television channels: a lowering of the cultural level, a debasement of the public, and an increasing advertising invasion. I will leave on one side the question of "national private radio stations." In the case of true local radio stations, centralized stations of a capitalist type easily carry the day. Freedom brings about concentration. A final relevant factor that has to be taken into account is the power of the transmitter. The more powerful the transmitter, and the larger the area it can cover, the more chance it has of being heard. Once again we find technique and finance triumphing over culture. Long live freedom!

The centralization produced by decentralization is my final example of a foreseeable reversal. I will look only at the technical and not the political aspect of this phenomenon. We know the theses of B. Lussato and his distinction between the macrocomputer ("the large cauldron") and the microcomputer ("the small cauldron").[5] He sees well the problems raised by the former but still seems to have illusions regarding the latter. The important thing is to disclose correctly that the computerization of society, the passage from the computer to telematics, brings with it, whether it is suitable or not, a concentration. But no matter what the qualities he finds for microcomputers, he cannot deny that we find the same results as in the case of independent local radio stations, and that large computers carry a great deal of information that they pass on to one another (correlation analyses). He says rightly that the true distinction does not lie in the different sizes of the machines but in their degree of centralization. Large computers are an advantage in important economic and financial operations, in a society in which there is already a concentration of powers, and in the manufacturing of everyday products. But these are the three decisive elements in our society. We cannot hope that microcomputers will ever become a means of decentralization, personalization, or invention. We have only to see how they are used (e.g., CB or Minitel) to realize that personal autonomy is of no interest whatever to the public. The technical instrument is a game, which allows us to divert ourselves and to make jokes (or frauds).

The microcomputer is not going to lead to freedom but to conformity within the technical system and to smoother acceptance of the system. It will acclimate us to the computerized world. I think that Lussato's distinction between "the large and the small cauldron" is false. The small serves the large. It is the toy which the technical society provides in order to make more acceptable the concentration that is taking place in every area of the computer system.

I prefer the probings and reservations of M. Ader,[6] who points out that the medium of computers remains united even though their usage permits a certain amount of decentralization. We see the effect of decentralization very well in banking, but it goes hand in hand with a national centralization of accounting. The debate is not so much

5. B. Lussato, Le Défi informatique (Fayard, 1981).
6. Martin Ader, Le Choc informatique (Denoel, 1984).

between organization and the computer, as it was a few years ago. The computer can adapt itself to organization but does not condition it. The economy and the efficiency of a system can justify centralization. But centralized instruments can have geographically scattered branches throughout a region, and the microcomputers in these branches can all be linked up in a network, so that centralization advances. Naturally, there can be advantages for individuals. At work a computer can increase efficiency, take care of routine tasks that make few demands, and make it possible to engage in higher types of employment. But Ader shows that there are also negative effects (accelerated pace, loss of competence, greater inflexibility, increase in mistakes). In particular there is a demand for sustained, intense, and exhausting attention that is hard to achieve, and then incontestable physiological ailments (vertebral and visual), and mental problems that are now seen to be connected with the automation of tasks in this field (sleeping disorders, personality disorders, depressive tendencies, etc.). The net result is a significant rate of absenteeism (A. Wisner).

I have given these last illustrations in order to show that if we want to achieve awareness of the internal contradictions in the technical system we have to take into full account the great multiplicity of facts. To put it simply, we might say that the more these instruments bring us together, the more we are distant and strangers. The telephone has done away with the habit of visiting people, of engaging in true human contact; it has ruined the art of writing letters, and sending keepsakes—lovers phone one another instead of writing. We see here a limitation. Technical instruments are very good for technical operations; the phone is excellent for making appointments or passing on brief instructions. But these instruments become demonic when they invade one's whole life and replace such human activities as talking with someone face to face.

Telematics accelerates the process of driving us apart (by the very fact of the rapidity of communication). Communities will finally have to be redesigned so as to make possible uniform networks. Real meetings will become rarer than ever. We will see one another only by way of machines.[7]

Lussato hopes that he can fight against this in favor of the

7. On the positive and negative effects of the computer, cf. Robert Castel, *La Gestion des risques* (Ed. de Minuit, 1981), who points to the danger of collective instrumentalization. The computerization of the management of populations at risk transforms our relation to illnesses and handicaps. It leads globally to a loss of personal identity rather than to a gain in administrative knowledge. Obfuscation results rather than an accentuation of concern.

microcomputer. But all that he proposes (including simplifying structures, restoring autonomy to city districts, opposing bigness in every field, repopulating small towns and villages) will be possible by the microcomputer's opening up the villages; the small, decentralized computer units will allow all this. These proposals are all good, but they do not take into account the global reality of society nor the fact that the microcomputer has entered into the technical system. This is the problem which computer scientists for the most part fail to see. They are convinced that the mere presence of their enormous system combining the macro- and microcomputer will by its very activity (the creation of networks, etc.) create a new society and bring urban, industrial, capitalist, political, and technical organization. This simplistic hypothesis is based on the fact that computer scientists are the first to submit to fascination with their technique. They completely ignore the solidity and rigor of the technical system. They do not realize that their technique will serve this system and even in modifying it will reinforce its distinctive logic.

2. Fragility

A second contradiction in the technical system arises out of its fragility. First we have the fragility of all big organizations. I have already discussed this in an earlier work *(The Technological System)*. The bigger an organization becomes, the more accident prone it is. The links among the different sectors are more numerous, and it is at such links that breaks occur. This is true no matter whether the organization is economic (multinational companies) or political (Napoleon's empire or the USSR). It naturally applies to the technical system, which grows unceasingly and absorbs more and more areas.

Let me deal with an objection to this growth of fragility. Technicians say that the bigger and more complex a system, the more chances it has of finding resources and compensations to make up for an accident. If one organism cannot deal with a matter in a crisis, another one will. To some extent this is true, but suppose that the system is not centralized, that it is complex but not complicated, and that internal communications function perfectly. This might mean that the system can respond well to an accident, but I would reply that in these circumstances the accidents are more numerous and more serious. Experience also shows that big organizations that are centralized are also with a few exceptions complicated and that information circulates in them very badly in spite of computers. Centralization also has its own weaknesses.[8] The more there is, the greater the impact of each incident on the rest as the center spreads it. Individual means of solving problems also fade through collective distribution. Solutions do not come by way of individual disentangling; no one can now find the right response. In a word, we need to denounce what is now called decentralization, the rather feeble lie of pseudo-decentralization in 1982-83, decentralization by the computer or electricity. Branches have to be set up to give an appearance of decentralization.

Julien Bok has excellently explained the trend toward centralization in his study of hierarchy and the principle of least difficulty.[9] He shows here that large groups have a hierarchical level that we must not exceed (eight echelons, he thinks, is the greatest number for maximum efficiency). "If an organization grows and keeps its structure of n hierarchical echelons, at first its power increases N times, but when the maximum is passed, its power decreases drastically. Should the number of echelons become N_c, the power will move toward zero. Internal quarrels will reduce the organization to total impotence." The reaction then will be to harden the constraints and increase the central power (as we well know). But this leads to even greater impotence. "When an organization has the maximum structure and increases its power, the introduction of another hierarchical echelon reduces individual production by half. . . . But even if individual production is reduced, an organization with many echelons and individuals will have greater total power if it corresponds to the optimum structure. Here lies the explanation of the trend toward gigantism. The sense of power is felt by individuals even at the bottom, and they find in it protection and reassurance. Finally, an organization sees its power grow in constant relation to the growth of discipline." We are very far here from the illusion of decentralization and autonomy. This is found where the thresholds of reversal and the fragility of big organizations meet.

How many times have we experienced fragility as a result of the

8. Lussato gives an interesting example of the effect of the over-big in the realm of computers. Around 1975 with the search for bigger and more powerful computers it was found that at a certain threshold static multiplies and causes aberrations. No matter what the combinations, this static cannot be eliminated. It is inherent in the size of the system. Beyond a certain threshold Grosh's law is no longer valid. An over-big system has an internal resonance that makes it uncertain. Once the threshold is crossed, growth becomes vertiginous as the size and complexity of the machine increase. It seems to me that this principle applies to the whole of our technical society. Static does not stop multiplying, and it is so diverse and numerous that neither politicians, economists, nor sociologists can deal with it.

9. J. Bok, "Le principle de moindre difficulté," in Colloque de Cerisy, L'Auto-organisation (Paris: Seuil, 1983).

weather. In France the winter of 1985-86 was not too severe. But railroad traffic was often interrupted. Many people were without electricity because cables snapped due to an accumulation of snow and ice. Icy roads interfered with cars and trucks and resulted in supply difficulties. There was nothing unusual about these problems. Similarly, hot and dry summers empty our reservoirs and cause many individual problems that we need not recount. Another feature of this fragility is that it is not only so common but it is also accelerating. The more perfect the system becomes, the more it has to function at the maximum of yield, speed, and efficiency. Therefore, this is also a factor in fragility (hence Formula 1 cars are much more fragile than standard cars).

To return to the question of weather, a little snow and ice seldom prevented people on horseback or in carts from going on their way! But high-speed trains and powerful automobiles are brought to a halt. If we consider all aspects, we see that minor failings or flaws have major negative consequences (cf. Seveso, the pollution of the Rhône, etc.). If an error was made with a nineteenth-century nonautomatic machine, it caused only a temporary stoppage, but an error in programming modern computers can have incalculable consequences. The speed of our machines makes them much more fragile and magnifies the consequences of the least mistake.

Though it might seem astonishing, the faster the functioning, the slower the response to an unexpected event.

Because of the interpenetration of technical constraints, we lose the possibility of responding to events in time. Materially, adaptations have become too slow. Techno-organizational integration paralyzes economic reactions. This is a relative paralysis in comparison with the speed of the development of environmental factors which we must reckon with if we are to keep our powerful, integrated instruments functioning. To this first result of techno-organizational integration we must add a second that is even more surprising, namely, the acceleration of events and the multiplication of disruptions. The combination of these two simultaneous results is what leads into an impasse. Beyond a certain point in integration the relation between the two reverses itself. Events explode as the mega-machine falls into paralysis. . . . The more numerous and diverse the variables in a technical system, the more events of which it is conscious and the constraints of which it must accept. The greater then is the probability that a variable will develop at any given instant. . . . As variables multiply, the average gap that separates fluctuations tends to diminish. . . . The pace of the events to which a response must be made tends to increase. . . . When a certain number of integrated instruments has come into use globally, a threshold is reached beyond which variables enter into synergy and snowball.[10]

That is Granstedt's remarkably perspicacious theoretical analysis of the fragility of the technical system. But the strange thing is that experience shows how accurate is this analysis of complexity and size. Hypothetically, one can conceive that the technical system might decentralize itself, decomplexify itself, reduce itself to fluid interconnected networks, and hence become both less fragile and less rigid. But experience shows that all technique is made up of more than networks, and that it proliferates in powers, means, and products, so that no simplification or synthesis seems to be possible in fact. For this reason it is all the more fragile. We have to remember this when we evaluate the power of terrorists. (It should be added that we do not have here a form of discouraging pessimism, for the noting of limits and the recognition of fragility are the most necessary forms today of a "positive pessimism.") The growth of technique augments the vulnerability or fragility of the socioeconomic system. This is a factor which reduces the global fruits of technical progress. As Giarini has well seen, vulnerability is the situation in a socioeconomic system whose functioning and survival can be challenged at any moment by accidental events due to human actions (errors, sabotage) or natural phenomena (earthquakes, landslides, etc.).

Naturally, we must not confuse the vulnerability of a system to disasters with the probability of such disasters. But we have to realize that the more a society within which such a system develops is troubled (like our own), and the more extended and complex it is, the more chance there is of disasters. In 1943 it was relatively difficult to blow up pylons for high-tension wires, but now the network is twenty times greater and it is x times easier to do it. Vulnerability goes hand in hand with greater uncertainty and lesser reliability. Insurance companies are right. The risks that are most readily insured are not the very rare ones, which involve large amounts in a restricted market and thus prevent the law of greater numbers from functioning. They are the very probable ones (like automobile accidents), for in these cases the companies know where they are. In the case of nuclear accidents they do not know at all.

A special study is needed of the fragility of computers.[11] With

10. See Granstedt, *L'Impasse industrielle* (Paris: Seuil, 1980), pp. 107-12.
11. Cf. Yves Lasfargues, "De la prime à la panne," *Le Monde*, Sept. 2, 1987.

the speed of production even a short breakdown can be disastrous. At Peugeot one hour's breakdown can mean the loss of 100 vehicles. Computers are reliable enough (only one breakdown per 1,000 hours), but in an average company that uses 200 computers this comes to one breakdown per day. Breakdowns go hand in hand with automation and complexity.

We must also take into account the following fact: The more concentrated a system is, the more it needs specialized subsystems. As we have already seen, the relation between these subsystems is the weakest link in a system, whether this be organized in traditional style or by a computerized network. In a specialized system, a single breakdown (e.g., in a computer) can stop everything. In this analysis of fragility we must remember fact as well as theory. A technical system does not develop in a pure universe. It faces many dangers: nationalization of affiliates, pollution, interruptions of supplies, political blackmail, etc. To reduce the vulnerability of the system there is needed a large measure of agreement upon political and economic organization, a reduction in violent opposition, and a social structure on which the technostructure can rely. Private and corporate interests must not quench devotion to the general interest. "A large and high-energy society is infinitely more vulnerable to social blackmail."[12]

A final aspect of fragility that we need to consider is the more rapid appearance today of new techniques that have to be integrated. This is an important factor in our world. It is simply an application of a rule that came to light in 1950, namely, that anything that technique can do ought to be done. So long as innovations were on a human scale and could be adjusted to existing techniques and structures, this was no great problem. But in recent years the explosion of new techniques has totally upset the techno-industrial as well as the bureacratic and political scene. We have not yet mastered the upheaval. We need to examine the problems that it causes in the economic and financial world. In reality most of the fragility from which we suffer comes from unlimited technical growth, regarding which there is, as we have seen, less and less wisdom in putting the real question of feasibility.

We cannot absorb indefinitely the repeated shocks of overwhelming techniques. The more novel and powerful a technique is, the more it disturbs the world, and the more, in doing so, it contributes to the fragility of the technical system. In response it is not enough to

12. Rufus E. Miles, *Awakening from the American Dream: The Social and Political Limits to Growth* (New York: Universe Books, 1976), a fine analysis of vulnerability.

reduce the dimensions, to reduce the scale, to make domestic robots. The scale does not affect the reduced impact of frequent innovations on the system. Indeed, the age of "small is beautiful" is now past. There is a new orientation to the grandiose, to the network of networks, and finally to a global network (cf. Bressand and Distler).

As I see it, these are the aspects of growing fragility and vulnerability in our technical system and therefore in our society as a whole. One might object, of course, that each time one of these phenomena comes to light, an attempt is made to deal with it, as with negative effects. But we have here a fragility that is hard to discern and a vulnerability that threatens the whole system, so that they are not easily isolated and nullified.

3. Compensations

A final internal contradiction in the system that we need to note, and that will lead us back to a point already studied under ambivalence, is that the technical system increases what one might call countercompensations, and especially scarcity. Dumouchel has made a fine study of this scarcity in the economic sphere.[13] He begins with an analysis of classical thinking about the economy. Scarcity is ambivalent; it may be either positive or negative. It may cause violence, but it is also the basis of trade. The fundamental fact is the parsimony of nature that drives us to progress. If scarcity is both the basis of the economy and the cause of violence, we must distinguish between relative scarcity and extreme scarcity. This is taken up by politicians. Reduction of growth leads to scarcity, which in turn engenders disorder and conflict.

Dumouchel insists, however, that we cannot be satisfied with earlier economic and political analyses of scarcity. Having studied it in primitive societies (following Sahlin), he concludes: "No quantity of goods or resources banishes it, no parsimony of nature defines it. Scarcity is woven into the web of interpersonal relations. The structuring of social space either permits or does not permit the appearance of scarcity. Scarcity is purely social. It exists only in the network of intersubjective exchanges which give rise to it. It is simply a matter of social organization."[14] Having interpreted the phenomena of scarcity in terms of sacrificial mimesis, he rightly shows that it institutes the

13. See Dumouchel, *L'Enfer des choses* (Paris: Seuil, 1979), Part 2.
14. Ibid. Naturally, Dumouchel does not deny an objective scarcity, e.g., due to famine.

modern world as the sacred did the primitive world. Modern scarcity is produced by excess of production, by the universalizing of mimetic rivalries. Social and technical changes today have brought scarcity by destroying the traditional obligations of solidarity and opposing the social consequences of human actions to the individual consequences. Scarcity rests on the renunciation of the obligations of solidarity, on the abandonment of individuals to their own fate. There is a general belief that conflicts and misery are due to scarcity and a general commitment to the pursuit of personal interests. In this kind of world the worst violence is supposedly rational.

Neirynck, too, has offered us a profound study of shortages in a society of abundance.[15] There is a reality of abundance that has ceased to be a chimera; it has become an incarnate myth. But it is not absolute; it is for a minority. (The gross national product for the 450 million people of Africa in 1980 was only half of that for France.) It is explained that the abundance of the developed countries is necessary for the Third World, and Neirynck formulates a new beatitude: "Blessed are the satiated, for we have to feed them in order to nourish the hungry." In other areas he shows that abundance is a trap, that essential services are deficient, compensated by forced overconsumption of useless services.

The inability to organize services in the technical system is not due to chance. This system is based on the excessive using up of resources of free energy. It is known how to increase the productivity of primary and secondary sectors by using more energy. But services do not benefit from an injection of energy—they deteriorate. "Neither education, medicine, nor banking is a true service according to consumer expectations when computer terminals are used and abused in teaching, diagnosing, and informing." Neirynck gives many examples of this shortage, and I think it is essential to regard them as compensations for our enhanced power, efficiency, and abundance.

The above studies lead on to what follows. There is an absolute scarcity (close to famine) and a relative scarcity (by comparison with the rest of the possible consumption). In both cases, however, scarcity is a matter of social organization. The most obvious example—the object both of studies and of inflammatory speeches—is the fact that the technical power which engenders wealth and abundance in the West has Third World poverty as a compensating factor. It would be absurd to fall at this point into a propagandist discourse accusing of

---

15. See Neirynck, *Le Huitième Jour de la Création. Introduction à l'entropologie* (Presses polytechniques romanes, 1986).

the West. Talk of this kind involves moral judgments. It implies that we Westerners are robbers or wicked people. But the phenomenon has nothing whatever to do with moral virtues or vices. It is no individual's fault if technical growth demands more and more raw materials which have to be readily accessible at a good price. By various processes which we shall examine, this leads to the impoverishment of the Third World, the unbalancing of the economies of the various countries in it, and the sociological disintegration of their society, that is, proletarization.[16] At issue is not just the simple fact—often denounced—that we do not pay enough to the Third World for raw materials. Even if we paid full price, this would solve nothing. The example of oil is plain. The incredible wealth of the oil emirs does not prevent proletarization in their countries. It has also grown very clearly in Nigeria.

In reality it is the technical system itself which produces the global proletarization (as well as the poverty) of the Third World. Here, then, is a compensation. Our abundance by way of technique entails this proletarization. (It is ridiculous to say merely that we are robbing Third World countries of their wealth and that this wealth is the basis of ours!) Yet this is not the only form of compensation. We shall have to point briefly to others. Thus we have on the one hand an abundance or superabundance of things, of consumption, of information, of tools, but on the other hand we live in real poverty as concerns land, air, water, and nature. On the one hand we have a superabundance of human contacts (but superficial and artificial contacts), and on the other hand a poverty in relation to animal contacts which, unconsciously felt, explains the love of townspeople for cats and dogs, since it is not possible to think of human beings without animal contacts. In spite of the optimists, we cannot live surrounded only by other human beings, in a desert of concrete and mechanical objects. Naturally, we have to be adaptable; we have to be able to live in different circumstances. But an adaptation of this kind means regression to the most elementary human stage. Scarcity of space is forced upon us whether we like it or not; the more people there are, the less space there is. As regards scarcity of air, all the specialists agree that it is not just city air that is polluted, but that the upper layer of the atmosphere now contains what are becoming dangerous levels of carbon. As regards water, both surface water and ground water, there is growing pollution of all accessible reserves and a progressive reduction of the amount of water available. As regards nature, we have lived thus far in an

16. I studied this ineluctable process in *Changer de révolution* (Paris: Seuil, 1982).

environment that was made for us, but we are now making our own environment without nature. This enormous transformation brings a superabundance for which the compensation is a basic scarcity, the unavoidable result being an economic seesawing, a seesawing of humanity itself, of which we are now seeing the first signs.

In this situation there occurs a kind of transfer, which is also one of the symptoms. Mere objects, machines, no longer count as such, for we cannot live in that kind of world. Barthes and Baudrillart showed long ago that we use symbols, and the system of objects shows that everything is different from what it now is. There is no true symbolizing in the primary sense. What we have is the creation of a fictional world in which our religious sense incarnates itself. Objects like television, computers, bikes, and rockets acquire a fabulous dimension by reason of the sense of their power, their ubiquity, their domination, the unlimited access that they give, their secret, which remains strange to us, and the sacred awe that we experience face to face with nuclear fission. This complex is typically religious. The religious and the sacred that we have chased out of nature are now transferred to objects. Be it noted that the transfer is not quite the same. We originally related our religious feelings to our natural environment. The tree, the fountain, the wind, the animal were the focus. We invested them with a formidable greatness and they became sacred. But the things that compose our human environment now play this role. We ourselves have not changed. We still relate our sense of the sacred to what constitutes our environment. We adore and use with joy and fear that which forms our environment, making sacrifice to it. It is the environment that has changed. But how far we are from the supposed dedivinization of the world! It is simply that the world we now know bears no relation to the human world which up to half a century ago seemed to be eternal.

PART II

DISCOURSE

If we want to uncover the vast enterprise of bluff into which technique has plunged us, it is obvious that we must begin by analyzing the basic elements of the discourse[1] that is common among technocrats, technologists, and technolaters (I will not say simple technicians or scientists). For it is this discourse that incessantly surrounds and envelops us. We are entering into the field of what Janicaud calls "the technicism of technodiscourse":

> Technodiscourse is a discourse which is not strictly technical or autonomous, a parasitic language which is based on technique, which helps to spread it, or which, for lack of anything better, makes any radical retreat, any specific questioning of the contemporary technical phenomenon, nearly impossible. . . . Every technique has its own vocabulary, codes, and listing of events, problems, and operating scenarios. This is not technodiscourse; it is not strictly scientific, philosophical, or poetic. . . . In large part it is a functionalizing of audiovisual language; it is advertising. It involves technocratic "thinking." It is the excitation of computer scientists à la Servan-Schreiber. It is politico-ideologico-audiovisual puffery regarding global competition, productivity, etc. If it abounds, is this not because it has a function within and on behalf of the technical world? It undoubtedly has a social and technical function. We have only to think for a moment what the technical world of the West would be like without advertising. . . . When we consider that society is being technicized, we see that technodiscourse stimulates

---

1. See Langdon Winner, "Myth Information in the High Tech Era," *Science, Technology and Society,* 4/6 (1984). This very judicious article shows to what extent advanced technique has become mythical.

123

and shapes the process. It plays the part of relaying information perfectly and accelerates the process of technicizing the whole planet. . . . It blocks access to an understanding of techno-science. . . . There takes place a work of autosymbolization which tends to recodify all reality in an informational, manipulable glaze.[2]

With these magisterial thoughts of Janicaud as an introduction, we can now try to analyze some specimens of technodiscourse, showing in each case to what extent it is totally fictional.

2. Janicaud, *La Puissance du rationnel*.

CHAPTER V

Humanism

All technodiscourse either is or seeks to be discourse about humanity, about human primacy and objectives. It does not merely seek to assure us of happiness, nor does it discuss power. (There is never any question of power in this pious talk.) Its theme is true human fulfilment, which it rates very highly. Nothing is more important than the human race. The more technocratic the author, the higher the rating of humanity. Julian Simon entitles his book *L'Homme, notre dernière chance* (Humanity, Our Last Chance).[1] This is admirable, but what follows is all about technique in a context designed to reassure us. Michel Poniatovski's study of new technologies has the subtitle *La chance de l'homme*.[2] This is a common theme. We have not yet achieved true humanity. Technique, especially the "new technologies," can now offer us a chance we could never hope for. All technical activity orients us to a greater humanity. It is not just a matter of supplementing but of achieving.

We are far distant here from the supplementing of the soul that Bergson thought to be necessary. The issue now is the realizing of human potentialities that the limits of society, morality, and the body have thus far negated. In the enthusiasm of unlimited discovery we can now overcome these barriers. It is self-evident that technique purely and simply serves humanity, and that the race is now on the verge of miraculous adventures. We are embarking on a new Odyssey. "There where Ulysses faced the wine-dark sea of Homer is a technological,

1. As noted earlier, the French title differs from the English: *The Ultimate Resource*.
2. Michel Poniatovski, *Les Technologies nouvelles: La chance de l'homme* (Plon, 1986).

economic, and strategic ocean which invites us to new and countless explorations," says A. Bressand, for naturally technocrats are no strangers to the humanities: they love to adorn their speeches with phrases from Pascal or Shakespeare. Their discourse not only glorifies the race; it refers to the classical humanities. There is no contradiction, only enlargement and greater profundity.

This humanist discourse submits all things to humanity, the measure of all things (yes indeed) and sovereign over all these marvels. The first man to walk on the moon fulfilled at last what had been the dream of the race from the very first. One might also note that all technical progress is said to correspond to a basic desire of the race from the very first. We always wanted to fly like birds and control fire and plumb the ocean depths, and if anyone throws prudent doubt on such assertions the answer is triumphant: We wanted it because we did it. In all the vast array of techniques exploding on every hand we human beings with our unique nature and genius and sovereign freedom are the ones who wanted it. It is our free will that has produced it all. And if we *were* free, we will now be even more so thanks to these aids. There can be no concessions on this point.

Human freedom supposedly increases with every technical advance. This is perfectly clear and simple. We can do what we could not do before; is that not freedom? For each wish we can now choose among a hundred objects that might meet it; is that not freedom? We have great labor-saving devices; is that not freedom?[3] Is it not freedom to escape the ancient biblical sentence that we must work and earn our bread by the sweat of our brow (Gen. 3:17-19)? We can now move easily and swiftly from one point to another. Is that not freedom? Each year we have new hopes for life. Is that not freedom? I could go on for a long time. No matter where we turn in what is said about technique, we find the same proclamation. And since it is always stated that freedom is our very essence, we ultimately find a certainty that we are now more human than ever before. Furthermore, everything is made for us in this purely technical world. If industrial cities were ugly, fetid, and unhealthy, there is now set before us the dream city: Technopolis. "Respect for the site and the high estimation of nature are clearly advanced in descriptions of the new scientific cities. In this green, technological, radiant matrix the people of tomorrow can be gathered and reborn. . . . Mastering the future by technology, finding a place for people in an

---

3. It is regarded as bad taste to point out that unemployment is the price, and who are more completely free than the unemployed?

environment of machines, and promoting harmonious sites . . . that is the project."[4]

> A new spirit of enterprise is appearing that is the opposite of the spirit of the 1950s, when the model in business and daily life was the "organization man" who was fully integrated into the techno-structure and suburbia. The new culture values risk taking. We have a right to make mistakes. Progress is only by experimentation. . . . Individualism is valued. . . . Individualism and a spirit of initiative are no longer associated with a conservative ideology. Modern-ization becomes a factor of consensus . . . , a loose antibureacratic structure, a spirit of enterprise, teamwork, a constant renewal of creation.[5]

It is surely a dream when a sociologist as informed as A. Touraine discovers that our society has become individualistic again, that the individual has primacy, and that technique and the individual are associated.[6] What more could we want for humanity?

The development of technique is for humanity alone. (We are not to include God, massive structures, or the state.) Everything is oriented to humanity and its happiness. Humanity is the measure of all things, even where there is excess. This is the first time that the philosophical ideal will be realized. Again, the whole person will come to expression with all its potentialities. Nothing is ruled out; there can be no repression. Even what was once regarded as an aberration, like homosexuality, will be accepted.[7] In other words, this technical in-dividualism fulfils humanism and achieves all that theologians and philosophers have advocated and demanded. We must not forget that theologians, on the basis of Genesis, have explained that in ourselves, by an act of the Creator, we have an inventive genius, and that like creation, we are as yet only possibilities posited by God. Creation is not yet perfect or fulfilled. It is we who according to the divine plan are to

---

4. See Yan de Kerorguen in "Technopolis," special number of *Autrement* (1985).

5. Ibid. The only problem is that Silicon Valley has become overpopulated with a very high density of population, and that a sub-proletariat inhabits the examples of technopolis in Japan.

6. See A. Touraine in A. Jacquard, ed., *Les Scientifiques parlent* (Paris: Hachette, 1987).

7. The only embarrassment is when total permissiveness proves to be destructive. If drugs give pleasure, why not use them? Boredom makes us subhuman and will kill us.

perfect it. It is we who are to fulfil possibilities and develop potentialities. We will do so thanks to technique.

If I understand the matter, for 500,000 years there have been no fully human people, only embryos. True human beings have emerged only during the last century. This is in keeping with one aspect of the thinking of Marx, who argued that up to 1880 humanity had had only a prehistory, and that true history would begin with the coming of the Socialist city. On different premises, but with the same result, technique has been regarded as the fulfilment of human nature since the early 18th century, when this nature was based on reason and even confused with it. Supposedly, we are completely rational and we express ourselves necessarily in rational works. Science is not enough, for it is not strictly a work; it is knowledge. We have to go a step further to achieve humanity. We have to master the irrationality of things and impose on them practical reason. It is not by chance that the discoveries of the 18th century were in the first instance technical discoveries before being "raised" to the level of science. At this stage again we must thus glorify the profound unity between humanity and technique. Uninterrupted humanistic discourse! All technique has human well-being as its end, allowing us fully to realize and express ourselves.

A question remains. This century of technique was also the century of the "Rights of Man" that are an integral part of modern humanism. I am always astonished that this expression commands universal assent and everyone finds it perfectly clear and self-evident. The French Revolution spoke of the "rights both of man and of citizen." The rights of citizen, I understand—in any given regime various rights are ascribed to each member of the body politic. For example, there is the right to property. This is clear. Jurists also speak of subjective rights, the rights of creditors, or the rights of parents, or the rights of minors, or the rights of suspects. This, too, is clear. But the "rights of man"? This implies that we have rights by nature. But what is human nature? And what does the word "rights" mean? Until the contrary is proved (which would be hard to do), the word "rights" is a legal term and it can have only a legal sense. This means on the one side that rights can be claimed and that there are sanctions against those who violate rights. But a right also has a very precise sense which it is the task of jurists to define. Each right has only one possible sense. But when we consider what has been included under the "rights of man," we have to ask what is the precise content of the right to happiness, the right to health, the right to life, the right to information, the right to leisure, or the right to education. Such rights have no definite content, and the same might be said about many other striking declarations.

Let us press on. If we are talking about the inherent rights of human nature, of simply being human, how is it that on the one hand we include among such rights the fact that government must not impose upon us an ideology or religion, yet on the other hand one of the rights is the right to participate in the truth, which is univocal? No matter whether this be Communist truth or Muslim truth! There is complete disagreement among people, and to me this seems to indicate that these famous rights are not intrinsic to human *nature*. This leads us to a more comprehensible legal conception. Thus the right to property is recognized by government and granted to those dependent on it. Furthermore, these famous rights are written in a charter. They are not an integral part of human nature. It is just that certain political forces agree to state in concert that they recognize that their citizens have certain rights. On the basis of this agreement the rights are then granted to the people. But it goes without saying that powers that do not grant or recognize these rights have to have a different view of humanity, society, and power, and of the relation among them. One can neither summon them nor force them to change this. I would also like to be told by jurists what sanctions can be taken against those who do not grant these rights. Who can bring to justice the delinquent powers, states, or societies, and before what tribunal? If no one knows how to answer this question, then the expression "Rights of Man" is without real content.

But readers might ask what this digression on rights has to do with technique. The answer is that rights are part of the humanism which is integral to discourse about technique. Furthermore, the idea of human rights appeared at the same time and in the same country as modern technique, and I do not think that there is much that is accidental in history, certainly not here. It was not by chance that human rights were proclaimed in the very place and at the very time that techniques burst out (after horrible wars). The "dignity of man"[8] was asserted at the very time when humanity was acquiring the greatest power. Humanity thus completed itself, and with all tranquillity of spirit the most humanistic of discourse became possible.

But before advancing further, let us digress again, looking at the title of a book that was renowned in its day: *L'Homme, le capital le plus précieux* (Humanity, the Most Valuable Capital). This title fits in well with both humanism and technique. In the whole process human beings are the main (capital) factor. Everything depends on them. They

---

8. This common term, which has to have a moral and perhaps a metaphysical sense, shows to what degree there can be no question of "right" in all this.

are also capital in another sense.[9] People are the only true riches, said Jean Bodin, who thought that the strength of a country lay primarily in the number and quality of its inhabitants. From an industrial standpoint this means that people are apparently the most important factor in production, on which all else depends. They are thus the most precious factor. They have to be protected and used wisely. What they can give must be obtained and not lost. Capitalist exploitation is wrong because it wastes people, making the conditions of life so impossible that they die young. What are we to say about capitalists who squander their capital in this way? To describe people as a resource, as capital, means that they must be treated well, like financial or mechanical capital. People are the most precious capital of all. This thesis is in line with the humanistic discourse of the age.

But what remains of this humanistic discourse, considered from any angle, when we look at the reality of the world since 1900? What we see is deadly exploitation, the armed invasion of the whole world by colonization, two monstrous world wars with millions more dead than ever previously, concentration camps, police states, a mad development of torture, blind terrorism, scores of local wars during the past fifty years, and finally an imbalance of wealth and poverty that makes a joke of the wealth of the nobility in comparison with the misery of their peasants.[10]

In other words, what we actually experience is the very opposite of humanistic discourse. We must apply here the law of interpretation to which I often refer. In a given society, the more people talk about a value or virtue or collective project, the more this is a sign of its absence. They talk about it *because* the reality is the opposite. If there is much talk about liberty, it is because the people do not have it. The darker the reality, the brighter the speech. But readers might ask what technique has to do with the horrors of the century that I listed. Technique is not to blame for concentration camps and acts of terrorism. No, it is

9. Gabriel Dessus showed twenty years ago what this evaluation of people implies in "De l'inéluctable mesure des incommensurables, et de ce qui peut s'ensuivre," *Revue française de recherche operationelle* (1964).

10. I will not speak of totalitarian ideologies since we find these earlier, e.g., in Islam and medieval Christianity. Nor am I saying that previous societies were *better* with their slavery, cannibalism, human sacrifices, etc. It is no use arguing that in earlier days people did not know about such horrors but now we do know about them with our marvellous means of communication, and they make a terrible impression. This is not true. Our marvellous means of communication tell us only what we are meant to know. It took ten years to know about Hitler's concentration camps and twenty years to know about the gulags, though we can now learn at once about police atrocities in South Africa.

not the direct or immediate cause. But it does make possible the broader scale of these disasters and it also leads up to the political decisions. It is because technique is always demanding more and more raw materials and makes possible better methods of control that there are police states. It is because technique provides more efficient means of killing that there are millions of dead. Nor is the making of these means of killing simply a political decision. Military techniques have to be developed in order to improve peacetime techniques. The one is linked to the other. Technique also gives such possibilities of action to political organisms that they are induced to become totalitarian. None of the atrocities of our age would have been possible without technique. Technique is not responsible? No, that is true; people alone are responsible. But look! People have put all their passion and hope and desire into the development of techniques and techniques alone. This is the result. And at all cost we must now try to hide the brutal reality and the link between this reality and technical growth. This is why technocrats engage in the humanistic discourse which is a perfect introduction to the technological bluff.

By way of a final word which sheds light on all that precedes, I may point out that the author of the work with the fine humanistic title: *L'Homme, le capital le plus précieux,* was Stalin, who gave practical expression to his discourse in the gulag.

CHAPTER VI

Is There a Technical Culture?

1. Imperatives and Hesitations

Technique is here. It is everywhere. It is not represented merely by some useful objects. The whole field of computers is a challenge to our thinking. Are we able to think about the technical phenomenon? Are we able to think this "thought" of the computer? Can we continue to nurture culture on the ancient soil, a culture that can be popular (but the people have changed) or lofty (but does it not then lose its grip on reality and take refuge in dreams and ideas)? We have always conceived of what we call culture within and on the basis of our material, social, everyday, concrete world. Technique is here, it is our environment. Can we abstract ourselves from it? That is our problem.

The projects of enthusiasts for technical culture—politicians, scientists, or technocrats, also professors and entrepreneurs—have, along with nuances which we shall note, three fundamental aspects: the acquisition of technological knowledge, the adapting of the young to the technical environment (not just manipulation, but making them comfortable in it), and finally the creation of a psychological mood that is favorable to technique, an openness to everything pertaining to it. These are the general orientations. But we must try to analyze the currents and motivations.

First, among the promoters of technical culture we may distinguish between the minimalists and the maximalists. The minimalists want to preserve all ancient culture with its orientation to art and literature, that is, a classical intellectual education, and add to it a knowledge and use of techniques, bringing them into the existing structure. The maximalists postulate the totality of technique and its demands, which are vast enough and which imply a global questioning

132

and the invention of a new culture on a technical basis and with a technical content which is both its object and its means.

Naturally, the maximalist trend is the most interesting to study. It is also winning the day. The triumph of the microcomputer has contributed to its expansion and the sense of its inevitability. The many tasks that this remarkable machine can perform and the multiplying of information that it makes possible have carried away its creators. After all, culture is simply the transmitting and organizing of information, and since everything is changing in this domain, there must also be a change of culture. A dazzling task!

The interposing of the technical object between us and the world takes on another sense. Whether in the domestic, educational, medical, or commercial field, the value of consumption lies less in the actual cost of what we buy than in that of the progress of knowledge. We are moving into an economy based on the production and circulation of immaterial goods, that is, knowledge and information. The important thing is the

reappropriation of the knowledge of which industrialization robbed us. . . . After many centuries of divorce between highly specialized technique and nonspecialized culture, the technopole now mediates between what some know and what others do. There is thus a return to the technical culture that has marked each stage of human development. . . . In this attempt at reappropriation we benefit from a second-degree technology whose goal is to make access more and more simple to technologies that are more and more complex. . . . These are the facilitating technologies. They restore to us access to knowledge, but they also reestablish contact with other people. These are the interceding technologies. Overcoming space, telecommunications enable us to talk to anyone on earth. Overcoming time, recordings (audio and visual) provide us with a living memory of humanity the past and the future.[1]

According to the above analysis, the twin axis of this technical culture is, first, the use of machines and the access to knowledge, and, second, communication with others. These are the two elements in any culture. But there is more to it than that. What the creation of a new culture finally implies may be learned from the remarkable theory of networks advanced by Bressand and Distler.[2]

1. V. Scardigli, *La Consommation, culture du quotidien* (Paris: PUF, 1983); idem, "Les cités scientifiques," in "Technopolis," special number of *Autrement* (1985).

2. A. Bressand and C. Distler, *Le Prochain Monde* (Paris: Seuil, 1986).

We are beginning to get used to the idea that we can no longer picture the world as made up of fixed things or objects but that we must think of everything, whether in physics, economics, or sociology, as in flux. But the new theory now thinks of things, not in terms of (transitory) flux, but in terms of a new organization of the whole world in the form of networks, which differ totally from the new worlds that ten years ago were imagined according to the model of fixed techniques. The logic of networks is closer to that of language than to that of territory. The distinction is a good one. As a result of the computer, everything now is not flux but a network of information. Every human activity is now set within a network or an intersection of many networks that are combining and will finally become a global network, it is thought, which will be the agent and expression of all possible combinations of networks. We have here a supple and constantly changing force which before our very eyes, without our being aware of it, is structuring a new world. The indispensable thing is to realize what is going on, to learn the new rules of the game, to create a new culture. If we succeed, we will be citizens of this new world. If we fail, we will be its slaves. "The computer plays a role in the production of knowledge in anthropology and sociology, since, by its artistic and generally creative use, it is bound to have an impact on culture, *to obstruct the perceptions* individuals and social groups have of themselves; in brief, to expand profoundly on the imaginary."[3]

But it is not merely the creative capacity in these areas that we have to take into account. We have to understand that there will be a transformation of all human activity. Microcomputers permit everywhere things "made to order" and hence mean the disappearance of products for the "masses" and a transformation of intelligence around the networks. In the economy, for example, we can think of an alliance between material production and immaterial information. This is a summons to the creativity which has to produce a new culture and which is absolutely necessary if we are to have the ability to use all the possibilities of computerization in daily life, at work and in friendly relations. Only a new technical culture will enable us not to be like savages transplanted into the modern world.

We have to translate technical invention into social practices and new modes of thinking which seek to make every relation transparent. This transparency is correlative to an information society and it brings all information and all knowledge within the reach of all (so

3. Pierre Levy, *Vers un 1984 informatique?* (1984); idem, *Les Présents de l'Univers informationnel* (Centre Georges Pompidou, 1985).

long as they have, in effect, a suitable culture). But this culture is not idealistic. It is in the process of making out of simple spectators (like ourselves) those who are competent to use complex machines, and then the "computer freaks," who are able, for example, actually to invent new possibilities and problems for microcomputers.[4]

But the culture which is being created is not a direct human product. It results from the interfacing of the group and the machine. This interface allows individuals to master problems of communication that presuppose the transmission and processing of information. Naturally, this culture cannot remain enclosed in a limited world, in the circle of a province. Transmission and processing take place everywhere with startling speed. A culture of networks is possible only on an international scale. Based on the universality of knowledge, which is accessible to all and in all places, and also on the speed of relations, this culture has to be international. Naturally, the visionaries of this network culture have only scorn for what has thus far been regarded as culture: an intellectual, nonpractical culture, the expression of an elite, an intellectualism of the parlor, a dusty university collection of outdated knowledge, an "ethereal virginity" of the intelligence which leads to the "fantasy of human relations, a relational society," and similar amiable notions. Everything that has thus far been produced in the form of culture must be scrapped. And if some people are obstinate, within ten years they will be just as illiterate as those who could not read or write in the 19th century. The ability to use the computer will replace reading and writing. The illiterates are those who do not adapt (Bressand and Distler).

At a more modest, more concrete level, the first step is the development of a technical culture, beginning with education. We will not stress the fact that "Technology" is becoming a separate discipline in our schools and that professional instruction is emerging from this "ghetto." In primary schools children are being initiated into science and technology, and colleges have systematic instruction in technology (in mechanisms, in automatism, in electronics, in management and administration, embracing the dimensions of conception, manufacture, and use). High schools have detailed instruction in industrial techniques and mechanics. As we foresaw in 1985, computers are being brought into the schools and colleges. Students will have to learn how to use computers, since these are now educational tools, aids to pedagogy. Curriculum restructuring is making technique central. His-

4. See Jean-Louis Gasse, *La Troisième Pomme: micro-informatique et révolution culturelle* (Paris: Hachette, 1985).

tory becomes the history of techniques, and living languages become "modern languages." But more important than curricular change is the educational transformation stressed by Beillerot which involves "an innovative apprenticeship mingling an ability to adapt to the unknown with the transmission of knowledge."[5]

Pedagogy (learning how to learn) is a central feature of this technicizing of instruction, which implies a culture of practical intelligence in place of reflective or critical intelligence. At issue here are social changes that professionalize and rationalize anterior actions. Instruction will now have a practical aim, inculcating the ability to do a job. Normally, then, teaching will be a more or less complex technique and those who teach will be technicians. The main function of (technical) knowledge is to be a major element in production. All branches of knowledge are to be productive. The best and most appropriate pedagogy, then, is the technique that prepares young people to enter the cycle of productive forces but that is also a force in ideological reproduction. It is no use appealing to Bourdieu. The double current of an appeal to the growth of productive forces and an appeal to the necessity of ideological reproduction leads inevitably to educational changes with a view to adapting education to social needs.

Beillerot emphasizes, however, that this raises serious problems. If knowledge goes hand in hand with the social economy, its organized transmission in every area responds to an imperative of the social structure. But we cannot rest content with the original form. What is obvious in our society is the rapid obsolescence of all knowledge.[6] And the knowledge transmitted is only a fraction of what is acquired. It is also impossible that the last to be acquired should be transmitted. Older knowledge has to be transmitted first, so that the knowledge of the schools has little in common with that of science. Finally, Beillerot raises a vital problem when he points out that in the consumption of knowledge there is a struggle to give it a use value so that it may function as an instrument of exchange. Teachers must have an ambiguous practice: at the same time form the mind, give access to a certain culture, and transmit practical knowledge that can quickly be put to use. This might be done in two ways. An effort might be made to bring traditional values into the setting and practice of techniques. This is

---

5. J. Beillerot, *La Société pédagogique* (Paris: PUF, 1982).
6. There are many examples. Half of our medical knowledge is out of date every five years. Ten to twelve percent of techno-scientific publications are out of date each year. It is estimated that human knowledge doubles every five years, and in 1985 as much scientific information was published as between 1550 and 1950.

the great illusion of certain humanists. They think that we can maintain the existing culture and superimpose techniques upon it. They think that we can infuse traditional values into technique.[7] The other way, which is more concrete and efficient, is that of popularizing science among both pupils and the public, who must be brought into technique at all costs. The object of this popularization is at the same time the circulation of knowledge and justification. The public must accept gigantic expenditures on science by being persuaded of its certain usefulness.

Roqueplo has shown that popularized knowledge is not the same as scientific knowledge and that knowledge in the schools (the first to be popularized) is determined by higher education.[8] Giving knowledge a form in which it can be exploited and understood is part of the work of research if this is to justify its funding. But popularization never transmits anything but discourse, and technical discourse is not scientific practice. Scientific knowledge cannot give birth to a culture, for it changes its nature when it transforms itself into discourse and forgets the conditions of its own production. Popularization cannot pretend to be anything other than a socializing of science that leads to a hierarchy of knowledge.

However that may be, there is much popular interest in science and techniques, primarily in medicine, then in the natural sciences, and next in sociology, with the atom, space, and biology running well behind according to a French survey in 1981. We find a good example of popularization in the magazine *Culture technique,* which is trying to reintroduce the technical sphere into culture. A special number under the direction of M. Jocelyn de Noblet (1983) was devoted to the USA. It first described the birth and evolution of the great technical innovations there, with reflection on the technocratic image, the tradition of "know-how." It then noted the changes that technique had brought into American society: business competitiveness, personal computers, combined research. But none of this had much to do with culture. The theme was simply the diffusion of knowledge and the socioeconomic consequences.

Much more important is the observation of a kind of mass technoculture, which J. Chesneaux has fully analyzed.[9] It is character-

7. Cf. the books of Jean Onimus. But J. Cazeneuve in effect raises the same issue in *L'Homme téléspectateur* (Denoel, 1983), when he asks why traditional values are not admitted into shows.

8. P. Roqueplo, *Le Partage du savoir: sciences, culture et vulgarisation* (Paris: Seuil, 1974).

9. J. Chesneaux, *De la modernité* (Maspero, 1983).

ized and diffused in many ways. There is linguistic change, the American myth (high tech, technology, patchwork, network, the look, the feeling, etc.). But it also follows a mechanized model: we program, we decode, we link up, we process messages; we make everything visual; even dance is mechanical; the setting is more important than the text; we are interactive in an interface, we are in a network. Psychological jargon is also prevalent: megalo-, schizo-, maso-, sado-. Relations either structure or destructure. We are always in relation: to money, to the body, to death. We are "situated." These peculiarities of vocabulary are signs of cultural change in which innovation is valued as such. The "culture of modernity," as Chesneaux rightly calls it, is a product of industry with the sanction of the state. It is the creation of a throwaway, consumer society. It reflects the domination of currents and circuits over objects in transnational space thanks to the computer, which with other means of communication is the great agent of this technoculture that functions by segmenting knowledge and by eliminating those data that are not regarded as useful in this very formalized context.

Knowledge is modeled on the stock of information, and its acquisition is dictated by the logic of the computer. This is what students must learn at school if they are not to be illiterate. But in spite of the promise of intercommunication, communication is always to and among people. This technoculture rests on the enormous mass of information, a saturation which makes it impossible to isolate, to assimilate, to establish relations and perspectives, in particular to grasp facts during their life. Information is splintered and ephemeral. The confusion in the use of words aggravates the ambiguity. Information in the true sense is the active investigation of data in terms of individual needs and interests. A mass of data as such, available in bulk, is of no significance, but this jumble is what is called "information" (J. Chesneaux).

These, then, are the main directions in the attempt either to bring technique into culture or to create a technoculture. I would note in summation that techniques have always played a part in culture.[10] But the problem is that size, number, speed of development, omnipresence, and omnicompetence now make it impossible to insert techniques into a stable culture. On the contrary, techniques are now encircling and swallowing up all that has constituted culture from the beginnings of human history. At any rate, we must take into account two basic considerations regarding the relation between culture and technique.

The first has its basis in the position of Gehlen (and of many

10. For a fuller discussion see Ellul, *Technological System.*

theologians of the same period), who wanted to rehabilitate technique, setting it among other cultural fields because it expresses one dimension of being human.[11] Gehlen had an anthropological view of technique. He made humanity and technique coextensive. Human beings as he saw it have been technicians from the very first. Technique is a mirror of humanity. We project ourselves into it and in it extend our nature artificially. The human organism offers the key to technique. This continuity of humanity and technique finds itself backed by an instinctual theory, namely, that we satisfy deep instinctual drives by engaging in it. This obviously means that there is coincidence between technique and culture. But the mistake of Gehlen and many other philosophers is to think of technique "in itself" without considering that the actual technical phenomenon today has nothing whatever in common with the techniques of earlier societies. An abstract ideology of technique bears no relation to knowledge of the present-day world. I have quoted this thesis only because it is the expression of a whole current of thinking.

The other consideration, that of Roqueplo, is much closer to reality.[12] For him technical culture means knowing well the milieu in which we live. Those who do not have this culture do not know their own milieu and are doubly alienated. They have no mastery over their environment, and they are permanently dependent on those who have this "knowledge." Technical culture means having the knowledge and the know-how which can give us personal mastery over our environment and control over the activities of those to whom we have to have recourse. The absence of this culture is a cause of generalized alienation. I will not insist upon the fact that the extension of techniques rules out an "omnitechnological culture" (in surgery, computers, electricity, mechanics, genetic engineering, chemistry, television, etc.). That is a mere fantasy. Much more serious is the fact that the practical technical knowledge to which Roqueplo refers has nothing whatever to do with culture. This raises, then, the central question: What can we call culture?

2. What Can We Call Culture?

Our only solid finding thus far is that all that we have encountered in discourse regarding technical culture really has nothing whatever to

---

11. For a good analysis and criticism cf. D. Janicaud, *La Puissance du rationnel.* Cf. also Geistler, op. cit.

12. P. Roqueplo, *Penser la technique* (Paris: Seuil, 1983).

do with culture (as we have seen, the proof lies in the ironical way in which technocrats speak about the bookish, Jesuit culture of the 19th century). In this context I will not attempt a definition of culture. I will simply offer two models that seem to me to correspond to the historical and sociological reality. The first model is that of E. Morin,[13] who begins by noting that culture plainly oscillates between an absolute sense and a residual sense, between an anthroposociological sense and an ethico-aesthetic sense. In both cases the first sense is closer to Anglo-Saxon thinking, the second to Mediterranean thinking. We must also think of culture as a system that enables us to communicate dialectically both an existential experience and a body of knowledge. Thus culture is neither a superstructure nor an infrastructure but a metabolic circuit that links the two. Morin then offers this essential definition: "Culture is an informational and organizational sphere which ensures and maintains human complexity, both individual and social, over and above the spontaneous complexity to which society would give rise were it deprived of this acquired informational and organizational capital."

The second model is that of an author who is not at all close to Morin, namely, Roland Barthes.[14] Barthes does not think we should define culture as an attempt to acquire more knowledge or even as the maintaining of a spiritual patrimony, but rather, along the lines of Nietzsche, as the unity of artistic style in all the vital manifestations of a people. This demands an increasingly stricter stripping away of all that is accessory and an investigation of the unity of style. We must voluntarily set aside many nuances, peculiarities, and possibilities in order to present the human enigma in its sparse essentials. We must introduce a style where life offers us only confused and disorderly riches; it is the unity of style that defines culture. Culture rests on a tragic sense of life that is forcefully mastered, and this mastery is culture, which does not seek the origin or end of things but the reason for them. This culture is not just that of an elite. It may be the work of artists, writers, and thinkers, but it can be born only if there is in the people a profound culture, a communion of style between life and art. Thus we are not to distinguish between an aristocratic culture and a popular culture. The former is impossible without the latter, and the latter feeds upon the former.

These two models show how far we are from the ridiculous image that the technologists construct of culture and to what degree

13. E. Morin, *Sociologie* (Paris: Fayard, 1984), pp. 109, 345.
14. R. Barthes, "Culture et Tragédie," *Le Monde*, April 4, 1986.

television is not an instrument of culture any more than the microcomputer. These two models are truly at the center of our problem. At the same time one might think that they rule out an essential source of culture, that is, popular culture, artisan culture, maritime culture, peasant culture, etc. For we have to remember that all culture has its source in the profound myths and rites of popular creativity. Yet the criticism would be incorrect, for both models presuppose this popular infrastructure and on this basis engage in a process of abstraction and of explication. Technical and computer activity, however, genuinely excludes all the real creativity and spontaneity of the stable group which is in contact with the harshness of life. We find such culture in embryo among rebellious, marginal, and racial groups, which appear for this reason to be dangerous.

3. A Technical Culture Is Impossible

A technical culture is essentially impossible.[15] To make it possible, as we have seen, technologists reduce it to an accumulation of knowledge. If it were nothing more, technical knowledge could certainly replace literary knowledge. But as Morin finely points out: "The crisis in the humanities lies first of all in the area of knowledge. The predominance of information over knowledge, and of knowledge over thought, has disintegrated true knowledge. The sciences have contributed greatly to this disintegration by extreme specialization. Science can only create an aggregate of operational knowledge. . . . By its relational and relativist character it saps the base of the humanities. . . . In developing objectivity, science develops a permanent duality between the subjective and the objective."[16] This is why developing technological instruction in place of the formation and transmission of culture is such a serious matter. Practical knowledge glues individuals to the concrete with no intellectual capacity apart from an operational one. The massive information transmitted by television falls into a sidereal void of nonculture and therefore has neither place nor meaning. Even if those who watch television retain all that they see, they finally know nothing and understand nothing because there is neither the intellectual means nor the cultural structure to give the information place or relation, and to make it count in the global scales. Beillerot makes this point when

15. The only interesting part of M. Henry's *La Barbarie* deals precisely with this contradiction between culture and technoscience.
16. E. Morin, *Sociologie*, p. 374.

in relation to education he differentiates true knowledge from a mere accumulation of facts: "We cannot reduce the spread of knowledge to increased consumption."[17]

Let us now look at a second aspect of this radical contradiction between technology and culture. We can deal quickly with one feature. What is called culture in our society is totally subordinate to the economic imperative. Children have to be prepared for jobs. This is a culture of utility. Instruction and education play a corresponding part. Culture is also linked today to machines. "The culture of modernity is a product of industry with the sanction of the state." Our cultural products or cultural industries obey increasingly the logic of economics: big runs, standardized manufacturing, lowering of costs, mass consumption, capitalist concentration, futuristic commercial methods, homogeneous products distributed worldwide. This may be seen most plainly in the subjection of television to advertising.

"The game of modernity brings heavy technical materials into play. . . . The avant-garde tries to escape this culture of passive consumption but identifies itself totally with new techniques. . . . Economic constraints are more and more influential. . . . Technical imperatives have priority . . . quantitative thresholds are merciless. [Even a good book will not be published if the editors think it will not sell, as I myself know.] Technical culture is all-powerful even though it is finally subject to the imperatives of profitability, competitiveness, and productivity."[18] I would add to these fine observations of Chesneaux a most judicious remark of Lussato to the effect that bad culture is chasing out good.[19] Everything in the cultural field is becoming the object of information, whether accounting or data processing. Lussato, though he is favorable to the computer, looks at the music or drama transmitted by the media, where in an overwhelming manner the very worst, the most vulgar, the most titillating, the most pornographic, is chasing out a culture of quality. But it is not just a matter of art and the intellectual world. Once the harmony of a city is broken, its inhabitants lose the habit of loving it; they soil and degrade it. But it is not possible to insert an added culture into a degraded environment. The result is "a shapeless mush of a divided culture, piled up in this cultural concentration camp."

To be sure, the subordination of culture to technique and to the economy does not mean that we ought to separate it off and make it

17. Beillerot, *La Société pédagogique.*
18. J. Chesneaux, *De la modernité*, pp. 116ff.
19. Bruno Lussato, *Le Défi informatique* (Paris: Fayard, 1981).

again an idealism detached from the real world. All true culture has integrated the economic life of its time, but it has not depended on it (in spite of Marx's generalization). It has transcended it. For example, harvest festivities relate to an economic event but they give this event a universal character, a higher sense, linking it to a social universe and finding for it an aesthetic form. This is but one of many examples of the relation between economics and culture. But today this is radically excluded, and this is the link with what Roqueplo has shown in terms of impossible discourse. The question that Roqueplo was asking pertained to the prime minister's decision in 1972 to move the accelerator of CERN (the European council of nuclear research) to Geneva. What did he need to know to make this decision when it was a matter of technical investment? "Was the knowledge needed for the decision the same as that deployed in the technology to which the decision related? . . . If various forms of knowledge were involved (economic, political), what language would ensure their convergence? But if there is no common language, how can those concerned talk to one another?"[20]

Now, as Roqueplo shows with various examples, the problem is that the interlocutors do not speak the same language. The prime minister will talk about such things as exports and diversification. The technical field is restricted to the functions that administrators desire for it. There is a technico-technical language, a politico-cultural language, and an economic language. Each involves a "local" knowledge that can communicate with knowledges used locally in its neighborhood. Technicians cannot have any other language or knowledge. There is no communication with a culture, no ability to master technique. Common discourse is impossible.

Whether we like it or not, for technique all language is algebraic.[21] This has come to light plainly with the development of the computer system. As Jousse says, technique's ambition is to make the whole world algebraic. But if algebra becomes the universal language

20. Roqueplo, *Penser la technique.*
21. In *Le Partage du savoir: science, culture, vulgarisation* (Paris: Seuil, 1974), Roqueplo demonstrates clearly that if scientific information is to be truly appropriated, it has to make itself understood in a context in which it is plainly pertinent. The great problem is that there are two languages, an analytical and a functional, an operational and a strategic. Does this twofold structure of language leave any room for a technical culture? What matters is that we should all have a functionally pertinent knowledge in our own fields that permits mastery over the relevant phenomenon. At issue, then, is a "local" knowledge. But how can there be a technical culture in these circumstances? There is only understanding between specialized groups of technicians.

into which all other languages must be translated, then there can be no other communication. Communication between people will be destroyed and no culture can be created, for culture has to rest on the specificity of a language. The technical system implies a universal use which will attach itself, without taking root, to the various cultures and civilizations. Culture cannot be universal, for human beings are not universal. We all have a place, a race, a past, a formation, a specific time. But surely there does not have to be an *annihilation* of all that has thus far been regarded as culture. Not at all! It is just that each culture is made obsolete. It lives on under the technical universal but no longer has any usefulness or meaning. We can still speak our own language. We can read the poets and great writers. But all this is simply an amiable dilettantism. The astonishing thing is that what is fully achieved with all this is the judgment of the nineteenth-century middle class that art, literature, the classics, poetry, etc., are simply a matter of pleasure and distraction and bear no relation to serious matters. They are pleasant whims when important matters have been settled.

But one might say that we are speaking only of older and outdated forms of culture. Technique is merely eliminating these older forms and will create a new culture. That may be, but there are some basic antinomies. We have pointed out already that technique is universal, whereas culture cannot be. We must now go further. Technique means speed, always greater speed. So much so that we now have to admit that there is no longer any uniform time; there is *our* time and there is *machine* time, which counts in billionths of a second and which cannot be correlated with our time.[22] Machines command machines in this time, linking "elementary particles" under human direction of the network. But all this is the very opposite of a culture. A culture gives meaning to time. It corresponds to human speed, which has not changed through all the generations.

We cannot manufacture a culture as we do computers. It builds itself up generation by generation with successive contributions, in successive stages, and by successive, matured, and integrated adaptations. Arising out of everyday life, it presupposes critical reflection on that life, on customs and relations (to the world, to people, to different races, to objects)—reflection that demands distancing oneself from this life in order to appreciate it and give it a cultural form. No matter what means we might employ, these various operations cannot be done fast. But how can we put technique at a distance so as to engage in critical reflection? Technique will have advanced and changed a dozen times

22. See A. Bressand and C. Distler, *Le Prochain Monde* (Paris: Seuil, 1986).

as we do it. It will not be the same. In thirty years philosophical works on it will have neither meaning nor value. And they are no more than books and philosophy. We are still very far from building up a culture.

Once technical operations are in billionths of a second and even the best machines are out of date in a few years, no distance, reflection, or criticism is possible. Technique also excludes distance and blinds us. Piveteau seems correct when he says "that extensive television viewing anesthetizes the reflective action of the conscience and inhibits speech. It makes speech a residual act. . . . It kills the aware and responsible adult and infantilizes."[23] These are not the conditions of the creation of a culture. Correlatively, Morin in an unusually fine statement says, "Ideologues are incapable of conceiving not only of society but also of technique. Scientists are incapable of thinking not only of humanity but also of science. Nullifying these great problems can result only in intellectual nullity."[24]

It is above all technique that nullifies the problems both by its speed and also by its organization in networks. The intellectual and cultural tragedy of the modern world is that we are in a technical milieu that does not allow reflection. We cannot look at the past and consider it. We cannot fix on an object and reflect on it. The technical object encompasses us even though we know nothing about it, and reflection is impossible. We ought to discuss intellectual and mental knowledge. We ought to make it a matter of experimentation. We ought to reflect upon it and incorporate it into the experience of life. But this process is no longer possible. It is ruled out by the very means that technique uses to spread culture, by the speed of information, by the confusion between the image and the reality, by what is called mass culture (which excludes any possibility of reflection), by the impossibility of communicating humanistic knowledge, everyday experience, and techno-scientific knowledge.

All culture presupposes coherence between reality and knowledge. But reality is determined today by various technical imperatives, and the knowledge accumulated in data banks is not accessible to me. "We have the spread of a new type of ignorance amid an accumulation of knowledge" (E. Morin). Furthermore, all culture is of necessity the culture of a group, of society. It implies a group dimension and interrelation between group members. But technique combats this on two levels. On the one hand everyday technique tends to increase loneli-

---

23. Piveteau, L'Extase de la télévision (INSEP, 1984).

24. E. Morin, Pour sortir du XXe siècle (Nathan, 1981); cf. idem, Sociologie (Paris: Fayard, 1984).

ness (through the means of communication) and makes it unnecessary to establish contacts. People drive alone in automobiles. The telephone only simulates meetings. Television isolates. The keyboard enables us to give orders without personal contact. We have a collection of individuals without interaction (except in terms of technique or by means of it). This fact rules out any possibility of culture.[25] The other side of the same phenomenon is the discovery that the new techniques form themselves into a network that can be self-sufficient and exclude us altogether. Networks are abstract, invisible, imperceptible. They impose themselves on real life and condition it. They evolve with a speed that is beyond us. They thus eliminate any possibility of culture, since any culture that might be set up cannot express human life or stability.

The network is a significant feature of the new world of technique. But the individual network is only one part of a larger network that links all networks and that is even more abstract and inaccessible. There can thus be no culture outside the network. But we have to have a very superficial and childish view of culture if we think that the images, the information, and the exchanges that the networks authorize have anything whatever to do with a culture. The very idea of a fluid, all-encompassing network rules out humanity's dominant position. Human beings are simply within the network. As one network combines with others, their reality reduces both human subjectivity and human independence. We can only rely on the networks, which even as they increase our power reduce our independence, since without them we can literally no longer do anything. Basically, the attempt to create a counterculture or subculture in the 1960s, an attempt which finally failed when it came up against the structures of our society, is no longer conceivable to the extent that technique functions in networks.[26]

25. See C. Lipovetsky, *L'Ère du Vide. Essai sur l'individualisme contemporain* (Paris: Gallimard, 1977).

26. One might add that a technical culture would have to arise out of an existing culture; it cannot come into being *ex nihilo*. In Europe the problem is that there no longer is any European culture. During the last fifty years art and literature have been destroyed (cf. Ellul, *L'Empire du non-sens* [Paris: PUF, 1980]). Instruction has replaced apprenticeship (which united the values of life and knowledge). There has been a univeralizing of cultures which has fragmented them; cf. especially the Americanizing of Europe. (I am not hostile to America but only to invasion by its worst elements.) The habits, customs, implicit languages, and rituals which made up European culture, along with art and literature, have been overthrown. Today the great concept of a multiracial France, with the Islamic invasion, is destroying the coherence of French culture.

Finally, we must follow Mirabail when in his study of telematics he shows that it is endangering all Western culture by generalizing what used to be the concern only of specific economic or social sectors.

> Telematics is systematizing the reorganization of knowledge, over-throwing ideas and methods of work, and bringing to an end the break between two worlds, but prefiguring dimly the existence of the second. *Nothing can really be said* about this second world [my italics]. . . . We run the risk of vague thinking that is incapable of defining either the meaning or the nature of the new social-cultural objects. . . . Culture becomes documentation. . . . Everything is potentially stored in the memories of computers. . . . If thought is in disorder today, it is because telematics contributes to a reorganization of everything that can be expressed, of all discourse that has meaning. It is revolutionizing language. . . . For what society or people? Why symbolic data processing? The obfuscation of questions is all the easier now that the system in place is a signifying system which by way of response points to the *objects* signified which the *new services* are. But nothing demonstrates that the intrusion of symbolic languages, of systems of signs which computerized systems are, really describes a space or a time for us, a world of meaning which is our symbolic history, even if we imagine the accurate covering of anthropological questions by technique, and the field of symbolic constellations of the imaginary by videotex on terminal screens.[27]

I have referred at length to Mirabail because no one could better describe the conflict between culture and technique and the impossibility of a technical culture.

Culture is necessarily humanistic or it does not exist at all. It is humanistic in the sense that humanity is its central theme and sole preoccupation. It is simply an expression of the human. It has human beings (and not what serves them) at its heart. This includes, of course, all that they put in the form of questions about the meaning of life, the possibility of reunion with ultimate being, the attempt to overcome human finitude, and all other questions that they have to ask and handle. But technique cannot deal with such things. It functions merely because it functions. It is self-reproductive. Each technical advance serves first to produce new techniques. It is itself the center of attention and allows of no questioning outside the mechanical sphere. It is not interested in what serves humanity. Its only interest is in itself. It is

27. Mirabail, *Les Cinquante mots clés de la télématique* (Privat, 1981).

self-justified and self-satisfying. It cannot occupy itself with the human
except to subordinate it and to subject it to the demands of its own
functioning. Culture exists only if it raises the question of meaning and
values. In the last analysis one might say that this is the central object
of all culture. But here we are at the opposite pole from all technique.
Technique is not at all concerned about the meaning of life, and it
rejects any relation to values. It cannot accept any value judgment, good
or bad, about its activities.[28] Its criteria of existence and functioning
are qualitatively different. It cannot give meaning to life nor give insight
into new values. On any approach we have to say that the terms *culture*
and *technology* are radically distinct. There can be no bridge between
them. To associate them is an abuse of meaning. It is nonsense. But this
does not hamper the authors of politico-technological discourse. They
want not only technical efficiency but even more the halo and glory
that centuries of spiritual and intellectual life have fashioned around
the word *culture*.

28. This is why all the studies of medical and scientific ethics are useless
and absurd. I will quote a saying which shuts off all discussion: "What is not
scientific is not ethical." This gem of wisdom comes from Jean Bernard.

CHAPTER VII

Human Mastery over Technique

It is plain that an essential aspect of technological discourse holds to the basic thesis that we enjoy full mastery over technique. Technique is supposedly a passive instrument. Human beings have created it, they know it, they can use it as they wish, they can arrest it or develop it. A computer does only what it is programmed to do (though this is increasingly doubtful), just as a car goes where the driver wants it to go—except in the case of accidents! I am not concerned about Three Mile Island or Chernobyl here, but rather about the multiplication of risks that are beyond our control. There were comparatively very few horse-and-carriage accidents in the 18th century. Today in France alone 1,000 people per month are killed in automobile accidents and 35,000 seriously injured. Whether we like it or not, the risks are greater and the number of accidents is also greater. Their number has become, as it were, part of the structure of society (involving aid, medical organization, pensions, insurance, care for the injured, medications, etc.). Each accident is an instance of "technique," represented by machines, escaping human control. At this simple level, therefore, we can say that human beings do not always control technique. And the more swift and powerful and gigantic technique becomes, the more serious and numerous are the instances when human control is absent.[1]

I will not take up again the question of the neutrality of technique. Since 1950 I have shown that it is not neutral. At that time this thesis was violently contested but it has now become almost a banality. It is generally admitted that science is not neutral, and even more so technique. But this means that we cannot treat technique like a single technical object. To compare a car or a television station to technique

---

1. See Patrick Lagadec, *La Civilisation du risque* (Paris: Seuil, 1981).

is ridiculous. To say, however, that technique is not neutral is not to say that it serves a particular interest (as the left often alleges). It is to say that it has its own weight, its own determinations, its own laws. As a system it evolves by imposing its own logic. But even though it is recognized not to be neutral, many people still continue to talk about our total mastery over it.

In the first place there are innumerable champions of a political mastery over technique. Roqueplo defends this thesis more skillfully than most, believing that politicians give the orders and technicians carry them out!

> The problem is to know whether the city that the technicians are building is viable. Technique has an impact on politics and has to be politically criticized. This is precisely the phenomenon we are beginning to witness. This political criticism of technique goes to the very heart of technique as it is practiced in what are called developed countries [its aggressiveness toward nature, the technocratic Jacobinism of its macroachievements, its artificial sophistication].... The emergence of ideology is supposedly the sign of a transformation affecting *the very reality* of technique, that is, what results from it.... Things that would have taken their own course yesterday no longer do so.... In these conditions there is room for new perspectives. [We can] socio-culturally break the web of facts in which culture has enclosed technique.... *Some technologies* enable us to escape from the impasses to which "autonomous technique" has led us.... The question is whether we are to continue in the same direction or to change course by inventing different technologies.[2]

The argument is, then, that criticism of existing technique is as such a changing of this technique, that today's technique is a product of our culture and a change of culture will mean a change of technique (cf. the preceding chapter), and that it is by technique that we will master technique—a common theme, especially among those who discuss the computer, which, as a second-degree technique, is supposedly an ideal instrument whereby to control technique and give it a new direction. The old question remains: Who is to take care of the caretaker? Now, the above theses belong integrally to a specific view of society. For many authors society is self-creating. It is the mainspring

2. P. Roqueplo, *Penser la technique* (Paris: Seuil, 1983). Cf. Lech Wzacher, "Towards a Democratisation of Technological Choices," *Science, Technology and Society* 1/3 (1981).

of autonomy. Thus science and technique will master themselves. An orderly decision is enough to establish social control. This view is accompanied by an invincible belief in the uncontested supremacy of human beings. When appeal is made to the "force of circumstances," the habitual response is that it is we who create circumstances.

I will not enter into a new discussion of human autonomy and freedom, for in this book I do not want to deal with philosophical questions. I know, however, that in the technical society and world, events and structures are constantly occurring that no one intended. We can obviously reduce technique to its usefulness and use in business, in what helps business function better. Technique then becomes a kind of patrimony that can be managed. Reduced in this way it comes under human control. It is a good tool—no more! One could keep going. But reduction of this type is quite untenable, for it ignores 99 percent of the technical phenomenon.[3]

Let us pursue the question of human mastery over technique. According to Salomon, technology is one social process among others. The technical world and the social world are not two different worlds. Technical change is a model for society and society for technical change. Technical innovation comes from within the economic and social system. It is a human work; supposedly, then, it is not beyond human control except as we allow it to be. A society is defined not so much by the technologies that it can create as by those that it chooses in preference to others. The process of selecting among techniques is economic, political, social. What we do with the fruits of technology depends on our values. There is no inevitability about technical change; neither its pace nor its direction is predetermined. We do not have here a process to which individuals and society have to react by adaptations to technical change. Technology itself is a social process in which individuals and groups always make the choices. This is Salomon's perspective.[4] We shall indeed see that choices have to be made; however, these choices are not an expression of values or freedom but of other pressures which we cannot avoid!

Salomon has a program for the "unfettering of Prometheus." We must spread to everybody the information that makes everybody able to choose. He is happy that there are institutions for the evaluating of techniques and their social and economic consequences. We need to make education uniform so as not to make a division between those

3. This fact renders completely useless an otherwise good book like that of J. Morin, *L'Excellence technologique* (Publi Union, 1985).
4. J.-J. Salomon, *Prométhée empêtré* (Paris: Pergamon, 1981).

who have techno-scientific knowledge and those who do not. We must learn to see that technology is not a preserve of technicians but a social process which like any other is not outside the control of everyone.

Scardigli also discusses freedom of choice and the human ability to control technique. In our world, he says, the individual is the basic social unit. After an excellent exposition of the question, he ends his book on an optimistic note. He thinks that there can be, and already is, an attempt to reappropriate techniques. He, too, is oriented to information and education. We benefit from second-degree technology, which gives us increasingly simple access to increasingly complex technologies. These are facilitating technologies. They give access to knowledge and offer equal contact with others. There exists an everyday technology which makes possible a reappropriation of power, a collective affirmation of diversity, a personal conquest of autonomy. To be able to choose is the result of gaining new access to knowledge. Technique entails a capacity for homogenization but consumers come up with effective responses. New techniques (of communication) must be submitted to users, who will decide on the content. We are always free to disconnect our phones and shut off our television sets. We are always free to take a trail back to nature.

I might cite other texts which reaffirm human freedom. If I have selected those of Salomon and Scardigli, it is precisely because these two in their works have very well described the threats of technique, its power relative to humanity and society, and the rigor of its development. But they are not prepared to yield to despair, and they thus take a leap into a kind of wishful thinking, of metaphysics, of belief. It is not possible, they think, that we should not be masters of the process. Naturally, other writers who do not have the same lucidity engage in wild outbursts of enthusiasm when they look at human sovereignty, at human mastery over this vast apparatus.

Let us consider for a moment how empty are some of the theses. Do we really think that public bodies can master techniques? It is amusing these days (July 1986) to be able to state that a very profound report presented to the government by a commission composed of state councillors and tax inspectors concludes that the French agency for the supervision of energy is quite useless in view of the absence of any real work or results, and that the same applies equally to the center for the study of systems and advanced technologies and the agency for the evaluation of research. In reality these agencies and commissions do not have mastery and critical reflection as their objective but simply the acceleration and growth of technique.

The truth is that attempts of this kind simply follow the line of the great hope of mastering technique in the 1960s and 1970s, of technology assessment.[5] E. Q. Daddario and his committee finally found the reply to the great hope. The program was clear: identify the potential interests of applications of research and technology, specify the means of utilization, discover the secondary and harmful effects before they become unavoidable, and inform the public of these effects so that the necessary measures can be taken against them. These were the four aims; one cannot say more. The principles of action were also good. Errors arise in the application of science, technique, and morality. The power of decision must be shared equally among the citizens. These fine suggestions led to debates, conflicts of influence, discussion of the terms used, and finally the establishment of the Office of Technology Assessment. But amid the debates the project became increasingly less ambitious and more oriented in fact to the growth of research and development. Above all, "technology assessment" became a process of self-justification, an attempt to sway public opinion, much the same as happened in the case of "human" and "public relations" after the war. The system of "public relations" is in fact a technical system which, since coming into existence, has never given evidence of even the least effective control, of even the least reduction in the technical enterprise be- cause of the risks, to which it has never alerted public opinion or the electorate. Once again we see here what ought to be done but what is not done because of the combination of the techno-system, economic interests, and political forces.

In reality, mastering a technique or techniques is becoming increasingly difficult, especially since we remain globally unaware of the problem. An interesting point to note, then, is that there is great concern to master technique when we run up against very common- place moral questions raised by it: bioethics, artificial insemination, in vitro fertilization, etc. In such matters it is worth creating commissions of ethics and control (which do nothing) and holding discussions (like the international colloquy on bioethics in April 1985) which set up norms and points of reference that are no more useful than the human rights charter, since in spite of all the good intentions these techniques are simply part of the totality of the technical system and cannot be controlled unless the whole be controlled.

It is an absolute rule, moreover, that any technique that exists

5. See Derian and Staropoli, *La Technologie incontrôlée?* (Paris: PUF, 1975). For a good bibliography cf. T. T. Liao, "Technology Assessment," *Science, Technology and Society* 2/6 (1982).

will be applied, as Chesneaux finely shows from two standpoints.[6] First, according to the new computerized mechanization: with machines with digital commands, no more slack periods (as Bressand and Distler also observe to the glory of technique), the operators have to obey the machines and their programs. The subordination of people to machines is just as great as before, but with the difference that people are no longer the glorious conductors of the orchestra; the machines are in sole control. Automated machines and industrial computers exert absolute power in real time and at a distance. Their control of real time is instantaneous and global. Furthermore, at a distance the machines control the operators no matter what the situation. Work is desynchronized (variable schedules) and delocalized (terminals at home). The modern machine is a hierarchical entity which precisely conditions both materials and people, disciplining both body and soul. The total work time can be reduced, but the internal pressures are more severe. The more advanced the technique, the more it speeds up production and imposes a restriction on delays.

On the other extreme, workers are losing power as automation destroys the unions. Automation-computerization means an expropriation of knowledge from the working class and therefore a reduction of its power in collective bargaining. Professional training, once the work of a lifetime, is now no more than knowing how to adapt to devices that are temporary and constantly changing. At the same time the system entails an increasing diversification of the labor force. The working class is splitting up into small fragments. The proper functioning of an automated tool might need the input of six distinct categories of workers who are dissociated in time and space and interests, with no possibility of communication among them: those who plan, those who program, those who install (often by subcontract), those who maintain (often by subcontract), those who engage in centralized production, some merely keeping watch over the functioning, others intervening in case of breakdown or malfunction, and finally, those who service and manage the centralized equipment.

In reality, there is no common expertise, in many cases no meeting at the place of work, no collective interest to protect. The whole working class is slipping toward precariousness, fluidity, becoming unskilled, and insecurity in employment. Forms of precarious work are increasing (interim agencies, subcontracting). All this is in the industrial sector of high technology, but there is also a large sector that has not been modernized and is very precarious, suffering from terrible

6. J. Chesneaux, *De la modernité* (Maspero, 1983).

economic disasters (shipyards, textiles, metallurgy). It is to the changing of employment patterns in industry that we must ascribe the unions' loss of membership and influence. The more industry is automated, the greater the loss will be. In France, with few exceptions, the unions did not realize what was happening in the 1970s. They carried on with the same aims and strategy as in the 1930s, and like the political parties they have suffered total dislocation. If I have stressed this point, it is because the unions might have hoped to become the supreme instrument for achieving mastery over technique. Joint worker-management control is the way. But computerization in big business, far from aiding joint worker-management control, has become a major obstacle because of working-class change. The unions no longer have any hope of mastering techniques, let alone redirecting technique.

Furthermore, if we return to Scardigli, how modest is the freedom which consists of choosing our car or taking a trail! (Real though this freedom may be, yet to those who know the real countryside in which there are no ski slopes or roads, how ridiculous it sounds!) As in the preceding chapter, the real debate is this: What do we mean by freedom, and what do we mean by mastery? On a ship it is the captain or skipper who is master. And what is proposed here is no more than the freedom of sailors to put on gloves or not in order to make the maneuvers that are ordered. Lagadec in much more concrete fashion shows what is the desirable direction. We have to recognize the reality of major risks. This entails a general awareness, a policy of prevention, the use only of techniques whose risks have been assessed, and an orientation to new attitudes and practices on the part of those concerned, when it has been shown how impotent industry, state, citizens, and experts really are. We have to take up the challenge. Lagadec is thus much more prudent than all the other authors who deal with technique, but he still thinks that mastery is possible.[7]

I will not proceed to criticize the innumerable writers (Toffler and Servan-Schreiber are popular examples) who exult in human power, for I could only repeat what I have already said, showing that no one can in fact master technique and drive the technical system.[8] Teaching half a million French children to use computers only forces

7. Lagadec has the courage to quote some official texts, such as a 1955 report which says that the most satisfactory solution to the problem of mental health regarding the use of nuclear energy is to raise a new generation which has been taught to accommodate itself to ignorance and uncertainty (La Civilisation du risque, III, p. 49).

8. Cf. Ellul, Technological Society; idem, Technological System.

them into the system and takes from them any critical power or global understanding. I will simply emphasize the incredible contradiction, which ought to leap to our eyes, between the supposed human mastery over technique and the commonly asserted ability of technique to do anything. Thus Castoriadis rightly notes: "the unwitting illusion that technique is virtually omnipotent—an illusion which dominates our age—rests on another idea which is not discussed but concealed: the idea of power."[9]

It is in effect a common basic conviction that technique can fulfil all our desire for power and that it is itself omnipotent. This is certainly an illusion, says Castoriadis, but it is all the more interesting for this reason. Here we have the absolute belief of the modern world which implies our absolute renunciation of mastery; we delegate power to technique. Thanks to it we have achieved an unequalled power. But the greater the power is, the harder it is to master it. A high-powered car is much more difficult to control than a normal car. If this is true of a single machine, it is even more true of a complex system which constantly becomes more powerful, more complex, more complicated, and apparently more rational. Even though the rationality be merely apparent, we renounce the critical use of reason precisely because all reason seems to be concentrated in technique and techniques. "Seeing the world as object or image, we establish the domination of science and technique, an essential configuration of which is the reign of the quantitative. In principle we can represent all this accurately, that is, calculate it."[10]

But beyond a certain degree of power we enter the sphere of the incalculable. "The incalculable is simply the unconditional promotion of power, unceasingly measured and reevaluated."[11] But precisely when we come up against almost absolute power, against the technique which can do unheard-of things, we cannot master it. We feel this unconsciously. We thus experience a terrible fear. We sense that power is closed to basic questioning. If technique can do anything (as we are all convinced), we cannot stand up to it. We are not its master. This obscure feeling explains the enthusiasm, the delirium, the frenzy which has greeted the computer and its many applications. At last we can calculate the incalculable. We can master power with greater power. Humanity has discovered this.

9. Castoriadis, "Développement et rationalité," in Le Mythe du développement par Candido Mendès (Paris: Seuil, 1977).
10. Janicaud, La puissance du rationnel.
11. Ibid.

But one thing is lacking. We have not become the masters of the computer. The means by which to achieve domination is on the side of that which we have to dominate. What all those who think they can master technique lack is a basic understanding that technique is simply power, that no one can master power, and that by its very nature power forbids all questioning and slips away from all attempts to seize it.

Those who want mastery over technique also fail to ask what is meant by mastery. Mastery is an ability to dispose *at will* one's potential. If it does not mean this, it is nothing. It has thus to be a nontechnical power that is greater than technique and that enables us at our pleasure to direct, brake, accelerate, or modulate the technical system. But none of those who arrogantly talk about mastering technique dare think in such terms. All that they put forward are timid propositions for the decentralizing of decisions or the creation of a monitoring commission that can simply regulate one small technical phenomenon (like television, but still only in its programming) and nothing more.

This is what Gabor understands very well when he analyzes the impossibility of mastery, recalling the three main features.[12] First, rational autonomy is less and less referred to those who command machines and corporations and is more and more dependent on the self-regulation of technical networks (as I pointed out above). Second, chronological acceleration does not allow of any controlling body or will or any vision of the final reality of technique (experience showing in effect that any attempt at mastery is always too late). Finally, the great extension of the scale of the social effects of technical progress, along with the diversity, is such that we cannot pass value judgments or appreciate what it would be good to achieve by technical means. For mastery it is necessary to know in what direction we are going, but the polydimensional nature of modern technique is such that an absence of possible orientation rules out all mastery.

As Janicaud notes, there is at the moment a basic reversal. The power of the rational, which was the origin of technique, since rationality made it possible to judge and measure technique, has become the rationality of power. Power (mechanical, quantitative) is at one and the same time both the objective and the justification. We systematically explore the possible. We stake everything upon it as the potential for development increases. Why research? Because it is fashioning tomorrow's power. But will there be a search for power ad infinitum? Janicaud quotes Rescher, who predicts a cessation of serious synthetic

12. Dennis Gabor, *Innovations: Scientific, Technological and Social* (Oxford: Oxford University Press, 1970).

research in favor of analytic discoveries, when power will be halted at its maximum point. But Rescher is making the same mistakes as I made ten years ago.

Power has in fact been halted at a certain point (e.g., an aspect of industrial mechanization), but it is an illusion to think that the "micro," the universe of chips, involves anything but a search for power. Power remains the same. It involves a new subordination of the individual to machines over which there is less and less control. The innumerable mistakes of computers in all branches of life today show how impotent we are face to face with the inerrant network. (I realize that I will be told that the mistakes are human errors, not computer errors; we want the machine to be powerful and infallible!)

Confronted by this reality, we are forced to take refuge in wishes and desires and hopeful programs, as I showed in *Changer de révolution*. As André Gorz wrote, "The inversion of tools is a basic condition for the transforming of society: the development of voluntary cooperation and the weakening of the sovereignty of communities and individuals presuppose the installation of instruments and methods of production which can be used and controlled at the level of the commune, which will generate the economic autonomy of local collectives, which will not destroy the environment, and which are compatible with the power that producers and consumers ought to have over production and products."[13] This is in line with the ideas of Murray Bookchin regarding the introduction of anarchist thinking into the technical system.[14] It is also in line with the excellent earlier analysis of Georges Friedmann in *La Puissance et la Sagesse*, or the even earlier concept of "conviviality" in Ivan Illich's *Tools for Conviviality*.

In conclusion, I will refer to Lagadec's program.[15] Lagadec would banish everything unreasonable (the opposite of the rational!). It is unreasonable to launch out into innovations when we do not know the risks, or to take from citizens their ability to judge. We must find new liberties, producing slight innovations, increasing the possibilities and means of choice. We must make decisions political again[16] and not leave power in the hands of experts. Security functions should be

13. A. Gorz, *Écologie et politique* (Paris: Seuil, 1978).

14. Murray Bookchin, *Post-Scarcity Anarchism* (Palo Alto: Ramparts Press, 1971).

15. Lagadec, *La Civilisation du risque*.

16. Lagadec obviously suffers from illusions at this point. On the one hand the political class would have to be destroyed, along with the ideologies of left and right, and on the other hand the famous power of choice would have to be renewed, an illusory goal in politics; cf. Sfez, *La Décision*.

discharged by high-level technicians and not by responsible political delegates. New institutional measures should be adopted to safeguard the independence of the areas of expertise. The objective of security should be defined by those concerned. Finally, we should open up the process of decision (though to some extent this is incompatible with making it political!). He recognizes that traditional democratic procedures are no longer suitable. The watchword "participation" is strangely illusory. But the idea of participation is taken from the arsenal of advocates of microcomputers and of those who believe that we can master technique. But even were there institutions for citizen participation in decision making, who would participate? It would be people who are manipulated, who have been shown what to do by the media, who receive enactments of news that they are incompetent to judge. A. Jacquard *(Les Scientifiques parlent),* too, argues a need for collective reflection in which there would be general criticism. But this, he says, involves an effort to inform. In response I contend that information is not enough. There has to be awareness and a critical spirit.

I concede that the participation of all people in all decision making would be the ideal solution. But this is an impossible solution. It is not even made conceivable by modern psychological and intellectual development (in which we are all specialists at something). With best wishes Lagadec calls for the reinvention of democracy. Who would not agree with his formulation that the alternative is either the abandoning or the reinventing of democracy? We have to come to terms with reason, he says, that is, with rigor, argumentation, criticism, debate. We can engage in an ultimate act of will, the refusal in spite of everything to yield. If we want to avoid the collective and personal loss, are we not finally forced to affirm our freedom?

Forty years ago Bernard Charbonneau said much the same, but no one would listen to him. Faced with the triumphalist affirmation according to which we are now summoned to master technique, we can measure the impossibility of the task, noting what would be required to accomplish it and how great an illusion it is to think that it is anything more than wishful thinking.

CHAPTER VIII

Rationality

I think we may take for granted what Janicaud has written about the power of the rational, into which I offered some insights in the last chapter. I will not repeat myself here but will look at another aspect. Rationality is part of the reassuring discourse and also of the proof of the inevitability of technique. Technique clearly results from science, which is purely rational. Deduced by rational operations, technique, too, is rational. But reason implies humanity. Rationality is what humanity and technique have in common. As technique is the product of a process of rationality which is specifically human, how can we say that there is any contradiction or that technique has eluded us and shown itself to be bad?

Have we not seen, indeed, an endless progression in rationality from the single, individual machine functioning rationally to a group of rational machines and then to a corresponding ordering of society which shows obvious progress as compared with the social incoherence of previous centuries? Since the eighteenth-century philosophers, the political world has tended toward rationality with democracy, and it is taking the risk of finding fulfilment in the rationality of the computer.

The rational has made strides in every sector of human and social life. This is reassuring, for we know what to expect from the rational. Technique, obeying rationality, becomes an expression and instrument of human reason in our society. When from time to time a nonrational event takes place in the technical system, it is always a one-time accident, an unfortunate mistake, a negligible fact. Even when such accidents become common (as in the case of automobiles), they can be rationalized by statistics and integrated into rational tables of prediction for insurance purposes.

Some years ago Gabriel Dessus concluded that we have to evaluate human life in terms of money.[1] Money is the only way to rationalize the acquisition or loss of human life. Technique demands this in order to be truly rational. We dare not say this too often, because it is so shocking, but it corresponds to reality. And what is so reassuring as the rational? It is reassuring because it is both understandable and certain. Implying the development of a series of linked operations, it can be fully grasped. If, then, the world is to be grasped (i.e., understood and mastered), it must be rational. All that we ask of people in this society must be rational. It is rational to consume more, to change immediately what is worn out, to acquire more information, to satisfy an increasing number of desires. Constant growth is rational for our economic system. We can take the ordinary actions of 99 percent of the population in a so-called advanced country and we shall find that the key to them is always rationality.

And does not this rationality assure us of technical mastery? Contrary to what I said earlier, are not those who have mastered dozens of machines of increasing complexity helping to forge a global mastery over technique? Is not this situation clearly much better than that which our ancestors could know, immersed as they were in a world of nature and then of society? For nature seems to be essentially irrational, made up of accidents, of incomprehensible renewals, of the unexpected, of a lack of obvious causality, of chance phenomena, and we are tempted to ascribe coherence to all these things by using myths, to detect the presence behind the phenomena of forces upon which we can act and which we can command instead of the phenomena themselves, such forces as gods or demons made after the human image. That was an early way to humanize and rationalize nature.

Nature perhaps has its own order, but this translates always into terrible disorder for us. We have only to look at the Lisbon earthquake and Voltaire's poem about it. Society, too, is irrational, with its kings by divine right, its dividing up into age groups (I do not mean the class distinctions of Marx), its functional classes, its clans and castes. And all this is explained and justified in part, and upheld in part, by the total irrationality of religion, with its gods whose reactions and favors and retributions and angers cannot be known or foreseen. If the centuries prior to the 18th were centuries of obscurantism, it was because nature itself is obscure, the basis of power is obscure, and death is obscure.

1. Gabriel Dessus, "De l'inéluctable mesure des incommensurables, et de ce qui peut s'ensuivre," *Revue française de recherche operationelle* (1964).

In face of this situation, how can we fail to see that the rational, as a victory over the irrational in thinking, conduct, and society thanks to techniques, is the great triumph of humanity? The eighteenth-century philosophers showed very well that reason brings light, that it dispels obscurity and the incomprehensible. Science and its minstrels, like Victor Hugo, confirmed this in the 19th century.

Today we have moved into the technical age, and what we say about rationality is no longer the same. It is no longer a matter of fighting against religious irrationality or the extremes of nature. It is a matter of projecting human power on the whole universe, on every culture, on nature in its totality, thanks first to the machine and then, gloriously, the whole array of techniques. It is a matter of showing that this power, this domination over the irrational, this subjugation to the rational, is a fulfilment of the very being of humanity. It is very remarkable indeed that the projection of the rational over the universe always leads back to humanity. As the technical system is rational, humanity has to be rational in turn. We begin with a philosophical conviction about human nature and we come back to an ethical command to identify ourselves with the universe that we have created. This explains the countless numbers of the abnormal, the marginal, the handicapped, and the maladjusted who must either be brought to the level of our world's rationality or put on one side, maintained by society but without having anything to do or being able to mix in society.

No previous society has had to set aside as much as 10 percent of its population. This is the result of the coming of technical rationality. It is also remarkable that opposite routes lead to the same affirmation of rationality. For example, what is rational behavior from an economic point of view? For some it is planning, mastering economic phenomena, calculating raw materials, hours of work, prices, production, needs, tools, etc., so as to fix at will the rhythm and growth of the economy. Rationality is the power to impose upon economic facts and dominate them. Yet the advocates of economic liberalism are no less rational. The economy as they see it functions according to its own laws, like nature (in physics, chemistry, or biology). Reason for them is discovering the rationality of these laws, whether of the economy or anything else. Economic life is rational in itself. The various elements combine according to what may be seen to be rational principles. Wealth that we can calculate rationally is the result. The vital thing is not to interfere with the free play of the laws. Progress is in understanding, not in domination. But there is no less rationality. Both economic systems think that in their different ways they are expressing supreme rationality. This fact is remarkably confirmed by the new

interpretation of a global system of communications which with innumerable interconnections will encircle the planet with a network.
The word *network* is a favorite one at the moment and we can foresee
the (unavoidable?) time when the many networks will finally be
agencies of a central brain.[2]

We are no longer dealing here with the supercomputer which
imposes its will by means of a centralized system, but with a complex
system of financial and information networks, of transport, and of
abstract shifts, which finally converge on the central brain, which is
bound to appear even with no centralized or authoritarian intervention.
This network will have to have a rational understanding of all that
happens and to give out its impulses, which are also rational. The
computer, which might lead to local initiatives, cannot avoid interconnections that inevitably mean centralization.

From another standpoint the triumph of rationality comes with
the computer and all the machines that derive from it, which have for
a long time been called thinking machines. We could not have said this
thirty years ago but today we are close to it. Naturally, a computer does
not function like a brain. But fifth-generation computers fully imitate
all the operations of the human intellect, including the association of
ideas, rectifications in the course of reasoning, and self-programming.
The following question arises very seriously. Since the 19th century
the machine has replaced physical motions, human muscle. Are we
now confronting a new step in the replacement of people by their own
creations, that is, the replacement of their intellectual operations?[3] On
the rational plane, we are indeed. But it is forgotten that our own
rational thinking is not the product of an organism separate from our
being. It is nourished by memory (as might be said of computers, since
they record what they have done and have recall) and by foresight (as
computers, too, are beginning to make forecasts that will influence their
functioning). But human memory and foresight differ from those of
computers. They include memories of joys and successes and failures,
and the foresight is mingled with hopes and fears.

2. Cf. J. de Rosnay, *Le Cerveau planétaire* (O. Orban, 1986). This is a masterpiece of illusions about networks, depicting the many innovations which make a
network function like the nervous system—a comparison in which I take a naughty
pleasure since I envisioned it myself prior to the coming of computers in 1950, in
Technological Society. A technological tornado, which will do little good!

3. Computer translation is an example. But there are limits. Only restricted
Translation is possible, and it needs to be reviewed by human translators. Cf. "La
Traduction automatique," *La Recherche* 152 (Feb. 1984); also Makoto Nagao, "La
Traduction automatique," *La Recherche* 150 (Dec. 1983).

In other words, except in algebra there is no such thing as purely rational human thinking. Even our most rigorous thinking is inevitably intermingled with opinions and sympathies and antipathies and feelings. How often our reasoning and knowledge reflect the causes we advocate! Our thinking is never pure. That of computers is always pure unless it is programmed to take into account a specific feature. Yet even though its thinking is rational, there is often an irrational factor in the way that one poses a problem (to the computer!) or in the choice of the problem that one poses.

Machines can certainly understand the intellectual problems that people pose. They have to break them down into simple questions, but when that is done they do much better than any of us. Yet no real comparison is possible. And we grant that problems are raised by what is called artificial intelligence.

An aspect of the technical universe which is not wholly without interest is the loose way in which terms are now used, as if laxity in vocabulary were compensating for rigor in technique. I have often cited the wrong use of the word *technology.* I might also refer to "star wars," which is a very inexact term for the strategic defense initiative. There is also the "conquest of space," which is a wild exaggeration, or the "mastery of energy." Above all we have this phrase "artificial intelligence," which dates back to 1952 and the very first computers, when P. de Latil was talking about "artificial thinking."

The phrase has come into general use, but strictly we can use it only to the extent that we carry out a reductionism of intelligence and obey a mechanism. On the one side there is a reference to systems experts, and I agree that here, in a limited number of cases which can be put in mathematical terms, the systems experts might imitate an intellectual operation. We shall see the limits of this later. But the abuse begins with the general assumption that computer operations can be compared with every intellect or that computers can do all that the brain does.

M. Arvonny defines artificial intelligence as any reproduction by a computer of the processes of the human brain. In practice, however, we see that the reference boils down to the systems expert, and that it is hard to isolate that which specifically constitutes artificial intelligence within the totality of the computer industry. If it is just a matter of recording the experience gained by experts and putting it at the disposal of all, there is nothing there that can be described as intelligence. So long as we recall that we are far from producing computers that talk and listen and reflect as we do, I will grant in exchange that computers can do some things much better than we can do ourselves. But it is going much too far to say that the computer has brought a (total) revolution in intelligence,[4] which would mean that our human intellect ought to bow to the new order, or that in the sphere of thinking everything is possible for computers, from writing a Proust novel on its own to having intuitions like those of Einstein. This was the thesis of Hubert A. Simon, who won a Nobel prize for economics.[5]

I will take up Simon's article, since it is fully in keeping with the thinking of those who talk about artificial intelligence. Simon has three main points. First, if we can devise computer programs that can display in their data banks facts and the underlying *patterns*, we shall be able to understand the process of discovery. And the computer will be able to make discoveries itself, for a discovery is not a true innovation. It is the result of a laborious process gradually enacted on the basis of immense documentation and through knowledge of prior models. There is no real intuition. Intuition is simply the ability to grasp instantly the significance of a given situation. The key to it is recognition. Experts recognize the indices that give them instant access to a vast body of knowledge. Computers are also capable of this recognition and are thus capable of intuition (note the reductionism). We have only to reduce the complex processes of thought to simple operations that computers can manage. Combining them, they will then reproduce more complex configurations.

Second, computers can be programmed to simulate human emotions. We have only to work out programs that include symbolic input similar to that which the nervous system receives in connection with emotions. The computer system will then have the same ability to deal with emotions as the human brain. Simon adduces the example of a computer that can simulate perfectly fear or anger or paranoia in response to certain stimuli. But can we really think that the thousands of different emotions, which are often hard to differentiate (friendship, sympathy, love, affection, etc.) but which make up the fabric of human feelings, can be analyzed so well and reduced to such simple input that they can be entered into computers? Here again we have an over-simplistic view of human reality. But according to Simon's conditions, a computer might write a Proust novel. With a sacred simplicity Simon argues that it is enough that the computer has as good a knowledge of French as Proust and that it can be furnished with analogous experiences to those that excite our emotions. That is all; a combination of French and emotions means that anyone might write *A la recherche du*

4. See "La Révolution de l'intelligence . . . ," *Science et Techniques* (1985).
5. Cf. the very important article in *Le Monde aujourd'hui* (April 1984).

temps perdu. But a point which gives me pause is that Proust *decided* to write this work. Can computers make the same decision? According to Simon we come up here against the "revolt of the robots." One day the robot that is given an order will reply: "No, I want to write a novel!" The decision of the artist or writer is a specific feature of the human intellect and it is beyond the reach of the computer.

Third, according to Simon computers can do anything in the intellectual realm. They can understand and interpret, since understanding is simply the process which uses a variety of things stocked in the memory in such a way as to interpret new data of experience— another example of reductionism. It is conceded that computers are not yet capable of nonverbal understanding, of noting in what is said the intonations of anger, joy, etc., but in principle this is said to be quite feasible. Finally, computers can organize their own programs as well as their execution. There exists a computer (though I have found no trace of it except in this article) which can study a problem, work out new ways of solving it, and then, if successful, analyze what it has done, and use its experience to modify its program so as to know how to solve similar problems it might meet in the future. In other words, it is capable of heuristics. We see, then, that with these declarations of the Nobel prizewinner we are well ahead of systems experts.

My main objection is that all these marvels are imitations of the operations of a brain that is taken out of the skull and that functions in a jar (certainly with nerve endings) full of physiological serum. Intelligence is more than the ability to assemble and use knowledge or to solve problems or to memorize. Intelligence is a total human activity. It is nourished by human relations, by accidents, by fatigue, by (non-simulated) joy, by the desire to write or to calculate, by the selection of knowledge for a particular project, by psychological obsessions, by the wish to please or hurt people. The intellect is not algebra. I also reject completely Simon's abstract view of intuition. The intellect can be excited or it can be bored. In other words, it is a function of living people. The computer may be able to simulate but it can never do more than simulate—it is not alive. The most perfect machine has nothing human about it. The simple fact that the computer can only simulate human intelligence shows that it is *not* intelligent.

Simon ultimately ignores three essential aspects of human intelligence. The first is imagination. Without imagination there is no intelligence. But the computer is not capable of imagination, especially in the dimension that Sartre and Castoriadis, for example, have given it. The second aspect is spontaneity. All intellectuals know that ideas spring up on matters to which they have devoted little attention. All at once a kind of evident truth illumines them. Thought has been secretly at work. Such ideas then give rise to strict intellectual exercises. This is an integral part of intelligence, but the computer is never grasped by impromptu ideas that arise out of dreams or street encounters or the play of colors or nostalgia or hope. The computer can never record such things.

A third feature of intelligence is its general grasp of a situation, relation, problem, etc. This grasp is indivisible. That is to say, one might, of course, analyze it, but adding up the units of understanding (or communication) that analysis yields in no way restores the comprehensiveness of the intellectual grasp. Five notes of music tapped out separately do not constitute a musical phrase. The computer can never restore the phrasing of general intelligence. As regards Proust, this leads us to a very simple question. Do we really think that the computer could write as Proust does about what is evoked in him by the "madeleine"? Undoubtedly, it is a matter of recollections, and these can be stored in the memory of a computer, but they are evoked by an experience that the computer cannot have, and they are ordered and linked and developed in a way that is quite beyond the computer. This is why it is absurd to talk about artificial intelligence.

Moving on from Simon's wild dream, we might refer to a very reasonable special number of *La Recherche* on this theme.[6] It is not now a question of Proust or Einstein. The reference of artificial intelligence is to systems experts, to games of strategy, to the simulation of medical reasoning, to recognition of writing (though this is contested),[7] to the understanding of speech, to the representing of existing knowledge, and to a certain capacity for learning. All that is remarkable enough. We also find in the articles an indication of two important limitations. It is only abstract thought that can be assimilated, interpreted, and imitated by the computer. The more thought refers to life, the less accessible it is. It is only on the level of theoretical knowledge that the computer can measure and simulate. It is paralyzed in the field of practical knowledge. In no operation, then, do we really have intelligence. Yet following current usage, and with a certain pride, this special number of *La Recherche* bears the title "Artificial Intelligence."[8]

6. *La Recherche* 170 (Oct. 1985).

7. The computer can understand and recognize a single letter as it is written by a given person and might thus read this person's writing. But different people write letters differently, and even the same people form letters in various ways, so how are computers going to read what they write?

8. Sometimes the situation is purely comical. An article in *Le Monde* dealing with adaptation to artificial intelligence raises only the health problems of those

In sum, human thought feeds on our experience of life as we register and interpret it. Imagination, fantasy, myth, intuition, and experience transform themselves into thought. Computers can imitate the human brain, but the human brain is not a separate entity—it is part of a body. The experiences of this body provoke the reactions of the brain and set the rational process going in one or another direction. To be sure, novels (and we must not forget that Italo Calvino had a story to this effect twenty years ago) can depict an amorous computer and pretend that it behaves much the same as we might do, but this is no more than amusing science fiction. Computers do not have the dreams or fears or desires that feed and stimulate human thought. This is why they may imitate one of the operations of the brain, but no more.

It is for this reason that computers cannot produce anything that might be called art except in the purely formal sense of putting one color or one note next to another. Genius lies in the heart. Songs of despair are the most beautiful of songs. Some immortal songs are pure sobs. Is that romanticism? Not at all! We have here the origin, development, and aspirations of all art. In spite of the claims of some modern painters and musicians, the computer has no place in art. We shall leave it at that.

To put it simply, the most perfect machine is purely rational, but human beings are not. They are not rational in their feelings and opinions and conduct, and they do not find it easy to live in a purely and exclusively rational milieu. Who of us has not been frightened by reading about utopias, whether of the 16th century or the 19th, with their perfectly mechanical organization, their closed world in which there are no accidents, but everything is foreseen and regulated? No one wants to live in them. And whenever people find themselves in an over-rational society, irrational behavior breaks out at once. Imperfect though it may be, our Western society is already infinitely too rational. We have to submit to rules and constraints and exercises in collective discipline, and we thus react against the excessive rationality. The more a society wants to be rational, the more we express irrational urges. This is exactly the problem that we raised earlier. How can we lead people to live well and happily in the rarefied air of rationality without external constraint or force?

---

working with computers. The computer is artificial intelligence and adapting to it is just a matter of better spectacles and seating. In contrast I would recommend the article by H. S. Dreyfus, "Why Computers May Never Think Like People," *Technology Review* (1986). The intelligence does not function merely in terms of facts, meaning does not lie merely in words, and intelligent action derives from knowing what we are; this is impossible to program.

Yet we must also consider the complementary aspect of this mistake in discussions of rationality. The universe that is constructed according to a rational design, with rational means, and with an ideology of rationality, leads to the astonishing result of such irrationality that I can even speak (see below) of the *unreason* of the technical society. We have here a kind of monster. Each piece is rational but the whole and its functioning are masterpieces of irrationality.[9]

We must demonstrate this in detail, for it is obviously not the view of technocrats, who are not in the habit of looking at a complex whole or taking note of significant details that are not in keeping with the general plan. We are well aware that the general plan enables us to shut out after a time facts that do not fit into it. Such facts are simply understood as aberrations or accidents. But they are precisely the facts which should claim our attention if we are to focus not on what we pretend is happening, but on what is really happening and what will tomorrow become the significant whole. These facts reveal a notable institutional, economic, and political tendency, all according to techniques, toward a striking irrationality.

The most rational of systems increases social maladjustment and retardation and multiplies the number of marginal people. Devices that are meant to give greater freedom engender maximum inevitability. Acceleration of the evolution of the system produces a worsening of crises. Multiplication of means leads to disappearance of ends. Growth of universal power augments social powerlessness. For individuals the means of power act as prostheses which weaken the natural use of functions.

The more technique advances, the more it is supposed to serve human progress, but the more in reality it brings human regress. The wealth created by technique is collective, fleeting (in need of constant renewal), and dependent. The more devices seem to be obvious and useful, the more they reduce the human condition to absurdity. The more progress there is, the more repetition there is. Change accelerates sameness. In a shadow theater we see constant transformations from the same to the same by miraculous means. The sameness consists of permanent change according to an immutable principle. There is a remarkable development of chaos and the irrational in a superfluity of rationality, order, and taxonomies.

But that is not the end of what I have to say about disorder, chaos,

9. On this point cf. B. Charbonneau, *Le Chaos et le Système* (1973), who shows plainly that the more perfect the system becomes, the more chaos develops around it and within it as a counterstroke.

and irrationality. We are now said to be in a stage of recuperation by which order will be born out of disorder and (in communications) information out of noise. This is a seductive thought and true enough physically. But I find it fits best in politics and for social life as a whole. The new science is based on this principle (cf. the works of E. Morin), but I am not so sure whether the same is true of technique. For in the technical domain, which is our own, rationality is necessarily operational and instrumental. It is the rationalizing which results in technical exploitation of the earth and which reduces all things (human beings included, as we have seen) to objects of calculation for representative thinking (Heidegger).

When it is justly argued that reason and rationality are (metaphysically) very different, two considerations result. First, although technique may derive from rational operations, the great discourse about rationality which is now commonplace has as its aim the vindication of technique by means of, as well as in the eyes of, a higher authority, that of human reason. We in the West who are trained in reason have no answer to this discourse with its intentional confusion between veridical reason and pragmatic reason.

But Dumouchel and Dupuy have well shown that the violence that is done under the sign of instrumental technical rationality is in truth, and more profoundly, the product of a constant tendency which simply has no means of expressing itself, "the omnipresence of mimesis which provokes dehumanization and the cold and indifferent rationalization of human relations."[10] This is why it is not evident to me that the new science can be an alibi for the new technique or can validate it. All that science can do is solve one of the essential human problems, that of adjustment to the chaos caused by the very rationality of technique. This is why I prefer to speak about the unreason or unreasonableness rather than the irrationality of technique.

My affirmation is that the rationality of technique and all human organization plunges us into a world of irrationality and that technical rationality is enclosed in a system of irrational forces. Hence it is not a matter of finding reassurance in discussions of rationality, which can never be more than excuses and vindications. Rationality has been able to justify all things (as it previously did religion): centralization, rationalization, Taylorism, and also concentration camps, which provide the cheapest labor in the world, force opponents to work for the regime, and when they die salvage their hair, teeth, etc.—a masterpiece of rationality! In technical operations pure rationality can lead to all kinds

10. P. Dumouchel and J.-P. Dupuy, *L'Enfer des choses*.

of aberrations. Yet it has its place in the justifying discourse which seems reasonable enough to the public when technicians or politicians address it. Since it is the decisive argument in administrative commissions, we have to take it very seriously.[11]

11. By way of transition to the next chapter see the notable article by J. Sheppard and T. Johnston, "Science and Rationality," *Science, Technology and Society* 2/3 (1982) and 2/4 (1982).

CHAPTER IX

A Sketch of the Ideologies of Science

1. Classical Ideology

There has always been an ideology of science, not in the strict sense of ideology that we find in politics, but in the way in which the public, especially the "cultured" public, with whom ideologies have their birth, has accepted and depicted science, attributing to it certain qualities and arriving at a general representation of it.[1] Those who share this ideology do not know science. They cannot follow its methods and experiments. They trust in the summary explanations that they are given. They also see certain results in the form of techniques, which at first they did not understand, thinking that these results were the direct fruit of scientific research. The automobile, the railroad, the vaccine for rabies are all applications of science.

The ideology which has developed is hardly shared by scientists themselves. Scientists plunge into the reality of scientific research and seldom have any general attitude to science. Nevertheless, when they break out of the frontiers of their own special field they are ready enough to accept the views that are passed on to them by a favorable and even enthusiastic public.

It seems to me that one could write a veritable history of the common ideology of science. I would distinguish five periods from

---

1. In discussing ideology I am using as sources for the present period various reports and specialized works after the manner of Bretonnoux, "La Perception du message télévisuel dans un groupe témoin" (thesis, University of Bordeaux, January 1985), and more generally I am drawing upon the daily press and television (which hide their bias from the public) and the statements of politicians, who accurately express the views of average citizens.

1850 to today, passing very rapidly over the first four, which will serve simply as a benchmark.

The first period is characterized by what is called scientism (it is usually stressed that true scientists did not share this characterization). Schematically, scientism amounts to the notion that science discovers and will discover all truth, truth being equated with the concrete reality of the world in which we live. This world is finite. It can be analyzed, understood, and explained. Reality obeys fixed laws which enable us to predict events, since these always repeat themselves in the same way. Science evolves unceasingly toward "more," so that finally it will grasp everything and exhaust all problems. There is no mystery; nothing is unknowable. The laws of nature are never broken, that is, the laws as science has established them. This idea of total knowledge was present even among disciplines that one could hardly call scientific in the strict sense. As late as 1930 I knew a historian who told me that when we have searched all the archives in France down to the last detail, we can know everything that happened there from the historical point of view in the 18th and 19th centuries.

Science, it was thought, is never wrong. It advances by accumulating certain knowledge. When a scientific theory proves to be inadequate, it is replaced by a better one, but there can never be a scientific error. There is only progress. In the educated world between 1860 and 1900 people lived in a kind of enthusiasm that was not without a certain intransigeance. In the name of science they had to shatter false ideas, religions, cultural traditions, myths. All products of the imagination in the ages of obscurity had to be replaced completely by the light of science.

But this enthusiasm waned around 1900, and a second period came between 1900 and 1918, bringing not strictly a new ideology but a weakening of the common scientism. Less was said about science; it was not extolled so much. I see two factors in this regard. The first is habit. The educated public had become accustomed to the marvels of science. They did not yet know the big discoveries that risked changing everything (e.g., those of Einstein), but they were less captive to the glorious future of knowledge and light and elucidation. Habit made them still convinced of the absolute value of science but without wanting to examine everything closely. Anatole France still has a scientistic view in his works, but if there is unbounded appreciation of science there is not the same readiness to contend for it. This is also the time of scholarly conflicts, the battle waged around the church in the name of scientism. Scientism is so deeply entrenched, however, that Roman Catholics do not dare attack science. Science still seems to be impregnable.

The other factor that brings with it a certain distancing of opinion and waning of interest is obviously World War I. The main preoccupation of the majority of the people is clearly not science but the war itself. Yet in the case of the conflict with the Roman Catholic Church, it is significant that science is not challenged as progress in weapons of destruction is noted. Cannon and machine guns and planes have nothing to do with science. No attention is paid even to the importance of technique. At issue are the political order (as though politics were totally responsible for what takes place) and the economic order (as though the war were merely the result of conflicts of economic interest, and the improved weapons were merely products of arms manufacturers and the great economic powers). No note is taken of the fact that capitalists can produce and sell only the results of technical research. Capitalist interests—and we see here the full influence of Marx's thinking—are held solely responsible for every product. In this period, then, the passion for science is less central and its radiant image less striking. Scientist ideology persists, but in a minor key.

After the war, we see another facet of the ideology. It bounces back in other forms. I would say that in 1900 truth was central. After 1920 happiness is central. Science will assure us of happiness. Note is taken of the spectacular progress in medicine and surgery. People profit from the great industrial machine. The products of science, by way of technique, make their way into every sphere of life. There is an acceleration and multiplication of means of transport. The prophecies of Hugo are frequently recalled to the effect that the material rapprochement of the nations would lead to mutual knowledge, which in turn would mean that they can no longer hate and make war. Consumer goods increase, and here again, thanks to science, the standard of living rises spectacularly in spite of the 1929 crash and unemployment. Once again science escapes responsibility for the problems, which are attributed to poor economic functioning. Up to about 1936 no one doubts that science is destined to ensure human happiness. Huxley brings to light a different aspect, but at the time he is not taken seriously except by a very small minority of young intellectuals who dabble in technical problems and offer no challenge to science.

It has to be said that attention remains fixed on the discoveries and the traditional approaches of science, and the great debates of physicists and mathematicians are ignored. The names of Planck and Heisenberg are vaguely familiar, but most people are totally incapable of knowing anything about their findings or the implications. For the majority, science is obviously good and oriented to human happiness. This view explains the desire during this period to describe as scientific certain disciplines which had previously belonged to the humanities. Hence we find a "scientific" sociology, psychology, or economics. These disciplines now use the methods of the natural sciences, and especially mathematics. It is not just a matter of rigor and clarity, but rather of the conviction that these sciences can help us to escape from the muddle of political decisions and lead the nations to a more rational organization that will guarantee happiness.

During this period, in spite of some movements in literature and art (surrealism, cubism, etc.) which discover the value of the irrational, the collective conviction is still that rationality is the path of progress and that progress will ineluctably translate into human happiness. The growth of consumption and ease of travel (automobiles) make possible the projection of a future in which human beings will be at once more free and more happy. More free, for the growth of consumption will liberate them from the bondage of scarcity and bring with it political modifications. This is the time when the theory develops that abundance of consumption will make dictatorships impossible and bring a trend toward democracy. Abundance certainly means wider choice and therefore more freedom. The automobile also allows people to travel at their own time and pleasure. At the same time the idea of happiness undergoes a change. It is no longer spiritual or idealistic. Tied to consumption, happiness is well-being. It involves the great discovery of comfort. And the general belief is that comfort, consumption, and freedom are all the fruits of applied science. In some circles a distinction begins to be made between basic science (which the average citizen regards as useless) and applied science, from which derive the marvels of modern society. With its roots in actual experience, the ideal of science is thus strongly positive.

We now come to the fourth stage. After 1945 the ideology of science undergoes a crisis, which has many elements. Above all, we might say that the discovery of penicillin does not compensate for the trauma of Hiroshima. We enter a long period of an ideology of doubt and defiance regarding science. It is odd to have to note the contradiction between the many scientific discoveries from 1945 to 1975, along with astounding economic growth (the glorious thirty years), and inversely, on the psychological and ideological plane, the attitude of withdrawal, and for the first time a lack of enthusiasm. Ideologically, this period is characterized differently depending on the level one examines. On every hand there is profound change. We see it first among scientists themselves. There develops among them a kind of doubt, so that one might almost speak of a crisis of science, which would be inaccurate as regards its development, the progress of its methods and results, but not wholly wrong ideologically, since scientists, too, obviously have their own ideology of science. Thus there is less talk of the equation of scientific discovery and truth. Truth is no longer a primary goal of science. I think we might analyze the ideological crisis in two ways.

First, the consequences of previous theoretical discoveries have developed among scientists. Thus account is now taken of phenomena that were earlier evaded (disorders, tornadoes, floods, noises in communications). Entropy is accepted. Scientifically there is recognition of a decisive phenomenon like feedback. It is admitted that observers, being in the same system as the object they observe, affect it by their very presence. The ancient notion of inseparability has been renewed. An extreme view is that facts as such do not exist; we ourselves construct them as we observe. Such things alone, of course, do not mean a crisis in science. On the contrary, there has been a great advance of knowledge, and science has scored new victories. That is true. At the same time scientists have come up against complex, if not complicated, phenomena and theories that do not have the beautiful simplicity of Newtonian physics. What seems to be characteristic is the greater number of books written by scientists to justify their work, to find for it different foundations from those of the 19th century, to relate this infinitely more complex science to a philosophy, to a general conception of the world, or to deduce from it a morality, a rule of conduct. For the most part such works have come out after 1970, but the important point is that they are a kind of response to an earlier disquiet in the scientific community.

This disquiet does not merely concern the new complexity but also the sense that the more riddles we solve and the more the field of research widens, the more numerous and difficult are the problems. Earlier the universe was thought to be finite, so that the number of problems was also finite. When every question had been answered, the goal of knowledge would be attained. But now the universe becomes more complex and varied the more we know. The quest, then, seems to be endless, and no final result seems to be possible in any branch of science. This is just as true in the "exact" sciences as in the humane "sciences." This is the age when doubt arises as to whether we can really apply rigorous mathematical methods. Human phenomena seem to be too complicated and elastic for them. The previous epoch comes under the accusation of reductionism. In history it used to be thought that a strict economic analysis would explain political movements, but now we have moved on to much more uncertain and fluid inquiries, for example, the history of attitudes. Various studies and challenges will produce attempts at explanation and generalization. One theory is that science advances, not by accumulation of knowledge, but by paradigm shifts that result in a scientific revolution. An effort is also made to relate science to a transcendent truth; it is no longer self-sufficient. An interesting point is that this effort does not come from philosophers or theologians (in spite of Teilhard), but from scientists themselves (e.g., the Princeton Gnosis). The conclusion is that there is no such thing as a scientific method, a nature of science, a criterion of what is scientific. No one can say what belongs to the category of science.[2]

But there is another source of disquiet that is less scientific and more ideological. During the war, from 1940 to 1945, weapons multiplied indefinitely. In addition to the atomic bomb, techniques deriving directly from scientific advances appeared in every field. Chemistry and biology were harnessed. Bacteriological warfare became possible, or warfare using defoliants. Science was everywhere, and it served every end. The great question that was raised was that of the close bond between science and technique, and since technique was put in the service of the powers, the judgment gained ground that science does not have clean hands. Scientists are no longer ascetic and objective seekers of truth but (perhaps involuntarily, yet unavoidably) the creators of weapons of war on the one hand, and on the other hand the creators of innumerable products (e.g., medications) whose effects we cannot accurately evaluate. There is no such thing as pure science. One can even say that pure mathematics does not exist.[3] Science is implicated in the totality of impure activities (politics, war, the police).

No doubt we are horrified at the scientific experiments conducted by Nazi scientists upon prisoners, but so long as the patients are willing we accept as normal the experimentation upon people in the areas of surgery, of certain new products, and of genetics. People can be manipulated, but not with a good conscience. These are the various elements which, as I see it, constitute during this period the premises of a crisis in the ideology of science, in the scientific world and milieu. Naturally, this does not in any way halt research. Yet some scientists (e.g., in America) are in fact proposing a moratorium on scientific research in order to try to get a clear view of the situation.

2. See A. Chalmers, "The Case against a Universal Scientific Method," *Science, Technology and Society* (1985).

3. See Didier Nordon, *Les Mathématiques pures n'existent pas* (Actes Sud, 1981). Since 1960 there has been an urgent need for a professional code of ethics in research. Most important is the Japanese plan (1987) for a big gathering of scientists to propose that research no longer be oriented to the power of production but to the frontiers of the human, the environment, resources, demography, etc.

Among the public there is still astonishment at the extraordinary discoveries of science since the early days of the "conquest of space." We have seen a wide distribution of the technical results of science (e.g., television). Yet as I noted at the outset, the astonishment goes hand in hand with very great fear. And the ideology which is developing expresses itself in the wide sales of Orwell's *1984* and science fiction. It may be noted that between 1960 and 1975 the stories in science fiction in the USA always end in disaster and are basically pessimistic. People in the West have a belief in the absolute power of science. This is now the central theme rather than truth or happiness. There is a conviction that science can do anything, but not in the sense of positive greatness or success for the human species. It can do all things for either good or evil.

There is thus spreading abroad the consoling conviction that everything depends upon the way in which we use science (e.g., political decisions), that is, within our own ability. But there is ambiguity about the value and the positive nature of science. The writings of the period give evidence of both unlimited confidence and widespread fear. Comparisons begin to abound with the language of Aesop. The only feature that is uncontested is the omnipotence of science. All things are now within reach. It is very important to note the difference between the convictions of scientists and those of average citizens. We see this especially in the ecological movement. Ecology at first was a scientific discipline. But nonscientists seized upon certain findings and hypotheses as a weapon with which to attack Western "civilization." It would take us too far afield to study here the roots of this political movement, the conflict of generations, the fear of losing touch with nature, the effects of 1968, etc. But the movement of refusal, of rejection of all the orientations resulting from science and its technical application could take on its true dimension only by basing itself on certain aspects of earlier ecological research. The success of the ecological movement between 1967 and 1975 manifests very well the ideology of fear facing technical power.

2. The New Ideology

After 1975 we find a complete reversal of ideology respecting science.[4] We note at the outset the strange phenomenon of the differ-

4. The simplistic ideology of science is not yet truly outdated. We even find it among young people; cf. Patrick Tort, *La Pensée hiérarchique et l'évolution* (Aubier

ence between the situation and the ideology. For at the very moment of crisis we encounter afresh a triumphalist ideology. But it is different from those that preceded; it is much more refined. To understand it we must first recall that the bond between science and technique has become much tighter, less fragile, and also bilateral. Science has often been depicted as sovereign and independent, producing secondarily and almost incidentally this or that orientation which has a technical consequence, so that technique is totally dependent on science. In fact, this view is completely impossible. In order to advance, science needs technical products that are increasingly extraordinary and sophisticated. Spatial discoveries are the fruit of space techniques, without which we would know no more than a century ago. The same is true in the microscopic sphere, in nuclear research, and in biology. Computer science is obviously decisive. It is thanks to more powerful computers that scientists can do the calculations that are necessary in their research. I was greatly mistaken in 1980 when I wrote that the speed of calculation and information that computers had then reached was such that no great profit could be foreseen from further advance. It is true that in ordinary use no more is needed, but scientists have embarked on gigantic calculations which demand much more efficient

---

Montaigne, 1984), which rightly takes issue with the sociobiology of Eden Wilson, but which is simplistic when it argues for a pure, ideal, uncompromising science and tries to show that no ideology can be engendered by science (only by ideology) and that science (that admirable goddess) is distinct in nature from ideology. Alas, this might be true in the pure air of metaphysics, but nowhere else. Happily by way of compensation we have the very moderate M. Le Lannou, whose article on the relation of science and life I value for its strictness of thought and lucid humanism ("Le décalage entre la science et la vie," *Le Monde,* Jan. 1984). This article shows how scientific discovery at once becomes a certainty which as soon as it is known becomes a peremptory argument in an ideological thesis before being supported by other facts (cf. DDT and the discovery of cereals by the so-called green revolution). Scholars and research scientists no longer belong to themselves, he says. They are in the service of causes that are so big and so right that they must be upheld even when in error. He recalls the disastrous experiment in Kazakhstan which rendered millions of acres useless because experts had not taken account of the peculiarities of the climate or the characteristics of the people. The same is true of the Aswan Dam, which had to be cost-effective regardless of the evil but foreseeable consequences. He also points out that in the Third World Brazil is the country in which agriculture is most advanced but famine is also most serious. Unexpected numbers and "means" replace slow statistics, abstract space replaces local realities, the ardor of research causes forgetfulness of the time when science did not compromise with ideology and the collective wisdom that is more measured but also more universal than the philanthropy of the hour was under the control of the diversity of people and places. Cf. the fine book by J. Klatzmann, *Nourrir dix milliards d'hommes?* (Paris: PUF, 1983).

computers and systems, and it is in response to this need that progress has been made.[5]

Technique furnishes scientific research not only with the means of exploration but also with countless new materials without which both experimentation and research itself would not be possible. But another dimension comes to light at this point. This equipment is enormously expensive. It is often beyond the resources of a single nation. Except for the USA and the USSR, nations have to join forces to provide for science the means of investigation that it demands, as in the case of the cyclotron. In general, states have to budget for these costs; sometimes they will bring in private financiers. But science has to reciprocate. If the state spends millions to put a laboratory in space or to install a giant accelerator or to construct an experimental reactor, it is obviously not for the love of science or of truth. Like any investment, the investment in science has to pay off. Expenditures of this kind must yield returns. By way of technique, then, science has to furnish results in economic terms. I would add that things do not really go that way. We do not provide science with the tools it needs on the condition that it yields satisfactory economic results. The movement is precisely the opposite.

One can see that certain scientific discoveries bring great possibilities of economic growth, in agriculture, management, and industry. This growth of return with lowering of costs is called development. Since 1950 scientific research has been linked to development: the famous R and D. The one conditions the other. Scientific research brings economic development, but that development is essential if research is to have what it needs. Since 1970, of course, and especially 1975, there have been doubts about this linkage, which many economists have contested. But we should not forget that public opinion is always formed well after the analyses of specialists. For the average politician, then, as well as for the average citizen, the expression "R and D" needs no discussion. It is supported also by researchers in the exact sciences, who know very well that if they cast doubt upon it they risk not being able to continue their work. Ideologically, we find here an agreement between scientists and public opinion (contrary to the general rule).

But science ceases to be free. It is henceforth polarized. It has an absolute duty to serve the national economy. Certainly we have not yet reached the time when politicians can fix a goal for scientists or order

5. See W. Mercouroff, "Quelle informatique pour la Science?" *La Recherche* 146 (Aug. 1983).

their field of research. But if the imperative is less restricted or detailed, it is none the less strict. It is a life-and-death question in the rivalry between nations and also for science itself. There is no longer any science for science's sake. We now have science for development's sake. We have only to look at the different budgets for different sectors of research to see this. And it is firmly implanted among the public by the media. We can see this from television, in which almost every news program has a sequence on the glory of combined science and technique.

We thus have a belief in science that is oriented to national greatness, not through patriotism, but through the simple need to get a slice of the pie. The link between science and technique is not viewed negatively as in the preceding period but very positively. The words have changed; we no longer hear about progress, for example. But the thinking remains the same; growth is the only way to emerge from the economic crisis, from unemployment, etc. This is commonly said by both the right and the left. But the only means of growth is competitiveness thanks to higher productivity, and only scientific research can ensure higher productivity. This is the simple logic which is constantly propagated. And it is all very ambiguous, since in effect, to develop the new consequences of scientific discoveries, new investment, new enterprises, and the recruitment of new personnel are needed.

Thus science itself has an economic effect. And the more science, as it deals with phenomena that were previously beyond our grasp or vision, becomes complex and incomprehensible, not merely for average people but for nonspecialists in the narrower sense, the more it becomes an object of positive and optimistic beliefs. Certainly fear is not absent, but it is nothing like as great as in the preceding period. For we are engulfed and convinced by the daily miracles around us. Hardly have we assimilated one advance before another is proposed which surpasses and annuls it. Am I not talking about technical products? Without doubt, but these are always the results of scientific work to the degree that the frontiers of science are much less clear-cut. Is design a science or not?

Physics immediately has technical applications, and the relation is even closer in computer science. This ideology rests on the double fact of immediate exploration linked to absolute incomprehensibility. In some measure, of course, it was always thus. We do not need to know how an automobile engine works, and even less the scientific principles behind its origin, to be able to drive the car. But the existing situation is much more radical. The more we advance, the more the products are complex and incomprehensible. A television station is more abstract than an automobile engine and a computer more complex than a television station. At the same time the use is much simpler. We will finally arrive at a robot that can respond to voice commands by way of the analysis of scientific data (e.g., in linguistics).

We thus live in a paradoxical situation which enforces a certain ideology. The machine that will be the easiest to manipulate (I refer, of course, to those in public use) will be the one that needs the greatest investment and the most elaborate research! In average thinking this confers a concrete legitimacy on science. Clearly, technique does not do it alone. We know that we are in the presence of more than technique. What we have here is a sort of complete renewal of life and society. We are directly affected by marvellous products that are offered to us by the most elaborate scientific research. We certainly do not know the calculations or hypotheses or details of this research, but we have a sense of the extreme closeness of everyone to the scientific world. Of course, the computer plays a big part at this point.

Science is no longer confined in distant, unknown laboratories. It is present among us. This happens to be considerably reinforced by the political impetus toward the public and the young. Children must not only be taught at once to use technical instruments (the computer) but also given a love of scientific research. The future depends on their being qualified in techno-science. Parents and children must be convinced of the preeminence of science. This is presented as a matter of destiny. Young people must be fashioned for tomorrow's society, which will inevitably be scientific and technical. It is simply not considered that precisely in making young people first and foremost into scientists, we are transforming a possibility or a trend into an inevitability.

But we must look at another vital aspect of this new ideology. More than ever, through the social transformation that it entails, science is becoming not merely the discovering of nature but the response to everything that disquiets or troubles us. There is thus developing an ideology of science that I would call a soteriology. The present-day ideology of science is an ideology of salvation. That is, science is our only recourse, and any negative aspects are strenuously contested. Science alone holds the future to our society. No matter what problem may arise the inevitable answer will always be: Science will take care of it. This is obvious in medicine. With the incredible discoveries in all areas during the last twenty years, the average person now thinks that medicine will have an answer to everything that attacks us. The media not only tell us of new remedies but offer explanations, no doubt greatly simplified, of the scientific research that led to them. The same applies to space research. Along with the spectacular aspects there are explanations of the themes of research, the conditions, and the hypotheses. All this gives to the average person some sense of the vastness of the domains in which science works. To be sure, even in the past some people thought that science was answering our questions. The great difference was that up to the beginning of the century the problems that people used to come up against were for the most part the traditional problems of human life from early days. We were used to these. We had some defenses and responses, not always effective but reassuring. Science was useful, but it was not alone.

Today the dimension, quantity, and quality of the problems that we meet with in our society have all completely changed. Famine is universal and well-known. There is economic disparity between the centers of growth and underdeveloped countries. The power of destruction is enormous. The speed of change is increasing. We must adapt to an incessantly changing environment. Traditional models of society, morality, humanity, art, culture, etc., are disappearing. It is easy to see that the old remedies for our problems and the problems of the world have lost their efficacy. Politics is totally outdated. Education needs to be completely revised so as to integrate the computer and television. Relations between people are ridiculously ineffectual. The older legal system which regulated society is absolutely useless face to face with the new problems posed by the new applications of science.

We can no longer have any confidence except in science itself. It is our only recourse. It alone can bring salvation for the individual and the race. It is science that calms the anguish of the anxious. It is science (not useless accords and treaties) that will respond to the threat of war. We get some idea of this with the reasoning that we now hear increasingly: "War has become impossible thanks to the multiplication of nuclear weapons of destruction. No one today dare risk unleashing a global war, knowing that a first attack would call forth an annihilating response. War is thus impossible." I will not contest the validity of this reasoning. I am simply saying that it is part of the ideological arsenal regarding science. With the system of the interstellar destruction of ballistic missiles, science can even provide an answer to the nuclear threat itself. It is not generally recognized that this second point cancels out the first. But coherence is of no concern in ideology.

In the same way it is evident that science (and not treaties) will solve the problems of pollution. Each time more profound scientific research is needed (as in the case of the catalytic converter on cars). But strides have already been made. The Great Lakes and some big rivers have been cleared of pollution. We know how to go about it; we have only to give free rein to scientific discovery. As already said, discoveries come so fast in every area that in a few years we can be sure that many threats will have been averted. Science is our absolute recourse. We thus find a remarkable situation. In earlier scientism, science thought that it could suppress God and prove his nonexistence. But this was true only of a minority. Today, this is not a goal. But with the development of the idea that science is our only recourse, we have reached the same end in the popular mind by another route. God serves no purpose in the situation in which we now are. Science, thanks to ideology, has now become divine as never before. This is precisely the greatest danger. Kaplan put it well when he said that the danger is not so much the biologizing of ideology as the ideologizing of biology.[6]

Naturally, all this is accompanied by the rejection of pessimistic opinions or alarming facts.[7] To some extent one might say that people today do not want to see or know. After the long period of anxiety there is now a period of calm. The nuclear threat is still there, but the marvels of new techniques, especially in entertainment, have obliterated it. Here is one reason why the ecological movement has suffered reverses and setbacks. (There are other reasons, of course, linked to inconsistencies within the movement.) We do not want to listen because we infinitely prefer the immediate advantages that science offers to the diffuse and distant threats. The positive propositions of the ecologists are obviously less effective than what science enables technique to achieve, and it also seems quaint to want to protect seals or whales. As for the threats to the human race, people basically do not believe them. They are not concerned if the Amazonian or African forests are disappearing, and acid rain does not bother them much. The public has a fantastic sense of impotence in the presence of gigantic threats. It thus refuses to think about them or to be informed about them. We can do nothing about them. We have confidence in science with all its promises for the future.

The world in which we live becomes increasingly a dream world as the society of the spectacle changes bit by bit into the society of the dream. It does so by the diffusion of spectacles of every kind into which the spectators must integrate themselves, but also by the dream of a science which is plunging us into an unknown and incomprehensible world. This will no longer be the world of machines. In that world we

6. Kaplan, in A. Jacquard, ed., *Les Scientifiques parlent* (Paris: Hachette, 1987).

7. Cf. P. Thuillier, "Les origines de l'antiscience," *La Recherche* (1986), and the important discussion of science and anti-science at Paris in 1982, J. Testart, *La Recherche* (1982).

had a place, we were at home. We were material subjects in a world of material objects. The new world is no longer the familiar world of prodigious electronic equipment. In that world we were in a setting that was astonishing from many standpoints but that was still accessible and could be assimilated.

What is changing in an incomprehensible way is the very structure of the society in which we find ourselves. This is a direct effect of science. But the average person has no awareness of it, does not know what it is about, cannot understand the change that is taking place, but is aware only of being on the threshold of a great mystery. In our society information is becoming the key to everything. It is more useful to produce and spread good information than material goods. The wealth of a society is measured now by information rather than by products. But all this is very hard to understand and to take in. We are moving into an unknown world, which will be organized in a very different way from that of the past five thousand years. We are moving into a society that will no longer be one of institutions, stable groups, and hierarchies, and in which we all have a clear place. As already explained, we are moving into a society of networks. It is on different points of these networks that we find ourselves situated, belonging to many networks at the same time. But it is very unsettling to be situated in a fluid and apparently unstable world about which we know nothing. It is no longer power that seems to be the primary qualification of science; it is this transition into a world that has nothing in common with anything that has gone before.

Yet the old social system is not dead. We continue to go through the rites of politics. We live in a bureaucracy. We still have a "normal" pedagogy that needs to be "adapted." The only trouble is that we do not know to what to adapt. Even specialists cannot understand exactly what this society of networks will consist of. If we are farsighted enough, we know that state structures and national boundaries will disappear. There are already some impressive facts. With the development of techo-science, economic calculations no longer have any value. In science we juggle with billions and the question is not one of profitability. Once a new avenue of scientific research is seen, work must be done in it without any knowledge whether it has the least justification economically. It is typical that artificial economic justification is found later (e.g., for research into the Concorde).

This ideology of a divine, soteriological science in association with a dream world is reinforced by what we anticipate and by what is about to come seemingly with no human direction and in obedience to none of the existing classical laws. Science is becoming capable both of absolute novelty and also of the regulation of a world, as is only proper for a deity. Like all deities, it has an oracular power. We ourselves can no longer will or decide. We leave this to the beneficent science in which we believe.

Nevertheless, in spite of all the progress, we still have an uneasy conscience. Should we halt research? J. Testart raises a crucial question: Should we continue genetic engineering in all directions, or should we pause to reflect?[8] Research scientists at once blasted the very thought, in particular Jean Bernard, their guru, who stated that in no circumstances must scientific research ever stop. Yet I think that Charbonneau was right when he stated that this is the only important question in our society.[9] Testart was claiming a logic of nondiscovery and an ethic of nonresearch. At much the same time in California Peter Hagelstein, a brilliant physicist doing laser research for "stars wars," announced that he was giving up his research work for reasons of conscience. We should stop pretending that research is neutral, that only its applications are good or bad. Can we find a single example of nonapplication of a discovery when it responds to an existing need or to one that it creates itself? It is prior to discoveries that ethical choices should be made. What genetic engineering makes possible is beyond our power to imagine, and it affects not merely one aspect of humanity but the totality of our being, of couples, and of the social order. There is the rise of frequent meeting between the ardor of the research scientist to explore all possibilities and the exorbitant or insignificant demand of this or that person, as we have seen already in the explosive growth of pregnancy terminations.

Charbonneau is right to emphasize the importance of dissociation between procreation and sexual pleasure. But contraceptives had already achieved this, and genetic engineering is only another way to achieve sexual pleasure without consequences and procreation without any relation to love. But Charbonneau also shows that society cannot be content to let things slide. Given the consequences, society, that is, the state, must regulate. It must say what is useful and what is not, and set up a model in the general interest, or society's idea of it. Faced with such a realization and such a warning, we can no longer submit to the incoherent results of science. If we do not judge in advance, once a process starts it will roll on to the end. We ought to have judged before splitting the atom. If we do not judge, the reaction

8. Cf. his article in *Le Monde*, Sept. 10, 1986; idem, *L'Oeuf transparent* (Flammarion, 1986).
9. B. Charbonneau, "La question," *Réforme* (Oct. 1986).

of refusal on the part of the human species that is powerless to know what it is doing, or what it is, will simply take the form of a frenzy against which the progress of a science that is as ignorant of ends as it is efficient in means will furnish increasingly powerful weapons.

This is incontestably the process. Scientists warn us of it. Alas, others have done so already. We recall Carrel and Rostand earlier. More recently there was Oppenheimer, who specifically refused to continue research leading to the H-bomb. Then in 1974 eleven American scientists issued an appeal inviting their colleagues to declare a moratorium on genetic engineering. But in 1975 150 specialists in California decided to suspend the moratorium. They tried to impose security regulations and limits on genetic experiments, but in this regard no one listened to them, any more than they had done to the famous committees on scientific ethics. Let us entertain no illusions. Scientists will not accept philosophical, theological, or ethical judgments. Science simply leaves by the wayside those scholars who have scruples of conscience. It goes its inexorable way until it produces the final catastrophe.

Excursus on Science and Faith

The fine book by Henri Atlan sheds a decisive light on the debate between science and faith (or religion, myth, mysticism).[10] He recalls that science has full power and reason in its own domain and in terms of its objectives, but that it cannot go further and ought not to pretend that it can reduce the whole of the human to itself alone nor to swallow up para-scientific phenomena, let alone to deny their existence. Scientists must recognize their limitations and must not try to achieve a synthesis of science and religion as do the works that he loathes (Princeton Gnosticism, the Taoism of physics, or the 1970 Colloquy of Cordoue). But the "spirituals" (if we are not to use the term *mystics*, which I do not like) should not in turn try to annex science in order to prove their validity (e.g., using Heisenberg's principle of uncertainty to prove free will). Each to his own! There is nothing new about that. But on the condition, however, that neither excludes the other; on the contrary, that dialogue is seen to be indispensable. The differences are irreducible, and Atlan sternly criticizes the reductionism of science. But relationship (dialectic) is essential. Thus science needs an ethics

10. Henri Atlan, *A tort et à raison, intercritique de la science et du mythe* (Paris: Seuil, 1986).

but is unable to found or build one. Faith and the thinking of faith obey different rules from scientific thinking, but they are still the rules of reason.

Atlan advances an excellent formulation when he ascribes questions to science and answers to spiritual disciplines, but questions and answers of a different order from those of science. I might reverse this and say that science gives the answers and spiritual disciplines put the questions, but they do not meet. Spiritual disciplines are weak on action and strong on explication. They enable us to found law, ethics, and the family. Science is strong on action (techniques) but, contrary to the belief of the last hundred years, weak on explication. The two human disciplines need to listen to one another and to accept the fact that there is another reality in addition to science, that ultimate reality cannot be grasped, but that we must approach it in different but equally valid ways. I am less happy with the idea that it is all a game: the game of scholars and the game of mystics, each following their own rules. I know that this conception has the useful aim of preventing us from taking ourselves too seriously, and that every approach is relative. But my reservation is that we can join in a game or not; whether we do so is not important. But science and spirituality are imperative human activities that we are not at liberty to opt out of as we please.

CHAPTER X

Experts

The final aspect of the discourse of technological bluff that I will take up is that of resorting to experts. Naturally, experts are indispensable in our society, but above all they are in the public eye and command public confidence. The word of the expert is what counts. After that word is spoken, there is no need to say more, for we, whoever we are, do not have the competence or training or information of the expert. Like technology assessment, experts stand in the way of the public. I am not accusing them of bad faith or machiavellianism; they do so unwittingly. Nor am I denying their competence; they are the best specialists in a given scientific matter. They are not always engaged in research, but they are often linked to research bodies and their role in a concrete matter is to offer their opinion. We find them at all levels. Government has its experts for every issue, and on an important question it will sometimes appoint a commission of experts whose reports are often published. Each public body will also have its experts.

Experts assess the feasibility of a project, its dangers, its secondary implications, its costs, etc. They will often give their advice in "impact studies." Experts also figure at trials, or on television, each with a special field. In fact, it now seems to be impossible to do anything without the advice and report of an expert.

Chesneaux puts it well when he says that at "a great remove from popular technoculture there is the technoculture of experts. The former fills a social void and is for leisure hours. The latter forms the very backbone of modernity. Experts live on the postulate of the objectivity and rationality that are the basis of the neutrality of scientific advice."[1]

1. Chesneaux, *De la modernité*, p. 130. Cf. also Helga Nowotny, "Experts and their Expertise," *Science, Technology and Society* 1/3 (1981); idem, "Experts in a

He points out that when the French government plan for advanced techniques was being drawn up, each of the seven sectors that had priority in stimulating growth was entrusted to a committee of experts. Then in the application of the plan, mastery was needed in such matters as management, norms of production, industrial psychology, marketing, etc., and these areas, too, were put in the hands of experts. Then society had to be guided so as to be able to assimilate the innovations (urban living, publicity, social control), and more experts were needed. The result in the three basic areas was a whole host of experts with innumerable institutes and centers. There was no alternative. To be able to follow so vast a program information had to be shared and programmed, and only the combined abilities of experts could do this. It was a matter of informational integration as each expert became more and more specialized.

The greater and more complex the context, the more extensive the integration,[2] both at the general level of society and within each unit of production. As the technical, social, financial, monetary, political, and ecological variables increase for each unit, posts have to be set up for observation, evaluation, and expert study. Governments can hardly decide anything without commissions of experts. This is one of the aspects that is neglected in criticisms of the governments in African countries.

Reciprocally, the public will not accept a new product or a change of networks unless experts succeed in passing on information about it. Experts prepare the papers for the government, and when a decision is made, they make the link to the technostructure that will apply it to the public. Within the public the groups that want a mature and enlightened opinion, unions and consumer groups, also have to call in their own experts.

What I said about Africa is not true everywhere. Some technical elements in Latin America have come out in support of the technocratic tendency that is inherent in the dynamics of power. The global tendency to transfer decision making from political sectors into the hands of technocrats may be seen throughout the continent. Brazilian technicians do not all fight for technological autonomy.[3] This brings to light the ambiguous role of experts, which we see when we look at union

---

Participatory Experiment," ibid., 2/2 (1982); cf. also studies of contradictions at the debate on nuclear energy at Autriche.

2. See Ingmar Granstedt, *L'Impasse industrielle* (Paris: Seuil, 1980), p. 123.

3. See A. Mattelart and H. Schmucler, *Communication and Information Technologies: Freedom of Choice for Latin America?* tr. David Buxton (Norwood, NJ: Ablex, 1985).

experts or the experts of ecological groups, etc. The great problem into which we are plunged is that we do not have only one category of experts, but at least two, often at odds with one another. We shall come back to this point.

Lagadec considers another difficulty. Experts are at home with things that go well and recur. Their primary ability is to describe phenomena, to experiment, to reproduce, to measure. Any major risk, being exceptional and aberrant, causes them problems. Accustomed to working with relatively stable states, they come up against limited situations that cannot be reproduced.[4] E. Morin takes up the same point and confirms it. Experts can solve problems for which there have already been solutions in the past. They are powerless when they come up against new problems. They come to see how poor are the general ideas that go hand in hand with their specialized vision.[5]

These two factors entail great difficulties. Experts owe their social status and credibility to the ideology of science. The public is certain that what science says is true. When there are serious accidents, the bodies responsible again call in experts. The experts assure us that there is no risk. But too many experiences are gradually spreading doubt about this and confirming what I wrote in an article about Chernobyl, namely, that what alarms the public is the uncertainty.[6] What experts say is not incontestable. But this fact means disaster for the system. Furthermore, experts, as we have said, are very specialized. But most of the problems posed by technique today have many facets and complex nuances that cannot be dissociated. We run up against the same difficulties here as in political economy when we try to take all the external factors into account (e.g., the costs of complexity and those of ecological troubles). We cannot assign one question to one expert and another to another, for they are all interrelated. We cannot have one study to find out how noxious dioxin is, another study to find out how it spreads, and another to find out the possibilities of evacuation.

Finally, experts are traditionally guardians of the objectivity of science, independent of opposing interests. They have a duty to speak the truth. But the public has now found out that experts, too, are part of the power game, that they belong to organizations, that they plead on behalf of these organizations. This is exactly in keeping with the technostructure as Galbraith analyzed it. Experts who belong to an

4. See P. Lagadec, *La Civilisation du risque* (Paris: Seuil, 1981), p. 148.
5. E. Morin, *Sociologie*, p. 64.
6. See J. Ellul, "L'Incertitude," *Sud-Ouest Dimanche*, June 1986.

organization always justify it, for it is from the organization that they receive their power and authority. Experts are thus an indispensable part of the technical system, for they are exponents of technological discourse. They tie everything together. They engage in technological discourse themselves, but they do not correspond to the image that they are trying to project.[7] According to the severe formulation of Morin, technocratic competence is that of experts whose general blindness enwraps their specialized lucidity.

I will take some simple examples from actual experience. There is first the great problem of experts contradicting one another. One expert says that a nuclear facility poses no real danger, but another, equally competent, draws attention to the many dangers. This was true when experts for the French nuclear power program came up against experts from the ecological movements. But we find it in other areas as well. I will not speak of experts at trials, whose role is really to present the scientific arguments in favor of the one party or the other. Living under the reign of experts, we actually live in a world of uncertainty. We need only refer to the discussion of weapons techniques (Twenty-Sixth Congress of the International Institute of Strategic Studies, October 1984). Everything is up in the air, especially as regards the strategy of nuclear deterrence. What is the real issue, the strategy of technique or the technique of strategy? None of the experts could say.

There are also debates among experts about problems of effectiveness regarding the nuclear weapons of the USSR and the USA, or

---

7. But the human expert is now relieved by what is called the expert system. In addition to individual experts there are now expert systems, a new computer program, which will reproduce the approach of the individual to problems in that individual's area of competence. These systems reason by inference from various items of knowledge supplied by specialists in relation to limited fields. A great mass of information is recorded so that the best possible response will be offered according to the knowledge available. These systems also record the experience of experts and their ways of reasoning. They equate knowledge with fact, and on this basis construct their arguments. There are two stages in the constructing of such systems, that of logical inference, and that of data base. The process of inference is that of expert reasoning as various applicable rules are considered and the best are chosen. There are also "meta-rules" supplied by experts who state their strategies of choice between the different possible rules. A new step in computers is needed for this, that of "the knowledge of knowledge." Experts need to engage in a long study of their own methods (excluding intuition) so as to provide precise and adequate rules of reasoning for the computer. These systems can function to some extent in medicine, geology, and engine failure, but not so well in artistic expertise. Cf. *La Recherche* 151 (1984); Tohru Moto-Oka, "Les ordinateurs de la cinquième génération," *La Recherche* 154 (1984); H. Gallaire, "Les systèmes experts," *La Recherche* 133 (May 1982).

in economics between those who support a particular measure and those who oppose it. Here again the task of experts is to supply scientific arguments for the one side or the other. I have often heard it said that only government experts are reliable. The others are troublemakers. We may simply note that this argument implies that an expert's ability is not scientific but depends on the government stamp. But this is contrary to the role that experts are supposed to play. There are many contradictions, too, in chemicals and pharmaceuticals. Experts in one country may decide that a medication is harmless when those in another expressly forbid it.

I could go on indefinitely. The implication is that scientific truth is not simple, that scientific-technical truth is even less simple, and that only approximate results are possible.[8] Let me give another example. For big construction projects in certain areas impact studies are required. What will be the effect on the total environment (plants, animals, water, climate, etc.)? At first such studies were meant to be decisive, but soon they came to be regarded as merely advisory. A case in which I was interested was a plan to build a harbor and marinas at the north end of a big lake in a very beautiful cove surrounded by steep dunes. An impact study was conducted by a very qualified expert selected by the government. This expert agreed to an imposed three-month limit for his study. (This was the first mistake, for if he was to evaluate all possible effects of the vegetation he needed to study a full year's cycle.) His report was not favorable. The water in the cove was not deep enough, there was not enough movement to dissipate pollution, the prevailing wind would wash filth toward this point, etc. On every count the project seemed to be totally inadvisable. Yet the report concluded that under these conditions it was still very feasible!

On another, very different matter, I myself was once on a ministry commission of experts. After two years of work on very complex issues, we began to draw up our report, but the minister told us that the report had already been completed by secretaries who were present at the meetings and took notes. The text was read to us, and it bore no relation whatever to our own conclusions. The experts rebelled

8. The danger of the expert was admirably brought to light by J. Klatzmann, who shows that a statistical figure advanced by an expert is repeated without control by a string of authors. He cites a hundred examples; one is convincing enough: an expert wrote that 50 million people died of hunger in 1981, whereas the total number of deceased in that year, from all causes (wars, sicknesses, age, catastrophe, etc.) was at most 46 million. There were perhaps 10 or 11 million who died of hunger. But the figures of the expert were repeated by all the journals (see *Nourrir dix milliards d'hommes?* [Paris: PUF, 1983]).

and made the minister back down, so that our report became the official one. But the minister then introduced legislation which embodied not our conclusions but those that he had dictated to his secretaries and that ran directly contrary to our advice.

Again, experts often come up against questions which have never arisen before and on which they have only fragmentary data, the situations being totally unexpected and unforeseeable. (An example is the problem of milk for infants among some African peoples; it took time to realize that this milk was veritable poison in these areas.) The result can be terrible blunders by some experts. I recall that thirty years ago when the development of the Third World was at issue, the experts (apart from René Dumont) concluded that we must industrialize the Third World as quickly as possible, replace traditional plants with stronger ones, and help the Third World to increase its supply of energy. Industrialization would be in terms of the developed world, and the same technology would have to be transferred. Under the heading of aid, many machines were thus sent which served no useful purpose (e.g., trucks which had no paved highways on which to travel). Industrial agriculture was also overdeveloped at the expense of subsistence crops. I could run off a whole list of what were later seen to be mistakes.

At the time, however, no one dared to say that first of all we must develop subsistence crops so as to provide in this way (and not by aid) enough basic food, or that we should not export our techniques but find simple technical means that do not use much energy and are adapted to the needs of the countries. To say such things was to run the risk of general abuse. The experts, the leftist rulers, and the elite of Third World countries would all say that such an attitude had only one real objective, that of maintaining Western superiority and keeping the Third World in colonial subjection. Only later were the blunders seen to be such, along with many others, such as giving food aid, which is disastrous for a country because it prevents the development of local agriculture. Efforts did begin later to produce "appropriate techniques," adapted to the conditions in each country.[9] But the invasion of the Third World by multinational corporations brought new conditions of exploitation. Thirty years too late it was seen that the problem was not that of aid and transfers according to a financial

9. We must not think that everything has been put right. Recently the advocates of "appropriate techniques" have been accused of preventing development and of preventing African countries from reaching the technical level of modern countries.

and commercial plan.[10] What was needed was the conception of a global economy into which the Third World countries would be integrated as participants and not just as suppliers of raw materials or labor.[11] But this is only to say that the task of experts can be very difficult and their efforts fumbling.

As a last example I will take Chernobyl. As in the case of Seveso, information about this event was very confusing. But I am not dealing here with the problem and difficulty of information, as is usually the case. I am dealing with the question of experts. We had on television an expert who explained that there was no need for alarm, that the "cloud" had no chance of reaching France, that its radioactivity (in becquerels) was weak, that the accident did not seem to be serious, etc. The aim was clearly to reassure the public. This is where my inquiry starts. Is it the role of experts to reassure the public or to speak the truth?[12] As the days passed it had to be acknowledged that the cloud was covering most of Europe (except France). Some people began to ask why the Germans were much more alarmed than the French, and the answer was that they had fixed the limits of tolerance to radioactivity much lower than France. But were the Germans and the Atomic Institute of Vienna right about this? Why, if it is a purely scientific matter, do some fix the threshold at one point and others at another point? Which experts are right? I also noted that the television experts used the term *becquerels* in measuring radioactivity. Since I had never heard this term before I suggested that it did not do much to enlighten listeners. I noted, too, that when the nuclear power program was adopted in France, the French experts had themselves disagreed about fixing the threshold of tolerance.

I was treated very roughly by the experts of the nuclear power authority, who told me that I was very ignorant not to know what a becquerel is, since it had been adopted as an international measure three years before. They also denied formally that there had ever been any variation. But I was not satisfied. Radioactivity had in fact been measured successively in four different ways, and it was now measured

10. The apparently fair notion that we should pay for Third World products at their true price, and not, as is customary, much below this, is in fact disastrous, for to pay more for cocoa, coffee, etc., is to encourage the local growers to produce more and to grow even fewer subsistence crops.

11. See Jean Touscoz, et al., *Transferts de technologie* (Paris: PUF, 1978).

12. As regards radioactivity there were incredible conflicts between experts after the accident at Windscale in Cumbria. Some (nuclear experts) said that there was nothing to fear, but others (biologists and the medical center at Seascale) said that there was everything to fear (*Le Monde*, Nov. 1983).

in becquerels. But since only experts had access to tables equating the different scales, to talk in terms of becquerels amounted to no more than disinformation. I argue that it is often the true role of experts to overwhelm people from the height of their scientific superiority and in reality to disinform them.

As regards the second point, the experts in this case had very short memories. At first, around 1952, they talked about a threshold of absolute safety. They then changed their vocabulary and began to speak about a threshold of tolerance, which is not the same thing at all. As regards the threshold of tolerance, between 1955 and 1962 it was changed no less than four times (according to government circulars), and raised each time. It is true that there were no more changes after 1962. But this was exactly in step with the adoption of the nuclear program. I might cite many other examples of dubious estimates on the part of experts.

What are we to think of all this? Certainly I would not say that experts are incompetent—I am convinced of their competence. I also cannot say that they are not serious. They are very respectable people and do not talk at random or take risks. Not for a moment would I say that they are not honest. They undoubtedly follow their conscience and not orders. But granted all this, and their appointments and qualifications by recognized bodies, we have to realize that they are in ambiguous situations. They are scientists who have their opinions, and contrary to what is often thought, modern science is less and less clear and univocal, and convictions derive often from opinions based on such and such experiments and findings. They have to talk about their science to politicians who cannot understand everything. Necessarily, then, they have to give their opinion. But they have also to talk to the public at large, which has little or no understanding at all. They have to help them to accept a project, to calm their fears, to interest them in a new project, etc. Their science is thus the platform from which they proclaim orientations that are accepted as truths and that will finally shape opinion. In these circumstances they cannot be fully credible to the critical mind. Engaging in technological discourse, they end up as its slaves and have to follow the common path of progress.

PART III

The Triumph of the Absurd

In this part we shall run up against many themes that have been dealt with in profound and learned works. I am familiar with a number of them. But I am not going to engage in a compilation or a new scientific work. I will deliberately adopt the attitude of an average citizen who puts questions that intellectuals always avoid, the simple questions of common sense. I know that scholars and philosophers take a poor view of common sense, yet I want to keep to this elementary level, for in the learned studies that I have read it is plain that common sense is deliberately set aside, that it is never even taken into account. Now as I see it, if we reject common sense, we open the gate to nonsense, to absurdities and fantasies. Common sense seems to me to be a necessary expression of reason, and it is reason that we must apply constantly as we evaluate what technique proposes and what technology declares to us.

We have seen already that it is not because technique makes a pretense of rationality that reason remains intact. Rationality has made great technical progress possible, but it is reason that has allowed us to survive, to live, and progressively to affirm our humanity. We have simply not to let ourselves be submerged by scientific knowledge nor reduced to a common sense that is mere ignorance, that will not listen or know, that is closed-minded. I think that it is this simplistic view of common sense that has led to its condemnation. People of common sense end up being those who, anchored in simple certainties, refuse to question them and wrap themselves in a mediocre middle-class ignorance. The common sense that I esteem is that of Flaubert's Bouvard and Pécuchet, who themselves, let us not forget, claimed to be followers of science!

CHAPTER XI

Technical Progress
and the Philosophy of the Absurd

In France and elsewhere, in the postwar years, the philosophy of the absurd developed. This was apparently a contradiction in terms: the absurd and wisdom seem to be incompatible. But for a long time philosophy has not corresponded to its etymological sense. The philosophy of the absurd developed along with existentialism and within it. It did not, of course, characterize existentialism as a whole, but was related to the existentialism of Jean-Paul Sartre. Its main orientation was that life and all activity and human thinking was absurd; they make no sense, nothing makes sense. To live is a pure fact. There is no meaning in what happens, nor are we to search for meaning or to attribute it. History makes no sense, it is going nowhere, it obeys no rules, it has no permanence. Good and evil do not exist; hence no morality is possible except a morality of ambiguity. Relations with others also have no meaning; in every respect they are completely impossible. One person is not and cannot be understood by another, and does not and cannot understand another. There is permanent misunderstanding. What we do is foolish to others even though perfectly reasonable to us. It is impossible to communicate. The glance of others is the worst thing to endure; hell is other people. We are in an absurd situation from which we cannot escape. Every attempt to escape is absurd. There is no fixed point of view from which to evaluate an event or an act. There is no supreme being to which to refer. Only what exists is real. But this, too, is as shifting and uncertain as water or sand. Nothing has form. We might view this as freedom; after all, it does not matter what we do. Doing this or that is of no significance. We are free to do the one thing or the other, for both are indifferent. Choices do not have to have reasons. They simply are.

All this leads, of course, to disjointed and contradictory behavior. In the relations between men and women, uncertainty reigns, for we have only to be "honest" with ourselves. Honesty with the self is the only thing that is not absurd. We have to be fully ourselves each instant. I may love a woman or a man and give myself up to this love. But I must be on guard lest this love become a habit, a good, a loyalty. I must look out for the moment when I cease to love with passion, force, exclusiveness, etc. At that moment I can honestly say that love is over and break it off. What about the other in all this? As stated already, we can have no true relation with the other.

It is just the same in politics. I belong to the social body and therefore whether I like it or not I am in politics. I cannot not be. Being honest with myself implies this. But there is no just politics (justice does not exist). There is no doctrine to which I may adhere. Political engagement is only for the moment. It is what I think I must do or defend at the moment. Hence I may change my political position according to circumstances, impressions, or emotions. Sartre constantly changed his (always momentous) statements; sometimes he could write contradictory articles only two weeks apart (e.g., on the crises in Hungary and Czechoslovakia). Sincerity and commitment are only for the moment. There can be successive forms of sincerity in every area of life. Naturally, it is argued that even science cannot give us certainty of any kind. At a roundtable discussion Sartre made the famous pronouncement to celebrated physicists: "As a philosopher I know much better what matter is than any physicists."

In this kind of wilderness where no orientation is possible there is only one reality: our human being. We must aid this, as the doctor does in *The Plague* by Camus. Medicine, of course, is also absurd, but it is the only activity one can select. Sartre, too, involves himself on behalf of the poor and the unfortunate, but not out of pity, charity, or virtue, only on the condition that his action has no value and gives him neither meaning nor justification. Nevertheless, if I do not do it, suicide is ultimately the only option. This, too, is an absurd act that is not obligatory. Everything and nothing are the same. Hamlet's question is no question. We can simply stay in the corner into which we are driven.

Naturally, this philosophy of the absurd means rejection of all previous philosophy, which tried to find meaning and points of reference for the appreciation of life and coherence of human thought. The philosophy of the absurd gives rise to a plentiful literature (novels and plays) which is both striking and passionate. Thus we have the plays of Camus (e.g., *Caligula*), those of Sartre *(Dirty Hands, Devil and the Good Lord)*, then those of Ionesco. But on the same basis there are two different developments. In literature, for example, we move on from the drama of the absurd to absurd drama. In Camus and Sartre we still have a plot and personal relations as in classical drama. These lead the audience to the conclusion that in effect life is absurd and has no meaning. But this was not enough. There was thus a movement to drama which would itself be absurd. The absurd would not just be demonstrated; an example of it would be given. The characters would now exchange meaningless and incoherent words with no beginning or end. Then onomatopoeia would be used. In Beckett we see the transition from the theater of the absurd to the absurd theater. There, characters are still speaking and acting, but their words are continuously overlaid by booming music. In May 1986 Paris even staged a play lasting for two hours in which the two characters exchanged sounds that did not make any discernible sense. In novels we have the school of the new novel (which is now said to be outmoded, but which still persists).[1] This has neither story nor characters, and also no punctuation, so that there is no understanding it without the key. One of the specialists says plainly that there is no meaning and that we are not to seek one, that there is no story, that ultimately there is no author, and that even the reader does not exist as a subject (I am not exaggerating). There are only structures and structural games.

The second development on the basis of the philosophy of the absurd is nihilism. Nothing has value. We must reduce everything to nothing. Nihilism can be explained as such (Cioran), or it can take a theoretical form which leads in art, for example, to the view that the artist must produce a non-work and ultimately no work at all. But the nihilism which derives from the philosophy of the absurd can also affect those who are not philosophers or artists, leading some to suicide (cf. suicide among young people) and others to terrorism (who join hands with the nineteenth-century Russian nihilists and whom Camus views with much sympathy and understanding in *Les justes*). Naturally, I am not saying that actual terrorists in the world today have even a slight knowledge of the philosophy of the absurd. My point is simply that this philosophy has penetrated much more deeply than we think and created a climate in society as a whole in which terrorism could develop. Nor must we forget that some terrorists (e.g., the Baader gang) were in fact intellectuals.

1. I studied all these phenomena in detail in *L'Empire du non-sens: L'Art et la société technicienne* (Paris: PUF, 1980).

The philosophy of the absurd flourished in a political climate which explains it in part: the Nazi occupation, fear of the Gestapo, the impotence felt by the resistance, and the discovery of the frightful atrocities in the concentration camps when the war ended. In these conditions, in this excess of evil, one can understand why philosophers could say that life is absurd and that there is no solution (not even with the liberation and its painful aftermath). There are no righteous causes; there is no good or evil. Human beings are atrocious. Caligula is representative, and Malraux has to fall back on the history of (ancient!) art to try to find meaning in humanity and its history.

Finally, it is possible that this philosophy of the absurd has had some influence on scientific thinking. This might seem to be incredible. Yet if we look at the hypotheses of the last twenty years in physics and biology (e.g., cybernetics, feedback, or the main concepts of communication), we will be surprised to find in research such notions as loops, vortexes, and turbulences; we have also heard about research on the odd form of a candle's flame. But what has all this to do with the absurd? Simply this, that there is a complete reversal in the understanding of this order of phenomena. Thus in communication, noise used to be a purely negative notion. It prevented information from being transmitted and received. But all that has changed. Noise is now an important if not a decisive factor in communication. It is itself information and it must be integrated into the theory of communication. Similarly, in physics there used to be a clear distinction between order and disorder (even though this meant abuse of the laws of thermodynamics). Disorder was simply perturbation. Like noise, it has a negative connotation. But all that has now changed. Disorder has become a positive phenomenon and it must be integrated into research in physics (not eliminated from it). Physicists are beginning to say that order can finally come only from disorder (as information does from noise).

As Henri Atlan says, then, physicists are situated "between the crystal and the smoke." Smoke is not an unimportant phenomenon. It has a form that obeys laws even though these are more obscure than those of the crystal. No doubt all this is true (I do not have the competence to judge) and it is certainly appealing. But if we leave the sphere of abstraction it becomes a vindication of disorder. And I understand this vindication very well. Physicists do not reach it, I think, by pure scientific research. As in the case of all scientific hypotheses, I believe that scientists, belonging to a society and culture, are inevitably influenced by them. We live today in a very ordered and coercive society (even though it is morally lax). We have to break free from it. We have to arrive at a new appraisal of disorder, to deal with order by putting disorder in the scales against it. I can understand this. But it means introducing the absurd, which is the surest form of disorder. And I fear that in economic theory the insertion of loops will not in fact justify a certain absurd economics.[2]

1. Technical Absurdity

Granted the above summary outline, I want to advocate the following thesis. Over against the philosophy of the absurd stood the sciences, which were not at all absurd. Technical expansion in particular seemed to be a model of rationality, strictness, efficiency, and exactitude. There was nothing absurd about it; quite the contrary! In combinations of techniques as well as in economic systems everything was reasonable and rational. Undoubtedly, one might say that in a too organized and systematic environment people were maladjusted and had incoherent reactions (e.g., that of violence). This came out in 1968 with an attack on the whole techno-economic setup. But it was the human reactions that were absurd. The technical system remained coherent.

Now the new thing in recent technical evolution as I see it is that the techniques developed in the last decade (principally in the area of computers, telematics) are themselves leading to absurdity. They produce and demand absurd behavior on our part; they put us in absurd economic situations. At its extreme, modern technique is linking up with the philosophy of the absurd in a way that was not foreseen. It might always be said that this is our own doing. Human beings are present in all these situations and developments. I repeat, however, that it is the flood of techniques that makes us absurd. To give a minor example, which must not be taken too seriously or used as a proof, some modern films carry authentic background noises (e.g., of a street or airplane) and these are so loud that though we see the characters talking we cannot really understand what they are saying because of the noise. We catch only a word or phrase, so that we cannot say any more that the noise creates information. Yet the words that are swallowed up by the noise, and concerning which we know only that they are spoken, do stir the imagination, and we imagine what the characters might have been saying.

In this context I am not going to engage in an abstract and

2. Cf. Henri Guitton, *De l'imperfection en économie* (Paris: Calmann-Lévy, 1979).

theoretical study of my thesis that modern technical growth induces the absurd. I will simply give some examples that will form concrete matter for reflection. A first example is the absurdity of the inexorable constraint of technical growth. Things are produced that we do not need, that serve no useful purpose. We produce them because technique makes them possible and we have to exploit the possibility. Inexorably and absurdly we have to follow this direction. In the same absurd and inexorable way we also use things that we do not need. Let me give three examples.

There has been a great propaganda campaign in France on behalf of the expansion of the telephone network. The number of people with telephones has doubled in the last ten years; there are now twenty million instruments in service. Unfortunately, the level of use is very low—the French do not use the telephone. Statistics show only an insignificant number of calls. Should development be arrested then? Not at all! Ignoring the statistics, the technicians decided that we must have twenty-five million telephones by 1985. This means one instrument per family. But it also means a further decline in average use. To make good the deficit someone then has the bright idea of creating situations in which people are obliged to use the phone. This is one of the most important motifs in the creation of the system for which there has been tremendous international propaganda, that is, the Teletel (an electronic telephone directory). This combines telephone, computer, and television (to promote this system free computer consoles might be provided). With this system it will be possible by phone to obtain a telephone number, the times of trains and planes, the price of goods on sale, cinema and television programs, etc. To force people to use the system the printing of directories, timetables, etc., will be stopped. Subscribers will then be forced to use the phone, the statistics of average use will improve, and inevitable technical progress will be justified. We have here an absurd series of developments which are all dictated by the compulsion to apply sophisticated technical instruments that we do not need.

Incidentally, throughout these pages I keep coming up against the expression "we do not need." I am not unaware of the innumerable discussions among psychologists and sociologists concerning natural, artificial, inward, and cultural needs, etc. I will not go into these abstract analyses. I do not deny that what was not originally a need (e.g., drinking iced drinks) may become just as natural a need as one dictated by physiology when we adapt the habit and keep it up long enough. I will restrict myself to some simple facts. When I learn that shops in the USSR are full of certain industrial products that no one is buying, I infer that there is no need for these products and little chance that a need will develop. The same applies to telephones in France. But once an advanced technical product is created, the important thing is to force consumers to use it even though they have no interest in it. Technical progress demands this. One might say that this depends on the people who make the decisions, and that after all they might take a different course. But this is not so. If our country is to remain among the most advanced, we have to follow the march of progress and invent even better gadgets, that is, gadgets that are even more absurd and useless, in order to stay ahead of competitors.

Electrical energy offers us a second example. After the war the word in France was that we must produce as much electricity as possible to replace coal. Great hydroelectric projects were undertaken. Every small stream in the Pyrenees and the Alps had its power station. After 1955 there was too much electricity. The stations could not run at full capacity and were not profitable. There was thus a tremendous campaign urging the people to use more electricity. Big buildings were put up heated by electricity and using enormous amounts. A tariff was introduced whereby the price went down with increased use. But suddenly around 1960 it became clear that the growth of use had become exponential and there was a need for more production. Thus the nuclear program was launched, but not without violent conflicts between the engineers and the ecologists, who were not all idealists but included physicists, biologists, economists, etc.

A study conducted by the Center of Economic Studies of the University of Grenoble in 1971 concluded that the price of a kilowatt hour from nuclear power stations was three times greater than that projected in studies (propaganda!) by Électricité de France, which had promised a lowering of rates, and that the program as it was envisioned would surpass actual needs by 1985. Naturally, no one paid any attention. The remarkable thing is that the commission on energy for Economic Plan IX concluded in 1983 that the nuclear program ought to be halted because production was already outstripping need and the price was of the order calculated by Grenoble in 1971. But the first reaction of some groups was not that the program should be halted but that a new campaign should be initiated to incite the people to use much more electricity even if only to use up what was being produced and not for any useful purpose.[3]

The absurdity here does not lie merely in the arbitrariness of the

3. We will return to this interaction in more detail when we study unreason below.

procedures but in the actual situation of total inability to see what is necessary and has meaning. I am not saying that those who set up the vast nuclear program were dishonest. Nor am I saying that the report on Plan IX was made by partisans. I regard its authors as good and honest technicians. My point is that they are simply unable to see exactly what is our situation, what will be the needs in two or three years, etc. The absurdity lies in the prediction itself.

A final example of the inexorable processes to which improvements give rise may be found in television and radio. I might deal with this on two levels. The first is that of the big existing systems, the "chains." My simple question is as follows. The networks have extraordinary equipment (enhanced by satellite television), which has to be used. They have to transmit; this is an imperative. They *must* transmit most of the day: news, shows, songs, discussions, interviews, films, advice on health, cooking, etc. They must transmit something fresh every day. Thus they are in a terrible bind—they must! It does not matter what they transmit so long as the screen is not empty. But it is impossible each day to find something genuinely true, beautiful, intelligent, and fresh, something worth showing and repeating. Hence the screens are full of inanities. It does not matter what is put on so long as the screen is not empty. Whether it causes the viewer to laugh or to shudder, it does not matter so long as it is new. People are thus brought in who may not have any real quality but who are more or less well known or recommended. The demands of viewers are easy. They do not want people of genius who are too much above most of them. They want honest mediocrity. It is within this basically unimportant range that the stations have the best chance of finding people to put on. The main thing is to have something fresh. That alone is what counts. When real intellectuals are found, they have to speak at a level that can only do them dishonor. A novelist who is a best-seller is a safer bet. But a new one is not always available.[4]

We thus have a deadly combination of equipment that makes demands and that also gives a hearing to people without quality. Every hour viewers see the poorest of shows on almost all stations. We need in this regard Kierkegaard's profound analysis of the masses and the crowd, which mean mediocrity, baseness, and falsehood. The media cater only to the masses and the crowd, and they have this conglom-

---

4. Cf. the excellent work of Jézéquel, Ledos, and Regnier, *Le Gâchis audiovisuel* (Éditions Ouvrières, 1987), who show that those who think they know the television public align themselves with the most stupid viewers. The obsession with ratings, aggravated by private television, enforces a deterioration in quality.

erate of individuals as their implication. The level is all the lower as the number reached is the greater, and the technical equipment demands that the number be constantly greater. This is my final example of the absurdity caused by technical equipment itself.

We referred above to similar experiences with what is called free radio in France. Freedom is badly served. Once again we see an instance of technical absurdity. We have the instrument, the equipment, but there is nothing to transmit. The same is true of CB radio, on which the messages are ridiculous and childish. Communication is being perfected. It has become swift, global, accurate, etc. But unfortunately there is nothing to communicate except what is banal and inane. Yet the equipment is there; we have to cater to it.

2. Economic Absurdity

There is no need to recall that economic life in our world is wholly organized in terms of techniques. As I see it, the global situation is as follows. We still have an economic model in terms of the industrial system in which the primary function of technique was to promote industry. The movement is from investment to mass production to mass consumption to mass returns or profits, which are then reinvested. One can follow the circuit in different ways. If a good part of the profits is reinvested, with salary reductions and eventual unemployment, we have a liberal and ultimately a Keynesian approach. If we focus on consumption, a great deal of cash has to be distributed among the public so that they can buy more, and this will result in increased production and then increased investment. This is a socialist approach. We increase wages, give good grants to the unemployed, and encourage investment by favorable interest rates (unless it is the state itself that is in charge of industry, which is meant to have a stimulating effect on the whole industrial complex).

These are the two positions in outline. I beg to be excused for stating what is so obvious, but it is precisely as things are reduced to the most simple and elementary level that their absurdity suddenly appears, as we shall see in a moment. For this whole system has come up against a new technical development which upsets the placid approaches. The new factor is productivity, that is, producing more with less work. This can make itself felt in two different ways. First, there is competitiveness with others, with rivals, or internationally with other industrial countries in the search for global markets. According to the liberal logic the best will win, and we should thus let the most efficient business squeeze out the others. But efficiency can take the form either of making the best use of equipment or of producing new goods, whether improved forms of older ones or totally new ones hitherto unknown. According to the socialist system there is no competition, so that there are fewer problems in the domestic market, with fewer failures but also less innovation and progress. Yet socialist countries are now inevitably competitive with capitalist countries on the international market. They cannot live in isolation but have to sell their products so as to achieve a balance of trade.

Thus far everything is familiar and in order. But what I have written is no longer wholly accurate. The technical changes of the last twenty years have altered things. First, automation and computerization have brought with them unimaginable possibilities of productivity, so that there can be no hope at all of absorbing in new work the unemployed who have been put out of work by the new machines. In the industrial sectors the possibilities of productivity have become almost infinite. This means that economically it is the machine and not human labor that produces value. Human labor has become increasingly unnecessary, and it is possible to imagine that in another twenty years we shall be moving toward absolute unemployment.[5]

Now the economic logic has not changed. On the one hand, firms must be as efficient as possible, and on the other hand it is hoped that the unemployment problem will be solved by an economic upturn and by new ventures. But the new ventures will have to employ as few people as possible to be competitive. Socialist countries will either have to withdraw from the circuit (which is becoming impossible) or follow the same logic. Pumping in additional money to revive the economy is no solution. But what about new products such as talking computers, tapes, domestic computers, flat-screen televisions, cars with automatic pilots, etc.? When we consider the nature of these inventions, however, we see that they demand hundreds of researchers and millions in investment, but we also see that they are only gadgets that do not really meet any kind of need, not even an urgent desire.[6]

I realize that a broad statement of this type will bring a reaction from any conscientious technician, yet when in the innumerable studies that I have read I compare the real usefulness of, for example, computers (and their derivatives, i.e., office automation, telematics, robotics)—their usefulness in keeping accounts, in aiding scientific

5. Cf. *Informatique et Emploi* by the Economic and Social Council, 1984. We shall take up this problem in more detail in chapter XVI below.
6. We shall examine the world of gadgets in chapter XIV below.

research, in inventory control, in memory, etc.—I am compelled to say that hundreds of the objects that are now set before the public are useless gadgets and that there would not be an adequate market if only useful goods were produced. Thus we have many objects that are only for amusement or to cause a little surprise. We are adding superfluity to superfluity, and it is solely in this domain that we are creating new goods. The very definition of political economics has been upset, but the reasoning is the same, as if nothing had happened. It is true that putting on the market one of these astonishing, sophisticated, magical toys gives a business an important advantage, but the market is soon saturated, interest in the little miracle evaporates, and something new has to be manufactured.

Similarly, production in the whole area of computers can give a country an impressive advantage (e.g., Japan from 1970 to 1981), but the whole situation is absurd. The country has an advantage only so long as it is the sole producer. Once six other industrialized countries begin to imitate Japan in the hope of achieving the same success, the basic question (which French politicians who have entered this path seem incapable of asking) has obviously to be posed: Who is going to buy all these things? Do we imagine that we French can capture the Japanese or American market? If all industrialized countries begin to produce the same goods, there is no hope of capturing on a long-term basis a profitable market with them. The only hope, then, is the Third World. But the Third World is not interested in these products and does not have the money to pay for them. The example of the Japanese, then, is a bad one. This is apparent once we look at the economic situation globally, at the world economy (it is a world economy, and not, as is wrongly said, an international economy).

We here come across a great contradiction which startles us. On the one hand are the economies of the developed countries, which function as I have said, and on the other hand the economy of Third World countries, which is going from bad to worse, since even the most necessary, immediate, and vital needs are not being met. On the one hand are economies which can function only as false needs are stimulated and gadgets created, and on the other hand an economy which cannot respond to famine and the minimal needs of civilization. The absurdity reaches a climax when specialists have only one remedy for the Third World, namely, to put it on the same track as ourselves, to bring it into the industrial circuit, and to aid it, as Rostow said, to break free from an economic standpoint. How absurd all this is when we see concretely the results of this system of ours!

We are truly in the presence of erratic economic thinking (and sadly, we have to say, economic practice). If our system functions as it does, it is because it has accepted integrally the primacy of technical innovation and the law by which it is technique that permits economic progress.[7] Obsession with technical innovation brings our system into a series of logical follies and puts it out of step with the economies of people in the Third World which are very diversified and fragile and which demand very accurately adjusted technical and human intervention. The idea that computers can enable the Third World to break free is crazy. But the course of technical primacy is leading us even further. We are now beginning to talk about countless gadgets. This means that we shall be talking about sheer waste.[8]

I am not referring here to the enormous daily waste which is often denounced and which easily scandalizes us (e.g., the excess food which is thrown out by restaurants even though it is perfectly healthy, or the crops that farmers destroy because they are not paid enough). I am talking about the waste ineluctably produced by technique (e.g., the regular replacement of equipment, cars, motors, refrigerators, television sets, etc., often because people have to have the latest model). The great law that we must not stop progress operates both at the individual level and at the national level. The best example is the constant replacement of weapons. There is no end to the production of more powerful and sophisticated armaments even though it is known that they will have to be replaced six years later.

What we have here is techno-economic absurdity in its purest form, for the goods that are produced are totally negative. If we use them, the result is negative because of the enormous destruction they will cause, and if we do not use them they serve no useful purpose except that when we discard them we can sell some of them to underdeveloped countries! I am well aware of the economic argument. They keep the wheels of industry turning and supply jobs. On this reasoning the pharaohs who built the pyramids were great economists!

But technique is responsible for other forms of waste as well. We will not take up again the wastage of raw materials. I am thinking more of the waste of air, water, space, and time. These are vital elements and dimensions of human life even though they have no economic value,

---

7. Has not a famous economist stated that the actual crisis, following a Kondratieff cycle, will resolve itself like prior crises by a major technical innovation around which the whole economy will function? As the automobile led us out of the crisis of 1930, the computer will lead us out of that of 1980.

8. I will study waste in chapter XV below.

and we are wasting them at a frantic pace. Absorbed in technique, we never have any time. Demographic growth within half a century will fill all the existing space on earth. I will not stress these points. The facts are well known and incontestable. But they are so serious that they are always carefully obscured.

A third form of waste is much less tragic but by no means negligible. I refer to certain spectacles which are purely technical and justified only by technical imperatives but which the West now regards as obligatory. Thus we have races for Formula 1 cars in which each model will cost millions and be used only once, not to speak of the enormous waste of fuel. The only justification that is given is that experimentation brings technical improvements, but the improvements in engines and tires are only for the racing cars themselves or for some sports cars that can travel at high speeds without danger, when the great need today is to reduce speed because of accidents. We might also speak of the remarkable improvements in giant trimarans which serve only to improve the techniques on boats that no more than the privileged few can afford. This technical justification is totally absurd. Nothing justifies the enormous expenditure on pure luxuries.

Another aspect of economic absurdity has to do with the size of the numbers which we juggle. Thus the American budget deficit ran to nearly $200 billion in 1983, or 6.6 percent of the gross national product. One has to ask whether there is any sense in trying to formulate an economic policy that has to include a deficit of this kind, especially as the social costs of technical progress, which will surely increase it, are not taken fully into account. Then there is the stupefying phenomenon of the indebtedness of almost every country in the world. How can we integrate this into a world economy? The sums owed by Third World countries are staggering. Latin America owes $300 billion and in a few years the debt for all developing countries will amount to $620 billion. The total charge will in fact be $700 billion, and there was a net loss of $200 billion in 1982. But these countries refuse to accept a policy of austerity and limit servicing their debt to a percentage of their exports (which in turn results in a limitation of their exports!).

We have to realize that these exorbitant costs and economic problems derive solely from the rapidity of technical growth (and not from this or that economic organization). Third World countries have to pay 67 percent of the value of their exports to service their external debts (most of which are due to the purchase of armaments). None of them will ever be able to pay off the whole debt. Either the creditors will have to write off the debt or 50 percent of Third World countries will default. There is no similarity here with the Marshall Plan, for this borrowed money does not serve in any way to equip the countries economically according to their level and for their benefit.

In these circumstances how can we envision a global market that will function "normally"? Third World countries that grew fabulously rich through oil (producers in the Near East) in ten years did not really know what to do with their money. This throws serious doubt upon the whole theory of breaking free economically. There is no longer any economic rationality or logic. I might almost say that there is enough money in circulation in the world; it surely ought to serve some useful purpose!

But almost all countries are crushed by the weight of their military expenditures (Third World countries as buyers, producing countries because they are launched on this endless course of the most efficient technique). In 1983 the USA spent some $600 billion on weapons, or five times its industrial investment.

I could pile up facts of this kind. They all go the same way and show that the thrust of techniques in all areas is toward an economics that is neither possible, predictable, rational, nor capable of global organization. The only chance that economists see is to proceed faster and faster in the rapid adoption of all techniques (no matter what their significance or use) so as to be sure of keeping ahead of others once they appear. But no one can construct an economy in these conditions. Those that pretend to be planned economies in socialist countries with authoritarian planning present an interesting study. On the one hand is the planned and controlled and governed sector which is backward, outmoded, and inefficient, and on the other hand is a sector in which free rein is given to technical development, in which there is research in weapons, rockets, space, and the nuclear field. But in this domain we find the same disorder, the same absence of coherence, foresight, and rationality. It is a remarkable fact that an excess of technique always leads to absurd situations, to an impasse from which it is impossible to see any exit. The facts are beyond the possibilities of human consciousness.

In face of these realities I take an opposite view to that of politicians (like Mitterrand), who believe that it is by the most efficient "advanced technology" that we shall achieve universal growth (I would say universal disorder!) and solve the problem of unemployment. If I am not deceived, that is also the view of one school in the USA, the high-tech Democrats who believe that high technology will lead us out of the economic impasse. What we have here, I think, is a mistake in diagnosis. We are not just in an economic impasse but in general disorder. Politicians and economists of that kind are really dreamers.

They "believe"; they have a kind of religion of a radiant technical future. They do not engage in reasonable reflection. This religion is simply a confirmation of absurdity.

3. Human Absurdity

I am dealing here with a question that seems to be properly philosophical. Do we become more human thanks to techniques? Will there be a kind of transformation of the human species? Will technique enable us to fulfil the ancestral human goal? This was the belief of, for example, Teilhard de Chardin. I have tried to show how at one point (modern art) technique has brought a radical break.[9] Using the most modern techniques in music, painting, sculpture, and architecture has resulted in products that might be called art (it is a matter of definition!) but that have nothing whatever in common with the aim of the race for the last five millennia in doing work that we think of as art, that is, works with meaning (that is to be given or found), with beauty, with harmony, lofty works that give happiness. Modern art is the exact opposite of all that. We will not say that technical art is not art, but that what is achieved, far from being a fulfillment of the human goal from antiquity, contradicts it and negates it.

But that said, the question remains: will we become more human?[10] I will give an example that I have often adduced because it seems to me to be very significant. In the 1960s, at the beginning of genetic engineering, a French journal asked a number of Nobel prizewinners (biologists, chemists, geneticists, etc.) about the new techniques and about the human model that one could hope to achieve by manipulations of the embryo. But not one of these important scientists could give any answer. Except for banalities like making people better or more intelligent, they could not say what human model seemed to them to be desirable.

In reality we do not really know what we are talking about when we say that we must become more human. (For Hitler it meant Aryan genetic selection.) We do not know what to do with these remarkable and marvelously efficacious methods. This means that although we do not want to create Frankensteins, we are creating we know not what. Yet that is not my present point. The immediate question that arises for

9. See Ellul, *L'Empire du non-sens*.
10. The most important book in this area appears to me to be Vance Packard, *The People Shapers* (Boston: Little, Brown and Co., 1977).

me in the presence of the effects of existing techniques on people (whether infant or adult), and their probable development, concerns the type of person that is created by the millions without the slightest genetic intervention. I would describe this person according to my own encounters as fascinated, suffering from hallucinations, and distracted.

People in our society, once obsessed with work, are now fascinated by the multiplication of images, the intensity of noise, and the spreading of information. In all three areas we have the effects of techniques on all of us even though we might not be too fond of television or shows. We cannot escape these effects. I think of the general growth of noise in all forms of modern music. (This is not at all the same question as that above.)[11]

It might be said that all this is not the fault of technique but of the people who use it and who turn up their sets to the maximum (when not at concerts). But this is the very thing that troubles me, namely, that listeners demand that they be offered an overwhelming music which shuts out all awareness. This is the element of fascination. The listeners are like those on drugs who cannot want anything else. And to me the worst and most significant aspect of the situation is the development of the Walkman. It is an absurdity that young people cannot live a single hour without this music which batters their skulls. They are so intoxicated by the noise which effaces all else that they need it even in trains or cars. They cannot escape its magnetism, which prevents them from being aware of the external world, from receiving other impressions, from living in the real world, and from breaking free from their obsession. This noise, and we cannot really call it anything else, is doubling the noise of the urban environment. There is general agreement as to the harmful effects of permanent noise (e.g., that of automobiles or machinery).[12] Some attempts are made to fight it. But music is now forcing upon us an equally powerful noise which is all the more obnoxious because it is a matter of choice.

The invasion of images is to the same effect, not just those on television or films, but those of advertising, which is not just the neutral and immobile advertising on posters but the mobile and active advertising of attractive animation. Images both capture the attention and dissipate it. We are grasped by a universe of ridiculous possibilities that are forced upon us. I am not saying that this advertising makes all

11. We will look in detail at the matter of fascination, and especially the influence of rock music, in chapter XIX below.

12. Cf. J.-C. Migneron, *Acoustique urbaine* (MLS, 1980), who insists especially on the serious nature of background noise.

buyers buy the product. That is not the question. The point is that the multiplication of invading images sets us in a wholly artificial world. In this world no reflection, choice, or deliberation is possible. Such advertising is not in any sense innocent. If it succeeds, this is because it takes the reality of people today, their wants and desires, into account. When we analyze the main themes, we see that there is violence. Many advertisements, urging us to be modern, portray what are basically images of aggressiveness, conquest, power, and violence. Then men and women are shown to be idyllically happy. There is also a strand of advertising which stresses friendship, conviviality, and intimacy, especially when pushing products which least promote such things.

Computers are now indispensable helpers to the degree that the very excess of information threatens to result in disinformation. Happily, information is now received, recorded, assimilated, and always at our disposal thanks to the memory of the system. This is a marvel, but it means that we ourselves are dispossessed. The computer is not just an instrument that serves to answer our own questions. It has its own function, and we have lost the power to choose what information to keep and combine in it. That operation is not at all the same as having things done by the computer. It is qualitatively different to the degree that in our own handling of information a subjective quality is present that is not there in the machine. This factor is what enables us to make decisions on the basis of information. I am not referring to solutions to a problem (which computers can give) but to decisions which cut the Gordian knot.

These examples—they are no more than that—should help us to see what I mean by saying that people are fascinated. The environment of noise and images is so invasive that we cannot continue to live in a distant, mediate, and reflective way, but only in an immediate, obvious, and hypnotically active way. We have here three features of absurdity in the existential sense.

4. Conclusion

As a result of these sketches of modern technique, I cannot say that human beings are absurd in themselves, nor that society is absurd in itself. This would mean adopting a metaphysical position. But we are certainly in the process of becoming absurd, even in a philosophical sense. This is a totally new experience in human history and we must try to probe its significance.

The first point that we can be sure of is that no philosophy of technique is possible. Nor is any technical culture, in spite of the great pretensions of some modern humanists. There can be no philosophy of technique because technique has nothing whatever to do with wisdom. On the contrary, it is solely an expression of pride. It makes excessiveness finally possible (as we have seen), an excessiveness that on the one hand rolls on without our wanting it or participating in it (there is a difference in this regard from Dionysiac excess, which was human, and also from all that Nietzsche wrote on the subject). On the other hand, technique attains a dimension so exorbitant that we cannot even record its products, let alone direct them. We need machines to record what other machines are doing. Only computers can record sounds and photographs from the planets. Only the complex machines of microphysicists can record the phenomena that calculations tell us exist. As Nils Bohr could say: "Matter, the real, is what my machines permit me to record."

Hence our own machines have truly replaced us. We cannot make a philosophy of them, for a philosophy implies limits and definitions and defined areas that technique will not allow. A philosophy constructed in terms of the technique of 1950 (e.g., that of Pierre-Maxime Schuhl) has no value or meaning at all in 1980. At root the problem is to some extent the same. On the one hand is the rapidity of change in techniques, on the other hand is qualitative change (e.g., with the transition from steam energy to electricity or from the industrial system to that of the computer system).

But if there can be neither a philosophy of technique nor a technical culture, what are the tendencies and orientations of people (intellectuals) as they embark on this venture? I think I see two obvious reactions: the search for compensation and the search for justification. As we are disoriented, ill at ease, and anguished by the constant upheavals in the environment due to technique, we look for compensations, which usually take the form of escape. I am not referring to the extreme and simplistic escape in drugs or drink, but to escape in religion, in the irrational, which seems to us to be all the more necessary as the world becomes more dangerous and incomprehensible. Along these lines we may refer to the explosive quasi-religious beliefs which are appearing everywhere, the belief in parapsychology, the return to a narrow religious mysticism (clearly discernible in Islam), the hope for an opening up of the world to unknown worlds from which will come meaning, happiness, and help—the extraterrestrials. The un-heard-of success of *E.T.* bears witness to this. The fact that so stupid a story could not only move the masses but also be taken seriously by intellectuals shows how disoriented we are and how far from any possibility of philosophy. The other aspect of our reaction is the search for justification. This does not usually take the form of a direct justification of the technical phenomenon or technical progress. It is usually indirect justification by way of politics or by an intellectualism which outbids the situation. The supreme justification, I would say, is that of absurdity (or nihilism). Nothing makes sense; nothing has value. Hence technical development has as much value as anything else. I have often heard this said during the 1980s.

To reach this conclusion we must deal with a more specifically philosophical question which arises directly out of what I have said. The self, the person, cannot constitute itself, or exist, or have a history, or freely become itself, unless it enters into the game of the possible and the necessary, of freedom and necessity. There is no individual, no human being, no self, if there is no freedom, no possibility. It is no good living if there is no margin of freedom on which the self can constitute itself. Conversely, freedom is not real unless the self comes up against a necessity or a group of necessities. The play between these two realities is what makes human existence possible.

We are caught in a web of determinations, but we are made so as to control and utilize them and to achieve freedom in this way. The self is already itself (necessity) but it has also to become itself (possibility). A self that is without possibility is desperate, and so is a self without necessity.[13] If possibility, in giving itself free rein, upsets and destroys necessity (e.g., in the case of general transgression of all physiological rules or social norms or values), so that the self bursts out without safeguarding a necessity to which to return and refer (as in nondirective pedagogy), we then have the despair of possibility. The self becomes an abstract possibility which flounders and exhausts itself in the possible without going anywhere.

On the other hand, if the self thinks only in terms of necessity, considering that everything is foreordained and ineluctably necessary, there is again despair, real despair. Kierkegaard uses a very simple comparison to show the link. The lack of possibility is like the babbling of an infant. Only the sounds are there. Necessity gives us the sounds, but only possibility gives us the words. There is no freedom except on the basis of necessity and in terms of necessity, but there is no reality (known and recognized by us) except in the struggle for freedom. This

---

13. On all this cf. Kierkegaard, *The Sickness unto Death*, ed. and tr. Howard V. Hong and Edna H. Hong (Princeton: Princeton University Press, 1980), Part One; and, of course, the debate on nature and culture.

is true both for the individual and for the social body. Being itself, it is necessary, but relative to the future, it is a possibility.

This dialectical game, however, has been fundamentally disrupted, or, I might say, destroyed, by the universalizing of technique on two levels. First, technique has become that which enables us to do things. It is both universal and absolute possibility. It enables us to walk on the moon (and to me this makes sense when we think of what "being on the moon" has meant from the standpoint of folklore and myth and popular appreciation). It makes possible speed, instantaneity, (false) immediacy, power, etc. It can achieve all that we imagine or desire. To come up against an obstacle seems to us to be a scandalous thing. We find it abnormal that there should be something that we cannot do, for example, cure cancer or create life ex nihilo. But this brings us under the radical judgment that if everything is possible then nothing is possible. Nothing is possible for the self because it is the object that is possible. As Castoriadis put it, absolute power is impotence.

"Thus possibility seems greater and greater to the self; more and more becomes possible because nothing becomes actual [with the reality of the self]. Eventually everything seems possible, but this is exactly the point at which the abyss swallows up the self. . . . The instant something appears to be possible, a new possibility appears, and finally these phantasmagoria follow one another in such rapid succession that it seems as if everything were possible, and this is exactly the final moment, the point at which the individual himself becomes a mirage."[14] There is no longer any reality (the question of modern physicists) because reality is a synthesis of the possible and the necessary, and there is no longer (in appearance, in illusion, in phantasmagoria) anything necessary. This is one of the reasons for our modern anguish.

But the opposite is also true. If technique makes everything possible, it becomes itself absolute necessity. I said thirty years ago that technique was our modern fate or destiny. I think that events have largely confirmed this. We cannot evade technique. It has laid hold of every domain and activity and reality. Nothing at all is beyond its grasp. It is *causa sui*. Ordinary common sense expresses this in the saying that we cannot stop progress. But this popular phrase has now become the last word in all consideration of these phenomena. When it is a question of dangers, costs, etc., at the end of the argument scientists and technicians close the debate by saying that we cannot stop progress.

There is, therefore, something absolute and incontestable

14. Ibid., p. 36.

against which nothing can be done, which we have simply to obey, namely, technical growth (for in our society progress obviously means growth). In other words, there is no other possibility for us. We have no freedom face-to-face with technique, for freedom here means the freedom to say yes or no. But who can say no to space probes or genetic engineering? It is here that we come up against an absolute determinism (not in our genes or culture!). This is the source of the basic despair of modern humanity. This is the key to it. We despair because we can do nothing, and we are vaguely aware of this even though we do not know it. This is the reason for drugs, as we well know, and for some aspects of hippies.

But we have not taken the final step. This happens when people, instead of being in anguish and despair, set about to justify the situation. In this case they take note of only one aspect of the phenomenon. We thus have those who theorize about the absolute freedom that technique gives us. These people deny (or refuse to see) the other side of the matter. They also overwhelm us with a totally inhuman responsibility. For if technique gives me sovereign freedom and I can truly do all things, then I am responsible for all things—for massacres in Argentina or Afghanistan, for Third World famines, etc. Suicide is the only option. But the same is true for those who want to look at the other aspect, the absolute determinism of history and a strictly mechanistic interpretation of politics and economics, with no possibility or intervention of an act of freedom. The weighty, imperative necessity of technique is here transformed into a "has to be." For this determinism has enough imagination to despair of possibility and enough possibility to discover impossibility. Conscience, then, disappears. There is only bondage in luxury or in misery, in conformism or in the concentration camp. Either way, philosophers, looking at things along these lines and moving on from existence to metaphysics, greatly worsens our human condition under technique.

One more step remains. What happens when instead of noting and justifying either liberation by technique or determinism, we take note of both at the same time? What happens when, without any dialectical relation between them, without any tension or conflict but only identity, we see that what might have liberated us has become our fate, or that what is in effect our destiny might be lived out and accepted by us as our liberation? What happens when we discern that in this technical milieu the possibility is the necessity and the necessity is our only possibility? This is real absurdity. But it is an absurdity from which we cannot escape. We are not dealing now with a philosophical thesis or an accidental example (for which examples to the contrary might be adduced) but with the very nub of the situation. We have rejoined by another route the philosophy of the absurd about which we spoke at the outset, but we are no longer in the realm of metaphysics. We have here a kind of ontology of the world fashioned by technique. On this basis the examples that I have given in the preceding pages are no longer isolated cases. They are truly exemplary and immune to contradiction. Such, in my view, is the significance of the absurd in the technical world.

CHAPTER XII

Unreason

It is obvious that we do not plunge into unreason for pleasure or out of vice. We always have reasons. There is a process which leads on from apparently sane and acceptable premises to unreasonable conduct and plans. This process is so coherent that if one denounces the premises themselves, the argument fails. We must first try to understand the principles behind this unreason before giving some examples.

1. Dissociation

As E. Morin has well seen, the whole problem is that of our modern way of thinking. In the realities of the world and society we dissociate and separate those things which can certainly be distinguished but which are in fact complementary and inseparable. Thus we separate the individual and society, or myth and reality, or tradition and innovation. Our mode of thinking is also reductionist and one-dimensional. We are prepared to see only one object, to reduce what we can see to a single dimension, to eliminate all difficulties and details and singularities. Formed by science, technique, and the media, our thinking is not global and complex, like reality—it is disabled.

It has become ridiculous to refer to the great human, humanist, or moral problems, for the thinking of the specialists of "technical culture" is incapable of either conceiving of them or stating them. This is connected with the fact that scientific thinking is more and more compartmentalized, formalized, and operationalized, and that technical thinking is wholly dedicated to functions and functionality. As the dominant thinking is so contemptible, so, naturally, is the rest. In principle the "technical culture" cannot think in terms of general

problems or reflect upon them. It cannot even think about itself. It is
for this reason that in scientific and technical circles, outside the areas
of specialization, we find the most shallow of general ideas and the
most summary of evaluations, as when scientists and technicians take
up positions on politics or economics. We find an advocacy that is as
pathetic as it is dramatic.[1]

This lack of thought, stemming from the severing of reality
according to the "arithmomorphic logic" denounced by Georgescu-
Roegen,[2] has two vital consequences: the irrationality of basic choices
and indefinite accumulation. In this astonishingly calculated world the
basic choices are made in accordance with a perfect logic, but they are
not thought out. To think them out in relation to alternatives other
mental features are needed than those set forth above. The conceptual
tools of a rationality that applies only to itself are profoundly inadequate, the more so because in our society the tools of scientific research
are subordinate to power, which gives its orders in a totally irrational
way. The congenital, structural weakness of techno-scientific thinking
is linked to the irrational imperative of power. Techno-scientists serve
military ends (we shall return to this matter). The technologizing of
discovery works in tandem with the interaction of science and technique according to the forecasts of research and development. Rationality is operationalized (the theory of systems, computerization, design), and all this destroys even the possibility of global and
introspective thinking. All this leads to indefinite accumulation which
can apparently never stop. There is expansion in every direction at the
same time. This makes everything incommensurable. Evaluation (let
alone judgment) is impossible. It is beyond the capacities of reason,
though this problem is never raised. On these grounds we need to
return to great and simple options. The unlimited progression that is
already part of the technical process[3] is even more accentuated by the
phenomena of economic and military rivalry. There is no limit to what
each nation wishes to achieve at the expense of others in these two

1. It is necessary to note, however, some attempt at generalization and
reflective thinking since the 1960s; cf. T. Kuhn, *The Structure of Scientific Revolutions*, 2nd ed. (Chicago: University of Chicago Press, 1970); Bernard d'Espagnat, *A
la recherche du réel* (Paris: Gauthier-Villars, 1981). The most grotesque of hundreds
of examples of scientists speaking out on political issues was that of Frédéric
Joliot-Curie on using bacteriological weapons in the Korean War.

2. Nicholas Georgescu-Roegen, *The Entropy Law and the Economic Process*
(Cambridge: Harvard University Press, 1971).

3. Cf. my study of the geometric progression of technique in *Technological
Society*.

areas, and governments act as if it were a matter of life and death. The search for power and productivity gives meaning to the social instrumentalizing of nature by technique. The social resources behind this search tend to make it indefinite, for competition has no need to come to a halt.

2. Paradigms

It seems to me that there are five main themes or paradigms at work in this descent of the technical world into unreason. The first paradigm is the desire to normalize everything. This is an older trend, though it used to be no more than a trend. The need is to create norms for everything, for the normalizing of the constitutive factors of society and humanity alone permits an integral application of techniques and at the same time universalization. Language must be normalized. This is especially necessary for computers, even to such details as punctuation and writing.[4] (I am not dreaming!) Only the normalization of meaning makes intellectual exchange possible and prevents misunderstandings. A normalized language alone can be of service, and this is the motive behind the search for normalization.

Aptitude for work, interchangeability of employment, and education all have to be normalized in the same way as bolts or weights. In the end everything will be normalized. But the norm, which supposedly corresponds to a normalizing human instinct, must never be imposed by fiat; it demands consensus.[5] Thus what was invented for the industrial normalizing of detached pieces now applies everywhere. It might be argued that this is not a serious matter. It is simply a sequel to the industrial age. I do not think that is quite true. We have here the spirit of general technicization. The pseudo-diversity of means, of media, and of computer creations, etc., is simply within an increasingly global normalization.[6]

The second paradigm of unreason is the obsession with change at all costs. This is the popular form of the myth of progress. Once we are in an era of progress we can never rest. The constant argument is

4. An interesting little example derives from the fact that I write my letters by hand. In the USA computers sort the mail. Some of my letters never arrive, and an engineer friend told me why. To the computer my 1 looks like a 7 and my 7 like a 2. I have to adjust my 1 and 7 to the way the computer writes them.

5. The Association française pour la normalisation (AFNOR) was formed in 1918, arguing that industrial normalization enabled us to win the war.

6. Cf. the important speech by the president of AFNOR, Feb. 12, 1975.

that to stop is to regress. This argument is always used against the theory of zero growth. But it means that we must always be doing something new and never be left behind. From this standpoint change is good in itself. We have to change all the objects of daily life. Nothing is made to last. Whatever lasts (even in human relations) belongs to a world that is totally past.

As we see, people are constantly changing partners, television sets, personal property, cars, etc. There is no reason for it; change imposes itself. Ideas change along with information. Governments change, art changes, and in art the new style must not borrow anything from the past, so that it becomes totally dislocated. In opera, for instance, we find lively debates about the staging of Wagner or, in 1985, *Aida.* An interesting phenomenon here is that we cannot change the music or the plot, yet change has to be made. Something new must be done or the operas cease to be of any interest. The solution is to change what is not fixed—the setting. Hence the setting becomes the important thing. It effaces the music. What has been changed is the interesting thing. We thus have grotesque baroque fantasies. Wagner was writing operas about the families of great modern industrialists. Intellectuals debate such novelties seriously, and their conclusion is always that the experience was interesting and revitalization was necessary.[7]

The third paradigm of unreason is growth at all costs. We all know this obsession in all areas. Growth is thought to be good in itself. No one asks why, or for what, or whether it is useful, or who it serves, or what will be done with the excess. The lack of interest in such questions is a sign of unreason. Growth is self-justified. It might have been foreseen that growing too many tomatoes here or peaches there or corn elsewhere would make it impossible to sell them, but so what? Everybody wants to grow at all costs and throw increasing quantities on the market and finally in the ditch. We find the same in every sphere. John Stuart Mill made a reasonable point when he argued that increasing production is an important goal only in backward countries. In more advanced countries the essential economic need is better distribution, one of the inevitable means being severer restriction as regards population. If the earth would lose much of the charm which it owes to things that the unlimited increase of wealth and population

---

7. A recent example was the staging of Hugo von Hofmannsthal's *Venise sauvée* at the Avignon Festival. According to the review by Michel Cournot in *Le Monde,* July 22, 1986, it was impossible either to see or to hear because a thick smoke filled the stage and the characters were at the back, speaking in false mechanical tones, so that one could not see who was speaking or to whom. What counted was the smoke!

takes from it simply to make a larger population possible, but not one that is better or happier, Mill hoped sincerely, for the good of posterity, that we would be content to remain stationary except as necessity did not permit.[8] But we did not listen to this wise advice. Growth is now for growth's sake. But growth is in contradiction with growth (e.g., that of population with that of the standard of living). We thus engage in a never-ending pursuit of which we are the witnesses and agents.

We need to ponder the truism that unlimited growth is not possible in a limited world. According to Jouvenel's calculations, if the number of vehicles continues its exponential growth, by the year 2000 all France will be covered by automobiles. It might be replied that that will never happen because we will have the sense to stop before it is too late. But it was between 1970 and 1975 that we ought to have perceived the negative effects of economic growth and the excessive growth of the costs of this growth. Then came the report of the Club of Rome and the deregulation of the international monetary system: a beginning of wisdom. But as the crisis ended this was swept away again and obsessional growth recommended. The human race has never given historical proof of the wisdom that is needed. We always want more, no matter what the damage or the costs.

It might also be replied that the world is not really finite or limited since there are millions of galaxies and we have only to colonize them to find plenty of room. I am not joking; some authors seriously make this point. But for the moment, in spite of satellites and the space lab, there is no conceivable possibility of an emigration of millions of people to Mars or Venus. In twenty years it will not be conceivable. Hence the limit remains, and it is unreason to think that we can ignore it.

The fourth paradigm of unreason is that of speed. Curiously, a contrast has sometimes been made between the industrial age when speed was demanded (Taylorism, etc.) and the technical age, when machines are certainly faster but they are said to save time for us, so that we can live and work at a more relaxed pace. Now it is true that many tasks which demand accelerated work can be performed today by machines. At the same time life globally is forced into increasing speed simply because of these machines. The problems of pace and rhythm have certainly not been solved. An anecdote will show how imperative speed is.

In 1983 M. Le Garrec settled some delicate matters by circular. When attacked, he replied that if he had followed regular procedures

8. John Stuart Mill, *Principles of Political Economy* (Toronto: University of Toronto Press, 1965).

and consulted all competent commissions and representative bodies, this would have caused serious delays and affected wages and measures against unemployment. That argument is typical of the technocratic mentality. Legal procedures are undoubtedly slow. Reconciling interests slows down the decision-making process. Government processes are slow. This is always the argument of dictators. Democracy is slow. That is why we have now to make a parody of democracy. Like Le Garrec, we have to settle things quickly without observing the procedures that restrict us.

Our judgments have to be as quick as our machines. But reason puts two questions. Is it a bad thing to adjust decisions and actions to our human speed? (Specialists in computers say that this is impossible!) Is it a bad thing to lose time on political and social relations so as to achieve genuine consensus by grass-roots negotiation rather than arriving at authoritarian decisions by way of the hierarchical echelons?

Everything must now be done fast, whether we refer to decision, to actions, to judgments, or to human relations. A few years ago I calculated that if the USSR were to launch a nuclear missile against France, we would be told at once of its departure. But it would arrive in *six minutes*. Six minutes for the president to learn of it and descend to the shelter to reach the red button! At best the president would have, therefore, no more than a single minute to decide whether to start a nuclear world war. This is an illustration of the human condition in the technical world. It is no longer a matter simply of *going* faster (as in *L'homme pressé* by P. Morand) but of *deciding* and *acting* faster. Parodying Galbraith's famous phrase, one might say that we have control over the machines; we now have to follow them!

The final paradigm of unreason is the implicit one that all judgment now depends on techniques. This is part of the autonomy of technique. No judgment is admissible that might hamper the progress of science or technique. There can be no independent moral judgment (this is bad!) or rational judgment (this is unreasonable!). As regards moral judgment, morality is very uncertain, and in a society like ours there are no longer any basic principles from which to derive the actual consequences. Furthermore, moralists seem to be out of touch with the age and only a few are ready to confront it (cf. the studies of Fourastié or Jankélévitch).[9] As regards rational judgment, the arguments can easily be turned around.

9. Yet there is a confused sense of the need. We see this from the many commissions on the morality of communication, on medical ethics, on the ethics of

The question of armaments provides the model of unreasonableness.[10] What sense does it make to manufacture and sell more and more powerful weapons? There are two questions here. Ten years ago a minister gave me an answer: France has to be armed and defended. For this defense it cannot risk being dependent on others. It must provide for its own needs. We thus make our own weapons and keep them up to date (the principle of competition). But this means that older ones have to be discarded. Again, once our industrial capacity is geared to produce weapons, we make more than we need. Hence we export those weapons that are outdated and new ones that are surplus to our own needs. This concludes the argument.[11] It all hangs together. Our whole problem with technique receives here a clear and unequivocal response. The irrational nature of the presuppositions (the entity of France, the threat hanging over it, its isolated capacity, etc.) is completely ignored.

A further question is whether it really makes sense to manufacture more than we need. But here we come under the double logic of industrial and technical functioning and the profit motive (which is present in socialist countries too!). One might also ask whether it makes sense to think that in the era of the superpowers, the USSR and the USA, France can provide for its own weapons and security. The idea of absolute sovereignty surely makes no sense in the age of multinational corporations and the structuring of a global rather than an international economy. (All governments today are having the same experience!) Again, is it reasonable to sell armaments to Third World countries? They squander their resources to buy weapons to the detriment of their true development. Yes, it is replied, but we cannot let the weapons rust, and there are also the demands of foreign trade. Even though the sale of arms is not the most important export, as often alleged, it represents 15 percent of exports, and we cannot accept a deficit of that size. But I press the question of what is reasonable. Is it reasonable to think that if we arm and overarm Third World countries they will not go to war? This is what we have seen happening over the last thirty years. And it is only a beginning. Eventually the Third World will inevitably turn on the developed countries. We are arming the Third World to a point of

---

research, of genetic engineering, etc., though naturally these lead to nothing concrete as regards either decisions or conclusions.

10. Cf. H. Eijkelhof and E. Boeker, "Weapons," *Science, Technology and Society* 2/1 (1982).

11. It may be noted in passing that for France the age of glory in this regard is perhaps over. Fewer weapons were exported in 1985 and almost none in 1986. But only four or five contracts had to be canceled to achieve this result.

recoil when we will pay the true price. My judgment here is not a moral one, but such is indeed unreason. In this whole area there has been for many years an inability to think reasonably.

3. The Main Areas of Unreason

Apart from armaments it seems to me that there are three principal areas in which unreason is almost axiomatic: pollution, the Third World, and nuclear technology. A characteristic of these three areas (as in many others) is a total failure to look ahead. This is very strange, since we have on the one hand the many forecasts to which we have referred already, but on the other the decisions taken by many public or private bodies, and by individuals, in which the future in no way affects the actual motivations. Theoretically it is calculated that in twenty years there will be so many inhabitants of the earth, that everything will be done by computer, etc. There is a kind of foresight that is based on what is approved in advance and that in no way challenges what is being done today. Even if forecasts run contrary to choices already made, the chosen course is blindly followed with no self-critical ability. We have seen this in the field of armaments. The same is true in the three other areas of unreason. Unreason, strictly speaking, is here a refusal to look ahead to the future. "After me the deluge" is our practice. One of the high French technocrats put it very elegantly when I talked with him about the pressing problem of atomic waste; he replied that after all, we can leave such problems for our children to solve.

As regards pollution, there is no more room for debate.[12] There is now general agreement about the danger of the accumulation of CO_2 in the atmosphere,[13] about acid rain, about the pollution of water tables throughout most of Europe, about the pollution caused by automobile exhaust, about the introducing of heavy metals into our air and water, about the pollution of our large lakes and inland or semi-inland seas (the Mediterranean and the Baltic), about the pollution even of the Atlantic, about the danger of the accumulation of chemical or nuclear poisons, about the dangers posed by the deactivating of nuclear power

12. Cf. Andrew I. Sors and David Coleman, eds., *Pollution Research Index: A Guide to World Research in Environmental Pollution*, 2nd ed. (Detroit: Gale, 1980).

13. CO_2 in the atmosphere increases by 0.5 percent per year. Doubling it would increase the temperature by 10° C. at the poles, upset the atmospheric circulation, perhaps make deserts of the temperate regions, and interfere with the productivity of the oceans.

stations, and about the almost unimaginable pollution of great rivers like the Rhine, not to speak of the pollution that is not so directly felt like that of noise, which is so crucial a matter for us in the West,[14] and the pollution caused by the density of human relations or the excess of information, though it might be argued that this is not really pollution.

Face to face with these countless facts, with these risks which we do not regard as major risks because they are not terrifying accidents, but which are still major risks over the long haul, nothing is done. The example of radioactivity is typical. It is said, of course, that the radioactivity emitted by X-ray examination or power stations or nuclear tests or even accidents like Three Mile Island and Chernobyl is small. Each time calculations are made to prove this point. But I have never seen it asked what happens when some rays are of average or even long duration as well as short. Let us take those of average duration (a few years). Obviously, the radioactivity they emit will build up. On each occasion the radioactivity may not be dangerous but the cumulative increase may become so. It might also be said that the amounts do not exceed those that people receive who live in granite areas (like Brittany or Auvergne). But some biologists have shown that in such areas there is a much bigger ratio of congenital malformations (e.g., goiters). Are we not exposing the total world population to these major risks over the long haul?

If we extend the question of pollution to certain aspects of ecology we again come across major risks. We recall that in ecology there are two different circles which do not coincide.[15] One group is mainly interested in ecological balance, and this means that everything in nature has a kind of sacred character. The destruction of rare animals and lands is a crime. Everything in nature is good; human intervention is always reprehensible. The other group is both narrower and broader. The important thing here is our own ecological niche. We risk destroying ourselves by destroying our ecological milieu. Our niche is an order of both nature and society. We are not concerned with nature alone. If we disturb its balance too much, we will have to bear the dreadful consequences.

Forests provide us with an example. Four-fifths of the world's forests are in Latin America, Siberia, and Equatorial Africa. Two of these

14. Already in 1975 there was a study of the danger of noise for workers. Cf. Laverrière, *Repenser ce bruit dans lequel nous baignons* (La Pensée universelle, 1982); and F. Caballero, *Essai sur la notion juridique de nuisance* (Pichon, 1981).

15. Of many works cf. especially the special number of *Science, Technology and Society* 3/1 (1983), entitled "The Finite Earth," an exhaustive study.

areas are poor in energy. Energy is expensive, and the forests offer a source that is close to hand. Throughout the world three hectares (almost 7.5 acres) of forest are cut down every second, or 288,000 hectares (720,000 acres) a day. Roads are being made through the Amazonian forest, and in addition it is the victim of mining and settlement schemes. If we carry on like this, the world's forests will be gone in fifty years. But trees mean oxygen (along with the surface of the sea, which is itself threatened by oil, etc.). Mad deforestation is leading to the loss of our source of oxygen and to the increase in deserts, which do not arise merely through drought but perhaps even more from the need for wood in these zones. Without realizing it, people are making deserts of their own border areas by burning wood.[16]

At the Nairobi Conference it was recalled that two billion people in rural areas use wood for heating and that four hundred million of these are in zones in which wood is scarce. In fact 53 percent of all Africa is rapidly becoming desert. One might speak of an ecological disaster which puts all of us at risk.[17] But we pursue in total thoughtlessness our triumphant ascent in space and computers. Who can tell us to stop producing dioxin or engaging in deforestation? These things are indispensable. We are sawing off the branch on which we are sitting. But we fail to see what we are doing.

There is taking place on a world scale what we have already denounced in battles against developing[18] the coast of Aquitaine. Take a forested area, with different kinds of undergrowth, a lake on one side and the sea on the other, and a developer comes along. How will he gain access? He puts roads through the forest. Picturesque developments are needed (e.g., marinas). Ditches are dug for water, gas, electricity, and telephones. The undergrowth gradually disappears. Lots are divided out. Advertising stresses the pure and tranquil lake, the marvellous air, the unlimited forest, the lonely beaches. Trees are cut down to make way for the many lots. Building goes on and on. The trees constantly fall. The water in the lake grows stale. The beach swarms with nudists. Nothing is left of what brought the people here. All that

16. I saw a small instance of this in my own town. In a forest close by, the council allowed gypsies to camp and they used the wood for heating. In less than five years the whole of the forest had gone and the earth was denuded.

17. See the detailed study in J. Gellibert's Bordeaux dissertation, "Le Choix de la biomasse comme énergie" (1986). Gellibert also adds that deforestation brings erosion. Thus in Nepal 240 million tons of arable land are washed away each year through deforestation.

18. Le Lannou calls this déménagement de territoire, "the emptying of the area" [in French a play on aménagement for "developing"—TRANS.].

was beautiful and healthy in nature is no more. Tourists have congre-
gated, living on a myth. Development has done a mighty work and the
promoter has reaped handsome profits. We have here an illustration of
the logic of technique combined with that of money and power.

I realize, of course, that some attention is being paid to pollution.
There is reaction to it on three levels. The first is to prevent it, for
example, by catalytic converters on cars, or the many ways of purifying
smoke or sewage. I concede that in most of the individual cases it is
quite possible to prevent pollution, but we have also to consider that
the public is slow to accept these measures and that there is much
negligence and a refusal by industry to apply them. Often, too, the
authorities fail to do anything, as in the case of noise. Thus it would be
easy enough to reduce the noise made by motorcycles, but no French
government is prepared to do anything that might provoke demonstrations by young people. Again, when purifying is costly, industries
threaten to raise their prices, which would make them no longer
competitive; thus adequate measures of prevention are abandoned.

The second reaction is curative. Some American lakes and some
rivers have been cleaned up.[19] But only in the USA is there enough
money to achieve this result, since the cost runs to millions of dollars.
The air in London, it seems, has become much cleaner over the last
hundred years. Well and good, but this is only a single instance among
thousands. Experimentation will certainly produce methods of eliminating organic wastes, which produce methane. China has taken
vigorous action since 1980. It is in this area that cleaning up is easiest,
but France has done very little. Here again there is need for a heroic
decision to make the cleaning up of pollution a primary objective of
politico-industrial action.

But this leads us to the third reaction, that of legal measures, the
passing of decrees and agreements to prevent pollution. Unfortunately,
I agree with Bressand and Distler that the law no longer has any place
in the comprehensive technical system! In a socio-technical situation

19. Cleaning up is always possible (except for nuclear pollution) but only
at the cost of much labor and expense. Lake Michigan used to be one of the most
polluted lakes in the world, but in 1975 seven treatment plants were set up,
including the biggest in the world, the West Sewage Treatment Plant. At a Water
Ways Control Center technicians were on constant watch for the slightest variations
of level in the sewers. But this was not enough. There was a need to drain off sewage
and rainwater, and so in 1980 a tunnel was built to a big reservoir to divert water
from the lake and make possible its recuperation. Positively, this provided additional
industrial activity, increasing the gross national product. Negatively, it cost vast
amounts of money. If cleaning up pollution is necessary, it is also extremely
wasteful!

it is totally outdated. We have only to see how impotent are the laws passed against noise or for the protection of fragile areas or for the purifying of sewage. We have only to see how poor has been the application of successive treaties on nuclear nonproliferation, or how impossible it has been to reach agreement to check the over-polluting of the Rhine. Thus in 1986 France decided not to apply a plan it had already accepted to bury the trailings from the Alsace potassium mines which were being dumped in the Rhine. The excuse was that another plan had to be studied. Similarly, the representatives and people of the upper Rhine fought a depollution plan so as to save money and jobs in the area. The cellulose of Strasbourg as well as the potassium mines are important sources of pollution. The Bonn Convention was signed in 1976 but not submitted to the Assembly for ratification until 1983, and it was then denounced in 1986. Twenty million cubic meters of silt laden with heavy metals arrive at the delta of the Rhine, along with fifteen million tons of salts, not to speak of nitrogen, detergents, phosphates, etc.

Now only four countries need to reach an agreement about the Rhine, but it is still impossible. What hope is there, then, for cleaning up the Mediterranean? No one even dreams about it. It is also impossible to arrive at an international agreement on acid rain. Even if an agreement were signed, who would observe it? Who would supervise its observance? What sanctions could there be against a delinquent country if that country were the USSR, West Germany, or France? We have already seen the general inability to stop nuclear proliferation. Hence we can have little hope. Pollution will continue to develop at the pace of technical growth.[20]

The Third World is our second place of unreason. There are warning signals, but no one heeds them. Here again I am speaking only of what is obvious. We all know the misery of the Third World. Developed countries with 30 percent of the world's population have 95 percent of its wealth. This is a well-known fact. What is not taken into account is that the misery is growing in spite of all that is said. It is growing on three levels. The first level is that of the runaway demographic growth in the Third World—a growth which cannot be accom-

---

20. Cf. Bertrand Goldschmidt, *The Atomic Complex* (American Nuclear Society, 1982); H. Blix, "Non-prolifération," *Le Monde*, April 1985; H. Laverrière, *Repenser ce bruit dans lequel nous baignons* (La Pensée universelle, 1982); F. Caballero, *Essai sur la notion juridique de nuisance* (Pichon, 1981); B. Charbonneau, *Nature du droit de la gestion des risques* (1978); idem, *Régime juridique de la lutte contre la pollution des eaux* (1985); idem, *Protection de l'environnement en droit de l'urbanisme* (1984); De Rosnay, *Biotechnologies et bio-industriels* (Paris: Seuil, 1979).

panied by equally rapid growth in the means of living. To say, as is sometimes done, that by cultivating all possible land with plants that have higher yields, and by distributing the resources equally, the standard of living might be raised, is both true and ridiculous, for within the actual structure of the Third World this is not even conceivable. The second level, an objective one, is the increasing impoverishment due to the using up of raw materials by our technique, the spread of transnational factories which require a labor force and transform peasants into an urban proletariat, and the reduction of local growers to poverty by competition and aid—things which are already obvious.[21] The third level, a subjective one, concerns poverty in comparison with the high standards of living in rich countries (as is now known) and depression in the Third World. Poor countries are miserable countries. But the technological gap between the rich and the poor is widening at an incredible speed.

From a human standpoint the Third World can never catch up with the developed world.[22] The well-known example of Southeast Asia is irrelevant because the reference is to Singapore, Taiwan, and South Korea and not to the true Southeast Asia of Cambodia, Vietnam, Laos, and Thailand, which are in an economic slump such as none of them has known for the last two centuries. Even Third World countries that have grown wealthy through oil do not know how to use their wealth because they do not have adequate economic and political structures. There is no widespread industry or development. What we find are government monopolies and a few localized instances of modernity.

Relations between the Third World and advanced countries are becoming closer and closer because of communications techniques and the need of the advanced world for all that the Third World can produce. These relations unavoidably engender violent feelings of frustration. In face of this, we can say that all the policies of advanced countries to deal with the situation have failed. Nowhere is there any rational cooperation between the two worlds and never has it been seen that the same policy cannot be applied everywhere. The problems of black Africa are not the same as those of North Africa, India, or Latin

21. Cf. my study of this question in *La Révolution nécessaire*, in which I explained that we should adapt industry to produce the necessary tools suitable to each country and its customs and modes of production. But if such tools were to be distributed free, this would means a severe drop in our standard of living, and who would accept that out of solidarity with the Third World?

22. See M. Kamenetzky, R. Maybury, and C. Weiss, "Scientific and Technological Dimension of Development," *Science, Technology and Society* 4/2 (1984)— a study of the possibilities and limits of development in the Third World.

America. But everywhere we see deterioration of the environment, and everywhere it is a mark of the failure of political aid and also of enormous debt. (We shall have to come back to the numbers.)

Naturally, everybody is concerned about so tangled a situation, and it is rightly seen that if this or that country stops payment on its debt there will be repercussions on the whole economy of developed countries as well. But no one dare go to the political extremes that the situation demands. Reasonable responses and decisions (which are not satisfying, rational "solutions") have to rest on a generosity, a spiritual (and not an economic) sense of solidarity, and a sense of economy (i.e., economizing) of which we do not seem to be capable.

The West implicitly refuses to give up its own extravagance and expansion of high tech. It tries instead to soothe its conscience (cf. J.-J. Servan-Schreiber) by arguing that it is precisely these factors that will enable the Third World to get out of the impasse. This is a technological bluff! But in view of this refusal, this total absence of all that is reasonable, what can we expect?

We could take things calmly so long as the Third World had no mobilizing ideology. An anticolonial revolt in this or that country was not too serious a matter. But today the Third World has a mobilizing ideology: Islam. Islam has every chance of succeeding in opposition to communism, which was imported from the West. Being of the West, communism is failing little by little in Latin America (except in Cuba and Nicaragua) and also in China, which understands that it must abandon communism if it wants to be the third great power. But Islam belongs to the Third World. With astonishing speed it is winning over black Africa and growing in Asia. It is a unifying, mobilizing, and combative ideology. We are now engaged in a true war waged by the Third World against the developed countries: a war expressed increasingly in terrorism but also in peaceful invasion.

Clearly, the Third World, even if it were to reunite all its forces, could not engage in a declared, frontal war on the battlefield. It could not engage in the trench warfare of 1914, nor in the war of movement of 1940, nor in the cold war of 1947, nor in economic war. It will never have enough military power nor economic superiority (as has been seen with oil). But it has two fantastic weapons: the fanatical devotion of its suicide squads and the guilty conscience of public opinion in the West concerning it. The strange thing is that Europe, although it cannot decide on the drastic reasonable measures that are needed to make the world livable, has a permanent guilty conscience. On the one hand, then, we have the Third World terrorism which can only grow worse and which cannot be stamped out so long as the terrorists are ready to sacrifice themselves.

When everything becomes dangerous in our world, we will be on our knees with no power to resist. On the other hand, there is inevitably a growing infiltration by immigrants (workers and others) whose misery attracts our sympathy but who also create among us strong nests of Third World militants. Intellectuals and the churches, for various reasons, are allies of these immigrants and try to open the gates to them more widely. Any measures taken by the authorities to stop them coming in or to control them run up against a hostile public opinion and hostile media. But their presence in Europe and the associated spread of Islam will undoubtedly lead to the disintegration of Western society as a whole. Because of the unreason that has been so evident among us for the last twenty years, in twenty-five years or so the West will find itself globally in exactly the same situation as the white minority in South Africa as it faces a black majority. This will have been the long-range effect of technicization on two levels, as we have shown.[23]

The final instance of unreason obviously lies in nuclear technology.[24] I will not take up what has become the classical argument that there is a link to the production of nuclear bombs, that there is the danger of a multiplication of atomic armaments, that nuclear power stations always involve risks, and that taking them out of service is difficult, as is also the disposal of wastes.[25] Instead, I will cite only one example of the danger.

23. On the Third World cf. especially J. de Ravignan, *La Faim, pourquoi?* (Syros, 1983); J.-C. Derian and Staropoli, *La Technologie incontrôlée?* (Paris: PUF, 1975); V. Cosmao, *Changing the World*, tr. John Drury (New York: Orbis, 1984); idem, *Un Monde en développement* (Éditions Ouvrières, 1984); F. Partant, *La fin du développement* (Paris: Maspero, 1982); J. Touscoz, et al., *Transferts de technologie* (Paris: PUF, 1978); A. Mattelart and H. Schmucler, *Communication and Information Technologies: Freedom of Choice for Latin America?* tr. David Buxton (Norwood, NJ: Ablex, 1985); John K. Galbraith, *The Nature of Mass Poverty* (Cambridge: Harvard University Press, 1979); Reports of the Independent Commission on International Humanitarian Questions: *La Déforestation et la Désertification* (Berger-Levrault, 1985); *Famine, mieux comprendre, mieux aider (reconstruire le monde rural en Afrique)* (Berger-Levrault, 1986); Ivan Illich, "Development: Its Three Dimensions," *Science, Technology and Society* 1/4 (1981); M. Moravcsik, "Mobilizing Science and Technology for Development," ibid.

24. See J. J. Romm, "Scénario pour un conflit nucléaire," *La Recherche* 149 (Nov. 1983).

25. M. Barrère, "Ou enfouir les déchets nucléaires," *La Recherche* 166 (May 1985), examines all the proposals, shows they will not work, and puts the wastes in two classes, the less and more highly radioactive. The issue of Aug. 1985 (no. 168) took up the question again, discussing the criterion of least risk rather than maximum safety in the selection of sites. Cf. also G. I. Rochlin, "Le stockage des déchets nucléaires," *La Recherche* 122 (May 1981); idem, "Que faire des déchets radioactifs?" *Journées scientifiques de l'École des mines* (1983).

A supergenerator is dangerous because it has no moderator. The rise in power is a thousand times faster than an ordinary nuclear power station.[26] A chain reaction can occur much more easily. The Super-Phoenix uses 5,000 tons of liquid sodium and 5.5 tons of plutonium, but a serious accident can be caused by only a fraction of these amounts. In the case of the Super-Phoenix a meltdown would release several thousand times more radioactive products than the first atomic bomb, especially of plutonium, which in the form of aerosol is fatal even in a dose of only one milligram. One might say that there has never been an accident of this gravity. But what may be said with certainty is that once there are many machines of this kind, a major accident is bound to happen (Puiseux).

Having said that, I will simply make two observations. The argument for the absolute necessity of the nuclear program in France was that it was the only way to achieve energy independence once all hydroelectric possibilities were exhausted. This argument presupposed an ever-increasing demand for energy. It was taken for granted that to be more civilized we must use more energy. The use per person became an indication of progress. In 1981 the average person in America was using twice as much as the average person in France, and the use in Holland and West Germany was also more than in France, though not in Japan, and certainly not in Ethiopia or India. The growth in energy use was supposedly related to human happiness, though in fact it often represented a loss of happiness. "We have here plain evidence of the irrationality of this kind of development that is said to be a model of reason and efficiency. American agriculture is a typical example, since it now uses more energy than it produces. Pimentel has shown how inefficient is the agriculture that is making such headlong progress. In 1940 in the United States 150 kilocalories of corn per hectare required 124 kilocalories of energy, but in 1970, 526 kilocalories of energy were required to produce 250 kilocalories of corn per hectare."[27]

26. Cf. the discussion organized at Paris in Sept. 1981 by the Groupe de Bellerive under the patronage of the French Academy of Sciences with the cooperation of the French Commission for UNESCO. Dr. Jochen Benecke, a German specialist on the safety of supergenerators, explained the concept of "hypothecality" put forward by Dr. E. Häfele, one of the fathers of the German supergenerator SNR 300. He stated that Super-Phoenix is a separate experiment and that the uncertainties are so great that populations are part of the experiment (Groupe de Bellerive, *Énergie et Société* [Paris: Pergamon, 1982], p. 488).

27. J.-C. Lavigne, *Impasses énergétiques: Défis du développement* (Éditions Ouvrières, 1983); he gives many other examples.

We must also consider the inconveniences (not to mention the possible accidents). A nuclear power station generates so much energy that a stoppage (not to speak of a dangerous accident) can paralyze life in a whole region. Knowing that the production of energy is concentrated in a few units poses the problem that all ideas of decentralization, local autonomy, etc., become futile. Since everything depends on one powerful center, the rest is a matter of mere words. Furthermore, this concentration carries the risk of great variations in a world in which there are so many imponderables. In addition, the complexity of the techniques, which are constantly being perfected, means that increasing recourse must be had to experts. The general public can no longer really understand or act in what has become an indispensable element in its life-style.

Inevitably, the nuclear program means centralization, as we have said, along with a regression from democracy. But investment in it is also long-term and expensive, while the life of the installations is short in relation to the initial costs, and there is no possibility of retreat if the enterprise seems to be a failure. Finally, this system of producing energy increases the gap between developed countries and the Third World. According to a German study it would take $5 trillion to provide adequate nuclear generators for the Third World. Poor countries are forced into an impossible choice between increasingly expensive oil (though the price has stabilized for the moment) and a crushing nuclear investment. Rich countries, however, can go nuclear and thus achieve a large measure of freedom from oil (which is produced in the Third World). These are commonsense arguments which are often used. But they do not carry the day against the great magico-technical achievement of the splitting of the atom.

I will simply insist on two final aspects. France congratulated itself on its nuclear program at the time of the first oil crisis. But thanks to this program it began to produce more energy than it consumed. As we stated earlier, the result was a publicity campaign to increase the use of electricity and a reduction of prices for higher use. But then came a lean period when the French were using too much energy and there was a need to economize and produce more. We thus have the topsy-turvy course: First too much, so a campaign for greater use; then greater use and too little, so increased production. At the moment we have reached saturation. We must now use more electricity so as to save other fuel; the nuclear generators will produce enough. But we cannot stop them or put them in reverse. We have to consume, whether there is a real need or not. Électricité de France is pushing operations that use the maximum of electricity, even though the use may be exorbitant and of no real benefit (e.g., the reaching of 20,000° C at the Centre de Recherche des Renaridères). What Électricité de France demands is absolute truth for the French government.[28]

At this point we come up against a basic reality for the whole technical process. If we do not increase use, what is to be done? True, we can export electricity. This has already begun, but the difficulty is that the export capacity is irregular. The importing countries want a regular supply, but this is impossible, since we have an excess in summer but just enough in winter, when the use is higher. A new solution is to store electricity (which hitherto seemed impossible) by electrolysis of water, hydrogen offering the most powerful chemical for the storing of energy. Thus a new technical process has come on the scene, but it is also a good example of the absence of mastery and the unreason in these areas.[29]

In my view, however, another potential danger of nuclear power stations is a more serious matter—the danger of attacks on them. It is obvious that terrorists (to use a more restrictive term) will finally make these their objectives of choice. I stated earlier that such attacks can come only with development. As terrorists perfect their methods, explosions at nuclear power plants will produce incommensurable fallout. Naturally, whole buildings would not explode in view of the solidity of their construction, but it would only be necessary to put a bomb in the right place for a disaster to result. It is said that security is tight and constant. But in view of the instances of security being breached, as in the case of the U.S. Marines in Lebanon, I have serious doubts whether the security is tight enough. Nuclear plants would need to be put in fortified zones with special police protection, and there would also need to be a network of police surveillance covering the whole population, since measures to discourage real terrorist nests are not enough. The multiplying of nuclear plants means a fantastic

28. At all levels Électricité de France says that choices regarding energy must be political and not technocratic. But the technocrats are also so convinced that their own solutions are right that any different choices are bound to be irrational and due to incompetence. Thus any decision that deviates from their advice is self-evidently suspected of invalidity; cf. Lucien Schwartz, "Syndicalisme, technocratie et politique," Le Monde, June 9, 1978. But accidents have a sobering effect; cf. M. Barrère, "Des fissures dans le programme nucléaire français?" La Recherche 107 (Jan. 1980). Le Monde, Sept. 30, 1987, carried a very interesting interview with the president of Électricité de France in which he recognized that they had been a bit too ambitious, that they would skip the year 1988, that they would not press on with supergenerators, and that the age of big projects was over.

29. Cf. the research into solar energy, J. Villermaux, "La chimie et l'énergie solaire (le problème du stockage)," La Recherche 149 (Nov. 1983).

multiplying of military risks. It is not possible to say that halting the
program would have been unreason because attacks were not then as
they are now. To continue the halted program today is unreason.[30]

4. Complementary Examples

I will select just a few other examples from the many available. There
has been much talk of the use of computers in farming,[31] especially in
raising cows. Each cow would have a collar attaching it to its little
computer, which would provide all the necessary information about
this particular animal: what it eats, how much, how much compared
to the average of previous days; its daily yield of milk, again in
comparison, etc. This is absolutely marvellous. Farmers can know
every detail about every beast in a herd in which even the worst cow
produces 8,000 liters a year. It is true that all the articles that refer to
this little masterpiece recognize that it is very expensive (about 20,000
francs) and that it will not be profitable unless one has a fair number
of cows, at least forty. Several farmers have sometimes combined to
make up a herd of a hundred, the number at which the computer
becomes really useful. There is thus a trend toward concentration, the
organizing of larger farms, etc. It is said that we want to keep up the
income of the average farmer, but in spite of what is said, many serious

30. I might list some of the points made in the excellent article by Paul Fabra,
"Le dogme de linfaillibilité nucléaire," *Le Monde,* Aug. 1986. Fabra argues that the
confidence of technicians and politicians who push the nuclear program in France
calls for criticism. They claim that an accident like Chernobyl is impossible in
France even though they allow that human error can always occur (and what other
error would there be?). No doubt they have good technical reasons for their
confidence. We do not have the competence to dispute this. But that is not the end
of the matter, as we see from the report on the Challenger disaster in the USA, which
led to the temporary suspension of the shuttle program. What led those who drew
up the report to question the confident NASA technicians was the deposition of the
engineer who expressed doubts about the safety of a crucial part of the shuttle. It is
easy to cast a stone at this engineer for not sticking to his guns, but he, too, was a
victim of the quasi-infallibility which previous successes had engendered at NASA
(as among technicians in the French nuclear power program). As regards the
profitability of the stations, at the lower price of oil no other producers of electricity
in the capitalist world are interested in additional nuclear power. Most nuclear
technicians argue that the low price of oil is temporary, but they may be wrong,
since OPEC seems incapable of enforcing an artificially high price so long as there
is no return of inflation.
31. See R. Bouchert, "Les fermes de l'an 2000 sophistication (Télématique
agricole . . .)," *La Recherche* 119 (Feb. 1981).

thinkers argue that we have too many farmers and that a mere 5 percent of the population could farm all France. Long live unemployment! In any case, by encouraging small farmers to incur the considerable expense of follies like computers on cows we are leading them into increasing debt and lack of control over their own vocation, since they will not be able to regulate the price of their products.

While we are on the subject of technique and agriculture, let us speak about another modern marvel: the system of forecasting yields. Thanks to satellites, we can get very clear images of fields, forests, etc. To interpret the pictures much preparatory work is needed in the comparison of various kinds at different distances. We also have to have electronic instruments to measure the temperature of a field or forest. The difference in temperature between the side of a leaf that faces the sun and that in the shade shows exactly how much water it needs or how severe is a drought. Finally, data are needed from observation of each field so as to discover, for example, the average size of corn leaves. Complex instruments, not hands, are used for this purpose. Bringing the three sets of data together we can forecast three months in advance what will be the yield of a given area and how much return the crop will bring. But in my view this wonderful operation is sheer unreason when we remember that a hailstorm can batter down a field of corn in ten minutes, or a tornado can flatten a field of wheat, or a violent wind with rain can blow down a field of oats. Such things can all happen, and often do, in a three-month period. Those fantastic instruments are, in fact, useless gadgets.

Need we return to the medicalizing of life?[32] Too many medications, too many doctor visits, too much hospitalization, illnesses due to excess medication—these are familiar themes on which there is much agreement. What worries me, however, is the prevailing notion that even the well are sick though they do not know it, so that we have the opposite of the situation in Balzac's *Le médecin de campagne.* Today we live in anxiety and fear. Our technical milieu makes us afraid of sickness. We resort to the doctor-sorcerer-magus. We stuff ourselves with tranquillizers, hypnotics, analgesics (I am no exception).

The important issue here is that of human quality. We today do not know how to suffer. We cannot tolerate the slightest pain. We cannot mobilize our own resources to combat anxiety or fear. We need

32. See Robert Castel, *La Gestion des risques* (Minuit, 1981). The computer has made medicine more impersonal, for medicine depends on the computer and statistics. Objectivity is justified in the name of efficiency, but it dehumanizes medical treatment.

help for the least little thing. The intensive development of medical prevention, which is useful in dealing with ailments like cancer and AIDS, helps to nurture this fixation, this obsession. Due to the excess of care, of medical and surgical methods, we cannot face up to ourselves and take responsibility for ourselves. When something goes wrong, we cry for help, we seek protection, shelter, tutelage. Overprotected, we joyfully hand ourselves over to others. We claim the right to health, but this right coincides with the right to rest, to vacations, so that the world today is full of people on the move.

Twenty years ago there was one migration a year, namely, for the summer vacation. But now there has to be a winter vacation. Indeed, there are approximately four holiday migrations per year. At Christmas 1985 some 500,000 automobiles left Paris. We have to recall that this number, which means about one and a half million people, presupposes an absolutely unreasonable expenditure (or waste) of money on vacations of at least a week in the mountains. To prevent traffic jams many police had to be mobilized and centrally located helicopters were also needed to direct things. The provision of these elaborate but useless services involved a secondary waste imposed upon society as a whole. People left the city en masse, traveled en masse, and arrived at the snows en masse. We have here total unreason in which individual unreason corresponds to the collective unreason of society. It is made possible and inevitable by the enormous technical machinery now at our disposal.

I might continue for a long time listing individual instances of unreasonable behavior. They all translate finally into the devaluation of the word and of language, which I have examined elsewhere.[33] This devaluation may be seen in various orientations like that to which I referred earlier, that is, people speaking in films or on television, but their words being lost in the background noise of the street or music or general conversation. This noise supposedly makes it authentic.[34] We recall the theory that information is born of noise. What information do we have here? Two kinds, but the background noise is what is significant. The human word has no importance at all. This is in keeping with the generalizing of unreason. But perhaps we should look more deeply. The point may be that in modern society, which is essentially the computerized world, our older classical language is no

33. See J. Ellul, *The Humiliation of the Word,* tr. Joyce Hanks (Grand Rapids: Eerdmans, 1985).
34. We find the same phenomenon in the modern theater in which the least important and most neglected thing is the text itself.

longer adequate. How can we talk about that for which we no longer have an adequate language because the methods of observation that condition it are no longer adequate? From the standpoint of information theory, what is at issue is a shadow zone constituted by the question of the creation of signification. From the standpoint of the postulate of determinism, what is at issue is a shadow zone projected by the question of the emergence of the new.[35] Face to face with this problem, may it be that the unreason of today is the reason of tomorrow? But what dangers do not then lurk on the horizon?

From these flashes of what I still call unreason, one may conclude that six dominant features are appearing secondarily in our society. I will simply list these and not develop them, for we shall come across them incidentally further on. The first is the almost total disappearance of ends that have been thought out and clearly conceived (at all levels). The second is the fading of human interest in the concrete sense. The third is the equation of the good and technical progress. The fourth is the combination of very complex multiple interests (classical economic, political, and social interests being largely outdated). The fifth is our inability to grasp a situation globally, or if we are outside a given situation, to take a longer or broader view. The last is our total inability to rectify our mistakes by analyzing the path we have taken and seeing the factors at work on it. We have looked at one aspect of this final point in discussing double feedback. But we are far from having examined the whole range of the absurd.

35. See Henri Atlan, "L'Emergence du nouveau et du sens," in L'Auto-organisation, Colloque de Cerisy (Paris: Seuil, 1983).

CHAPTER XIII

The Costs: The New Relation between Technique and Political Economy

We would do well to recall for a moment the successive relations between technique and economics.[1] In my first studies of technique I noted two periods in its relation to economics. In the first period it was taken for granted that economic life was determinative. This is what provoked and permitted the development of techniques. The market and producers created a demand. Machines had to be made to meet their needs and to achieve the ends that accumulated capital made it possible to project. Industry controlled technique. The transition came as little by little commercial capital became industrial investment. Technique at this stage was subordinate, governed by economic growth. It simply followed.

But little by little, in what I see to be the second and last stage of the evolution, technique became more independent and autonomous. It first began to develop in many areas that were not economic (e.g., in the case of techniques that are useful in scientific research). Then it began to govern itself rather than responding to economic appeals or needs. It thus became multiform, complex, and rapidly changing, obeying its own imperatives. In other words, a technical innovation came about because the conjunction of ten, twenty, or fifty other combined techniques made it possible. With the possibility came the fulfilment because there would be some research scientist who by experiment or theoretical calculation arrived at the combination of existing techniques and produced the innovation. Whether the innovation has economic value or interest was now of no significance. It was

1. I am taking up again here what I said about economic absurdity above.

243

not a determinative factor. The innovation was there; its application could always be discovered. The great principle came to light that when a technique exists it has to be applied. This is the stage which, since 1950, I have called that of the autonomy of technique.

Reversing the previous situation, technique now became a driving force for the economy. The creation of new techniques brought new possibilities of production, distribution, and consumption. The economy followed technique, entering upon the new paths opened up by it. Technique was not, of course, fully autonomous. It had to take into account certain economic limits. It was already creating goods whose economic usefulness was not immediately apparent. But if it did not meet needs that were felt or foreseeable, it brought to light latent needs, and also, notwithstanding the debates about this, the creation of artificial needs. It could be argued that since we kept up and eventually led in the game with much satisfaction, the needs were not as artificial as was said, that no distinction between natural and artificial needs can be upheld, and that after a period a need which is created by advertising but which becomes habitual is just as natural as those that are generally listed as such.

All this went hand in hand with a raising of living standards, and economists took note of the changes, being forced step-by-step to find more complex models for the effects of technique on all sectors of the economy, including its structures. But this work was possible only as the economy itself adopted the process, methods, and tools of technique and became more scientific, certainly, but also more technical. The economy in action became the possibility of the constantly accelerated progression of technique. Economic science became the mode of managing and incorporating all the possibilities of techniques in a harmonious whole. Yet between 1955 and 1975 I began to see a certain discrepancy in this picture.

I thus arrived at a new relation between technique and the economy. Although technique undoubtedly created values, it also caused much distortion between the various sectors of economic life. There were also so many technical innovations that it seemed impossible to put them all into practice in the form of goods for sale on the market. It was not that there was saturation. This has often been announced for one product or another, but it has never happened. The impression arose, however, that economic reality and economic science were being outstripped by the enormous proliferation, that they could not rationalize it into a coherent economic whole, that the liberal, Keynesian, and socialist models were all inadequate, and that the new ones which were trying to take all the facts into account were increasingly abstract and of relatively little use to a political economy. In other words, the relation between technique and the economy seemed to be changing. In its origin and creativity technique was still autonomous, but its concrete results were much less assured. The economy was now serving as a buffer and limit for technical enterprises. As always, there was the possible and the impossible. But these were now measured by economic rather than technical capacity. Technical growth became a function of economic possibilities. This brought us to the fourth stage in the relation between technique and the economy, which I call the stage of the wisdom of political economics.

Technique incessantly accelerates its innovative activity. But it has ceased to be autonomous vis-à-vis the economy, the economic reality, and the functioning of this economy. Economists are increasingly aware of this. Confronted by technical immoderation, by apparently unlimited proliferation, by the seriousness of the impact of techniques on the environment, by their overturning of the priority that elementary classical economies accorded to labor, raw materials, etc., political economics has begun to run into what seem to be insurmountable difficulties and therefore to raise some questions. The first two, which are familiar ones and which I shall not stress, are as follows. First, we have discovered that not everything is possible. Technique undoubtedly enables us to do this and that, but investment is needed which is of dubious profitability and which is so great as to make many projects impossible. Thus it is impossible to make the most complex and costly surgery available to everybody and at the same time indefinitely to create more hospital beds.[2] It is not possible to multiply satellites and at the same time to guarantee help for victims in every area. I could give many more examples of choices that have to be made.

For economic reasons techniques are no longer indefinite creators of value. Technique is still the most important creative factor but it is far from meeting all its own demands. This leads us to the second well-known problem, namely, that for a long time too simplistic a view has been taken of the actual costs of producing goods; we must now take a more complex view. The more techniques progress, the more they involve such difficulties as pollution, the exhausting of raw materials that cannot be renewed, social troubles, potential dangers, etc. We have to take into account the compensations that must be paid, the precautions that must be taken, and the substitutes that must be sought if we are to estimate the real cost of products. A technique that

2. Cf. "The Soaring Cost of Health Care," *Science, Technology and Society* 6/1 (1986), a study of health costs in the USA and the choices involved.

results in great risks of toxicity, for example, means that institutions have to be set up (e.g., specialized hospitals). This is the well-known question of the internalizing of external factors. It is a very difficult matter, for where are we to stop? What risk is negligible? What part of a problem cannot be imputed to a particular technique?

We cannot answer such questions concretely. But the very fact that economists have to consider such problems shows that they feel an obligation not merely to be scientific but to exercise a wisdom and moderation that stand in contrast to the immoderation of technique. The situation will be aggravated, it seems to me, by the coming to light of facts that economists will have to take into account. I might mention simply the fabulous growth of expenditures in all countries due to the development of techniques, directly because this becomes increasingly expensive both in feasibility studies and in onerous costs of production, which are now exceeding all bounds.

Let us look first at the larger figures before taking detailed examples. The French budget now amounts to more than one trillion francs and the foreign trade deficit is 360 billion francs; public debt increases every year: 614 billion in 1982, 900 billion in 1984, one trillion in 1985, 1.1 trillion in 1986, representing 20 percent of the gross national product. Interest alone on the debt amounts to $100 billion a year. The American budget deficit was $220 billion in 1984 (6.6 percent of the gross national product). Foreign debt in France was 20 billion francs in 1980, 75 billion in 1982. The U.S. Federal Reserve reports an "explosion" (Bourguinat) of trading deals on the various exchanges of up to $200 billion a day. We have already seen that the indebtedness of black Africa amounts to $110 billion. Cuba alone (an interesting point) is in debt to the tune of $3 billion. The debts of the Third World amounted to $1 trillion in 1985, with interest of $140 billion, which would absorb all the disposable revenues of the countries concerned. This foreign debt increased by $620 billion in five years, and the countries as a whole would have to contribute 67 percent of their revenues to pay off the debt.[3]

Throughout the world we find stupefying expenditures that no one can control. Let us take some specific examples. The Channel Tunnel is projected to cost 15 billion francs. Computers to assure safety

---

3. See F. Partant, "Campagne internationale sur l'endettement du tiers monde," *Champs du monde*, June-Sept. 1986. The system is kept going only by new borrowing. Rescheduling is not enough; Mexico had to be rescued afresh two years after rescheduling. We are now in the second stage of the debt crisis (H. Bourguinat). Oil is less costly, but everywhere there is waste.

for trains cost 1.5 billion francs. An enormous infrastructure in the desert for satellite retransmission between Paris and Dakar cost 800 million francs. The Villetts complex cost 5.5 billion, and one minute of computer images retransmitted by television costs 1 million francs. In 1985 advertising expenses in France (including clips) amounted to 70 billion. In a truly comic aberration, staging the opera *Aida* cost 25 million in 1984. The estimate for the high-speed train to the Atlantic was 12 billion, but the price had doubled by 1982. The Ariane space program (Ariane V carring a satellite of 5 tons in high orbit and 15 tons in low orbit) was estimated at 18 billion in 1986. Columbus (a kind of space lab that can function autonomously or hook up with the American space lab) will cost 20 billion. Hermes, a space shuttle, will cost 18 billion. The space program as a whole will amount to 56 billion. But we have to press on with it if we are to keep up. There was rejoicing when in 1985 Ariane brought in 7 billion francs worth of orders for launchings (including an American satellite), but we cannot deduct the 7 billion from the 18 billion cost, since each launching has expenses as well. Furthermore, after several failures of Ariane, the number of orders diminished. In 1982 Électricité de France had a debt of 8 billion. A single launching of the U.S. space shuttle costs $250 million. I am not trying to give a full picture or to heap up statistics. I simply want to show how enormous are the sums of money manipulated. We are playing about with billions, and we have to ask what this entails.

But first we need to recall firmly that these gigantic expenditures have their origin solely in the demands of technique, its production and application. There are individualized techniques (including those needed for advertising, to which we shall return later). There are also collective techniques (aviation, armaments, space, etc.). Undoubtedly, most of the costs relate to the human element: salaries, pensions, subsidies, various services, insurance for damages, etc. Yet most of these expenditures derive secondarily from technique or from the real or factitious needs stimulated by the technical milieu. From another standpoint, I realize that when a budget of 1,000 billion francs is split up among various departments and then among various services and branches, it is finally reduced to a human scale and there is nothing extraordinary about it. Yet this does not mean that government expenditures are not of an order that we find it impossible to grasp. We can neither imagine nor comprehend them. We must also ask whether it makes any sense to engage in economic forecasting when it would need only two or three Third World debtor countries to refuse payment and the whole international financial system, from banks to states, would collapse. Third World debtors can now cause global bankruptcy. How, then, can we forecast a regularly functioning economic world market?

We have referred already to armaments, but what do they mean from an economic standpoint? The factories are running and there is employment. But all countries are cursed by the weight of military expenditures: Third World countries as buyers, producer countries in the technical search for greater effectiveness, so that new weapons lose their value in less than five years and have to be replaced. Thus in 1983 military expenditures in the USA were five times greater than industrial investment ($600 billion). With figures of this order (which make sense only to a computer) nothing is really represented and anything is "possible." One trillion today, why not two trillion tomorrow? Some will say that a limit is imposed by the gross national product and the taxpayers. But I do not think so at all. At root the revenue from taxes is less and less important and will continue to be less and less important because what we have here is an abstract juggling. The proof is that all countries are ultimately in debt internationally and that the billions on paper change nothing. There is complete dissociation between the monetary and the real.[4]

We have to see how the process of abstraction unfolds. First, notes are covered by gold reserves, which do not circulate. Then it is allowed that only 10 percent need to be covered. Next, gold is dropped as a standard. Money does not rest on a gold reserve but on the value of the goods produced in a given system. Money represents the calculated value of the concrete products of human labor. The value here is the economic value of exchange. The goods produced are a monetary measure only as they are sold. It is on this point that the whole theory of Marx rests. The creation of goods is no longer primary. Naturally, they still exist, but they are no longer the essential thing on their material side. More and more we have the consumption of services, immaterial elements, images, sounds, collective goods (of a different kind from earlier collective goods) which serve as infrastructure and which are funded publicly. These are not objects that the public can lay hands on to buy or sell. They have no exchange value. An Ariane rocket, a satellite, a cardiac transplant, a cyclotron: none of these things has exchange value. Can we even say that any of them has "value"? The

---

4. Cf. especially H. Bourguinat, Les Vertiges de la finance internationale (Economica, 1987), which contrasts the financial economy and the real economy. The new international finance no longer has anything in common with classical finance, due to the worldwide application of financial markets, all kinds of financial innovations, the globalization of the financial function, and computerization.

only value is that which results from the expenditure made to obtain them.

In reality, however, almost everything which costs, which turns out to be very expensive, has a use value (airplanes, roads, etc.). But Marx showed that use value does not enter into the calculation of economic value (for we cannot count as economic value that which farmers produce in their gardens to eat themselves). At the same time all these high sums are due to things we all pay for. We pay for the consumer society, for modern agriculture, for the unheard-of waste of paper pulp, for polluted rivers, for degrading the soil, for breakthroughs in leading high-tech communications (which weaken the postal service), for individual cars (the absurd and disastrous competition in making sophisticated and dangerous products), for Japanese motorbikes and video recorders with the imbalance in foreign trade, for out-of-season foods and Third World products that are the happy result of the green revolution. We pay for all these things in money, which no longer represents anything, and also with imbalances, or harmful effects on the collective scale.[5]

Furthermore, in these enormous budgets we have also to take into account the negative costs (reparations and compensations for pollution and nuisances). In addition, two other factors are of increasing importance. The first is the disposing of wastes. This problem is very complex, for if there are wastes at every stage from extracting raw materials to the finished product, the wastes are not of the same kind. The environment can recycle some wastes, but this recycling takes time. The important point is not the growing quantity but the lengthening of the time needed for recycling. The waste products of modern industry are less and less capable of being reintegrated into the natural cycle (e.g., plastics apart from biodegradables). We ourselves, then, must either eliminate or recycle waste products. Simply controlling pollution from wastes cost the USA $17 billion in 1985. Some wastes can be put to use again (e.g., those that can reenter the biomass, which produces energy). Others can be recycled by very complex operations (nuclear waste) which in turn give rise to fresh wastes! Thus it is not possible to reduce to numbers the cost of the operations needed to deal with the millions of tons of daily industrial and domestic waste, especially as there is great variation from country to country. But we certainly have here costs that are unavoidable with the growth of the population and of industrial production.

The other great source of expenditure in our world, of totally

5. For details, see J. Chesneaux, De la modernité (Maspero, 1983).

unproductive expenditure representing a net loss from a monetary standpoint, is the covering of growing risks. I do not refer merely to the enormous costs of insuring against automobile accidents but to making provision for risks of all kinds, for example, droughts or floods in a region, which make it necessary for public aid to be given to farmers. Today we have to insure against an increasing number of risks. We thus have to distinguish the level of the acceptability of risk from that of the coverage. Insurance has to determine what compensation can be paid for a known risk when damage has occurred. What it guarantees is sanctioned by contract. The covering of risks due to technical change cannot be the result of statistical calculation. Nor is there any level of acceptability of risks as such. It all depends on for whom, at what time, and in what conditions. Each risk is a special case. But how can we estimate the level of risk that a society is prepared to accept? No calculations or data can provide an answer. An interesting phenomenon is that after many failures to launch satellites by rocket, and in view of the costs of ever more perfect satellites (the last destroyed by the failure of Ariane cost $200 million), the promoters wanted to insure them, but because of the number and the cost the insurance companies were increasingly hesitant to agree.

Unexpected risks of all kinds have become such that they form a system of their own. The elements that compose them and the chain of events that leads to them are unforeseeable. "To ask when a risk is worth running is from all appearances a question that lies outside the technical sphere in the true sense."[6] This means that risks are properly a matter for political debate, with competence on the one side and general participation on the other. "There are only two ways of approaching the matter of the acceptability of risks. We may compare the risks of a new project to those that exist in nature and talk about a normal level of acceptability (though this strictly technical approach reaches a limit in its hypothetical nature, there being no science certain enough to reduce contingency). Or we may search for alternatives to the project and thus abandon the technical debate and deal with the rationality of the project in terms of political choices."

This is enlightening, but I think that Salomon is wrong when he sets the technical and the political in opposition. The politicians are always inclined to accept the conclusions of the technicians, the experts. Furthermore, we have always to envision new and unexpected

6. J.-J. Salomon studies the interesting case of the Asilomar Conference which moved on from the utilizing of risky techniques to the problem of controlling the modalities and objectives of scientific research.

risks. Seveso, Bhopal, and recent transformer accidents were all unexpected. So is another possible risk. Now circling the earth are at least 2,500 satellites launched since 1970, along with another 15,000 objects. One or other of these falls back to earth from time to time, though we do not see them, since the engine finally explodes and they break into small pieces which burn up as they enter the atmosphere. Nevertheless, after the accident in Canada in 1971 most nations have stopped using nuclear-powered engines. But the USSR continues to do so, and its satellites, when they go out of service, break up into two or three large pieces of one or two tons each, which cannot burn up in the atmosphere and thus fall to the earth. These satellites have nuclear engines of 50 kilos weight which do not burn up and the fragments of which are very radioactive. But how can we insure against risks of this kind in these conditions? What guarantee can there be? Here is a new reason for growing expenditures that have no useful value.

At one and the same time, then, we are in a society of absolute buying and selling and in a society of technical risks and increasingly abstract goods which cannot guarantee money and which have no value in themselves. These goods and services are grafted at a second or third stage on operations that are strictly economic. The only measure of these fantastic costs, of this exorbitant creation of abstract money, is the actual activity of growth, the hope of progressing, on the presupposition of going faster and faster. It does not matter much whether what is produced is useful. We have to produce, for this is the activity that gives money its value. The activity has to be constantly accelerated to justify the expenditures. This process is wholly in keeping with the thrust of technique. But that also means that an "enterprise," with its structures and products, is not made to last. The pace at which modern enterprises are changing is accelerating, their expectation of life is shortening, and capitalism (though not the social side!) is adapting, so that it gives the impression of extraordinary economic vitality. New forms, like technopoles, are being invented, though these are already obsolete (as Silicon Valley soon will be). We have to realize that one part of the enormous capital expended is destined to disappear without a trace. Let us make no mistake—the economic flow from the industrialized countries to the Third World consists of capital that will be lost. In the long run three-quarters of the earth will become economically sterile.

Naturally, in these circumstances the fine old rural economy of the small farmer no longer exists. This is openly stated on television. When the building of a nuclear power station at Mezos in Landes (southwestern France) was announced, the commentator added that this would be the only economic center in the region. That which had made up the life of many thousands of farmers up to that point no longer existed. It was no longer worthy to be called an economy. Nor did the natural environment count for anything. Construction had to go on at all costs and France had to be subdivided. I am not saying that provision does not have to be made for the growing population, nor that run-down houses should not be replaced. I am simply saying that the rural areas are being devastated as the urban surface incessantly increases. This is supposed to be a sign of progress.

In the extraordinary growth of expenditures, we discern a sign of the competition among nations. We find less and less internal competition between companies, for this would imply budgets in which money corresponds to value produced and sold. But competition among nations need not pay heed to this factor, since the state can always turn out the billions needed. As the nobility previously had castles and jewels for prestige and security, states today have big budgets. To launch the first satellite, to be the first to land on the moon—these are goals for the state no matter what the expense in terms of money that does not correspond to anything. And the rival state techniques can serve any end, producing goods and services that can be sold, feeding the competition in aircraft, rockets, armaments, etc.

One has to find an opening for possible sales. Technical production is not in terms of utility but of the possibility of international clients. Only wealthy countries, of course, can play this game. The amount of capital and the complexity of economic interrelations mean that there is a constant trend toward a wholly abstract system. Capitalism is logically moving in that direction. Profit alone can be the common denominator. In the long run a business does not have to produce goods that make a profit; they simply have to make a profit (Gellibert).[7] Transnational corporations have become so complex that in the long run we do not know whether they are making a profit. Abstract sums are spent, are reproduced, and are concentrated. The same is true in large state enterprises, in which it is not obvious when hundreds of millions disappear. The very size means abstraction.

To take only one of many examples, it is interesting to note that Imperial Chemical Industry has made much more money by financial transactions on international exchanges than it has from sales of its chemical products in all its factories. Gellibert refers not only to this instance but also to the joke about the Rotterdam oil market: "If there

7. J. Gellibert, "Le Choix de la biomasse comme énergie" (dissertation, University of Bordeaux, 1986).

is a market at Rotterdam it is not a market, and if it is a market it is not at Rotterdam!" Representatives of the big corporations and oil traders negotiate constantly by telephone and telex. In these negotiations cargoes change hands four or five times en route and tankers change their destinations, the result being what the trade knows as "paper barrels," that is, profits with no corresponding barrels of oil (Gellibert).

I will be told that speculation has always existed. I agree, but we still need to refer back to the important law (which people in general refuse to apply!) that quantitative growth entails a qualitative leap. The billions of abstract dollars and the vast amount of equally abstract production entail such a leap. This is inevitable when the national market has given way to the international market and the latter to a world market, when one lives on a cable-television planet in which economies are strictly interdependent and universal, and networks of communications and politico-social structures are under the universal despotism of this market. All this is precisely the fruit of technical growth. Cable television, communications, universalization—these are techniques. It is these techniques that are so expensive. They constantly demand new investment as the competition among nations forces upon them techniques that are more advanced than can be applied.

As the example given above shows, this situation has truly absurd consequences. We recall the rise in oil prices in 1973-74, which produced a tragic crisis in the economies of all the developed countries. Technical progress has permitted a progressive reduction in the price of this main source of energy. But then the collapse of oil prices in 1985-86 brought a new economic crisis. This crisis now affected Third World producers of oil (Nigeria, Mexico). It was also a crisis within OPEC, with disagreements among the members. It was a crisis for Great Britain, which no longer had its projected oil revenues. It was a crisis for the international oil companies, which had stocks that they had to sell at less than they had expected. If this world crisis was not as obvious as the earlier one, it was no less profound. I also ask in passing whether, if it were possible to stop the traffic in cocaine, a similar world crisis would not result.

To conclude, I will state this time my agreement with the financial analysis of Bressand and Distler when they show that finally there is no longer any financial substance (the sums themselves no longer count) but only networks and the flow of money. Money does not have to be "saved" any more. It is made to circulate indefinitely, in the abstract. This means among other things that there can be no real financial regulation because, more and more, strategies of evasion are applied: to apply to what has become an immaterial domain, that of finance, the unlimited possibilities of innovation and strategic activity offered by the networks. The above authors give some striking examples of the way to get around strict financial regulation. Billions of dollars, which seem out of all reason to us, circulate (abstractly) in an incredible manner. The total international oil flow represents the exchanging of something like $250 billion in fifteen days on the New York stock exchange. But of the $500 billion per month only $35 billion correspond to commercial operations. The rest (about 95 percent) represent purely financial operations that are abstract and correspond to no concrete value. The degree of abstraction keeps increasing, for example, with special drawing rights or substitute accounts (which ought to affect the syntax of the monetary networks).[8]

Increasingly, then, we cease to have a world in which money plays the classical role and one can differentiate the government of people from the administration of things. Henceforth, thanks to and because of technique, it is the creation, management, and syntax of networks that now offer to the world of business and government the real fields of action. An enterprise will now define itself as an intersection of networks. Thus we no longer have a vertical or horizontal integration of businesses but a system of networks (which makes possible the complex creation of transnational corporations). This applies in the sphere of production, in the modes of access to multiple techniques, as well as in financing. The concept of the network is often used in large firms. To issue a loan, for example, there has to be a syndicate of banks under a single head, then the investment of the "paper" by the banks through their individual networks. It is the combination of networks that makes the launching of a new company possible: networks of risk capital, of evaluation, of counsel, of mobilization of savings, etc. Only computerized equipment makes it possible to do this on the necessary scale. All financial and exchange markets are necessarily electronic markets. They no longer engage in the money-changing operations of classical tradition.[9] Like paper money, all these transactions represent only the Brownian movement of flow

8. Cf. especially Bressand and Distler, *Le Prochain Monde* (Paris: Seuil, 1986), pp. 187ff., 195. Bourguinat, *Les Vertiges de la finance internationale*, emphasizes that the financial problem is increasingly difficult. There is a striking contrast, he says, between the number and scope of plans to reduce debt and the beginning of their execution. But if that helps to impede a solution to the crisis or to bring back inflation, that in turn can lead to protectionism and all the greater imbalance.

9. For a good example of this transformation cf. Bressand and Distler, *Le Prochain Monde*, p. 219.

and counterflow, for the market is no longer a place of meeting but a complex of services, rules, and electronic infrastructures. This electronic market can instantly bring buyers and sellers together, bypassing the traditional intermediaries.

Thus everything is faster and faster. One instability replaces another. All can participate thanks to their own microcomputers, buying and selling without going through a stockbroker. (Democratization, it is said, but to whose profit?) Furthermore, the development from the credit card to the microprocessor enables all of us to enter into the exchange networks and of fully computerized payments (the use of plastic money). By means of the terminal we can also manage our bank accounts (which we still need to have!).

But this raises the central question: Do we not have here a transformation of economic normality? Previously we needed a normal economic stability, with some controlled instability. When instability became widespread, we called it a crisis. But today we have come to consider as normal the widespread instability of the constantly shifting flow, and the need for stability is abnormal, since stability contradicts the brilliant march of techniques. This march cannot tolerate the slowness of traditional economic operations (though stabilization may have value for the Third World!). According to Distler, the mistake is to begin with a postulate of saturation of growth in the North and to reason in terms of reconversion and redistribution for the South. (If this is a mistake, it confirms what I said about the widening of the technological gap.) This said, I take up again the thesis of a conflict between technical acceleration and the inertia of the traditional economy, which in both thought and operation is not on the same wavelength. Hence this economy, rediscovering reality, will have to freeze the dance of techniques, by doing its real accounts, or else technique, completely gaining the upper hand, will lead economics into a world of abstraction, of unknown territory, and of crazy figures. This is our true dilemma today, it seems, though few pay heed to it.

Politicians certainly ignore the basic question: Are not our new, leading techniques the origin of the economic crisis (the very opposite of what politicians believe)? This hypothesis (which is more than a hypothesis, for it is beginning to be demonstrated) was upheld between the wars by economists like Kondratieff and Schumpeter. The essential thesis of the latter has resurfaced today, namely, that technical progress is the main dynamic factor in economic development, but that it is a destabilizing factor by reason of its timing, its rapid spread, and its many disruptive applications. All the big technical innovations call into question whole sectors of traditional economic activity. New materials displace the older textiles and steel. The arrival of new products explains some of the stagnation from 1970 to 1980. Replacements do not occur without crises. Furthermore, for some new products (e.g., electronics) there is at first a great demand for labor, but then with innovations (the fruit of an accumulation of knowledge in various disciplines) there is a tendency to economize on labor, and revolutions take place in specific sectors. The introduction of a new technique in an expanding system of production also causes structural crises of adaptation. Techniques do not allow harmonious economic development; quite the contrary. Policies of economic revival (often by the promotion of new techniques!) are no help so long as there is no understanding of the effects of technical evolution and the way in which it relates to income and investment policies.[10]

10. See C. Freeman, "Les technologies modernes sont-elles à l'origine de la crise économique?" *La Recherche* 125 (Sept. 1981).

CHAPTER XIV

What Use? The World of Gadgets

When I was twenty years old, the question that I am asking in this chapter—"What use?"—was for me the horror of horrors. In the period from 1925 to 1935 only the abominable middle class, which was materialistic in fact though idealistic in word, incessantly asked and repeated this question. Everywhere I came up against their triumphant logic: poetry, art—what good are they? They could admire Picasso because his work was beginning to sell at high prices. But history and Latin—what use are they? We must certainly learn to count; that makes sense. But change society? To what end? This was the question of the philistine. The important thing then was to find a place in society. Being red or leftist not only served no useful purpose; it was definitely bad. "What use?" was the absurdly triumphant question of a middle class which no longer believed either in values or in God, which had a positive but no less limited spirit, and which regarded the making of money as the only useful activity. Anything that might contribute to this was useful; the rest was of no account.

By way of reaction I was one of the generation that was greatly influenced by the surrealists and by Gide, who proclaimed the value of gratuitous acts, which were of value simply because they served no useful purpose and had neither origin nor goal but *were* simply because they *were*. All this prepared the ground for existentialism and the philosophy of the absurd. But at that time we had to crush the dreadful mentality of vulgar utilitarianism. A little later, however, I myself had to put the same question but from a different angle. The first time the issue was faster trains and automobiles. When meeting a man who had driven 100 kilometers per hour (in 1928) and cut fifteen minutes off some journey, I asked him what he had done with the fifteen minutes that he had saved. He looked at me with astonishment. Then when I heard that the Concorde had cut four hours off the flight across the Atlantic, or that the high-speed train had saved two hours on the run from Paris to Lyons, I asked again what people were doing with the time that they saved. Were they beginning a symphony or a sonnet, or thinking up a new experiment in chemistry? Were they simply enjoying the freedom of a stroll with no particular goal but in all the joy of liberty? But no one was ever able to give me any answer. They had perhaps taken a drink but in effect they had done nothing and had no vital experience. They had simply filled up the time in an empty and insignificant way. Or else they had profited by the time saved, as when a busy executive might squeeze three interviews into a heavy schedule, thus hastening the day of his or her heart attack. And always with the anxious proviso to the very end of the journey that the airplane or train had to arrive on time!

The time saved is empty time. I am not denying that on rare occasions speed might be of use, for example, to save an injured person, or to rejoin a loved one, or to go back to one's family, or for the sake of peace in a decisive meeting. But how few are the times when it is really necessary to save time. The truth is that going fast has become a value on its own, as is now acknowledged. What we have here is *L'homme pressé*, as P. Morand so well describes him, but not really "pressed" by anything. The media extol every gain in speed as a success, and the public accepts it as such. But experience shows that the more time we save, the less we have. The faster we go, the more harassed we are. What use is it? Fundamentally, none. I know that I will be told that we need to have all these means at our disposal and to go as fast as we can because modern life is harried. But there is a mistake here, for modern life is harried just because we have the telephone, the telex, the plane, etc. Without these devices it would be no more harried than it was a century ago when we could all walk at the same pace. "You are denying progress then?" Not at all; what I am denying is that *this* is progress!

Another incident that made me raise the question "What use?" was that of the first great massacres of the peasants, the kulaks, in the USSR. I asked many friends who were close to communism what purpose was served by killing these peasants, who had no real idea what was happening in their country. The embarrassed reply was always the same: They were capitalists. But all the evidence we had showed that this was not so. They were counterrevolutionaries, then, who did not want their land to be collectivized. This was half true, but did it justify mass killings? Did the reign of social justice, equity, peace, and freedom have to be inaugurated by massacres of which we knew both the horror and the extent in spite of censorship? At that time I was still an innocent in politics, but even though I was taking up the despised bourgeois question I could not stop myself from putting it. It is along the same lines that face-to-face with the fabulous technical progress of our day I put the same banal and vulgar question: What is the use of this immense mobilization of intelligence, money, means, and energies? Of what use *in truth?* The immediate utility is plain enough—dishwashers and robots save us time. It always comes back to that. But it is quickly seen that we are invaded not merely by objects, as Perec showed, but by innumerable working gadgets. There are individual gadgets and collective gadgets and gadgets of society as a whole, so that when I criticize them I stir up great, scandalized protests. But before taking a look at gadgets we must first give a summary sketch of our needs. For if objects are useful and correspond to true and original needs, they are not mere gadgets.

1. Needs

The search for happiness is not new. It is written into the U.S. Constitution, which is significant because the modern world opened with that declaration. But long ago I showed that there is a great difference between the ideology of happiness (or the utopia) found among past millenarians and our own ideology of happiness. It is a matter of means. Previously, the means did not exist to make people happy. The quest for happiness was thus an individual matter, a matter of culture, spirituality, asceticism, and choice of life-style. But for the last two centuries we have had the (technical) means to put happiness within the reach of all. Yet this is not, of course, the same thing. Happiness now consists of meeting needs, assuring well-being, gaining wealth and also culture and knowledge. It is not an inner state but an act of consumption. Above all, it is a response to needs. Though it may be a commonplace, it is worth recalling that we must make a distinction (contested) between basic, primary, natural needs and new, secondary, artificial needs. The lively argument against this distinction is that so-called natural needs are in fact modeled on a given culture. All needs, it is said, are cultural, so that a supposedly artificial need, when absolutely anchored in a culture (like the need for a car), is just as pressing as a "natural" need. As a rule, technical growth, in countries in which it has occurred, has enabled us to respond to the natural needs of food, drink, clothing, protection against heat and cold, shelter from bad weather. I know that there will be reaction to this simple statement, for it is not true that in our world everybody finds these natural needs met. My reply is also simple. The difference from past centuries is that then, if there was famine, people had to accept it as fate and do the best they could to survive, but today famine, or the existence of the Fourth World, is a scandal that we must immediately halt.

The difference in attitude brings to light the extraordinary change brought about by technique as regards access to happiness through the satisfying of primary needs. But the same technical explosion incessantly produces new needs.[1] That is the difficulty. Happiness is harder to achieve because of the acceleration of the production of new and different needs which become the more intense as primary needs are met. A young person wants a Walkman or a Honda because he or she does not lack food. The new needs are also multiplying. There are needs to compensate for the destruction of the traditional order: expenditures on nature, communication, cars, social life, leisure, sport. There are the needs of desire which are triggered by technical progress and which are in rhythm with the proposed objects: the desire for pleasure, for leisure, for longevity, for health. Though these needs are abstract and exist only because there are technical instruments, they are the constituents of happiness. There is need of music for the Walkman, need of the computer, need of the telephone, etc. E. Morin rightly says that the progress of technical and industrial development is the constant creation of new needs.[2] In other words, it transforms and extends our notion of well-being. The transformation is more by quantitative increase than by qualitative modification. Yet there are also qualitative modifications: an expansion of the consumption of the make-believe (cinema, television), and expansion of leisure in which eros plays a big part, so that life is eroticized.

Finally, there is also the creation of a need to compensate in the form of techniques of well-being (jogging, dieting, yoga, camping, etc.). In this case needs lead to well-being. This is not done only for consumption; one is searching for a "better life" (according to the banal slogan), or for being "in shape" (pushed by advertising). But there is a close connection between the better life or being in shape and technique, for we have to be in shape to be able to work. An ethic of production finds inner expression in the concern for a better life.

1. Several authors are aware of this fact even though they are not critical of technique. Thus M. Mirabail, *Les Cinquante mots clés de la télématique* (Privat, 1981), writes: "The offer of services anticipates a demand for them. Technology thus induces new needs." I would find this hard to contest!

2. There is even a supposed "need for communication between computers." We must weigh the terms here. The fact that computers can communicate with one another directly has become a "need"!

In sum, the problem of true and false needs, with the academic and stereotyped response, is outdated. Merchandising has been made necessary by urbanization and made possible by technical progress. Its corollary is the transformation of needs. Basic needs are swamped by others. Our life-style is made up of parts each of which is an object of marketing studies for the adjusting of needs and technical products. But are they still needs? asks Scardigli.

Sometimes the needs seem to have been dormant for a long time. Thus 15,000 French couples asked for in vitro fertilization. The need could not be there before the operation was possible, but from that moment there was an explosion of desire. Consumers who are caught up in the system of "the production of objects, which leads to influence, which leads to the production of a need, which in turn leads to the production of secondary needs" are constantly on the watch for new objects. They demand them again and again. They rush for every novelty that will perhaps respond to their desires. But the matter is more complex than that. As the framework of life is constantly modified, needs, too, are constantly modified. To live in this society we are under pressures, but these in turn create needs, and the products meant to respond to these needs create new pressures. We absolutely have to have an automobile or television. Television replaces the missing collective culture that was created by a living group. The automobile enables us to leave the city and drive the freeway believing that it is the country. Technicians never stop asking what is the new latent, unwitting, potential need that might be satisfied.

Roqueplo gives an excellent example from a business report. Someone asked about the next need in plastics. The embarrassed reply was that there was no need but there would have to be one. Not to have a need is abnormal. Yet this situation is more common than we think. There is a growing market for the new media (cable television, video recorder, videodisk, videotext), but those who resist are still numerous.[3] Pressure has thus to be exerted on these rebels. They have to be consumers even though they have no need. If no pressure is exerted, finding buyers will not be easy. A 1985 report showed that dishwashers and freezers had not caught on in France: 95 percent of the households had refrigerators, 91 percent television, 72 percent cars, but only 30 percent freezers and 20 percent dishwashers.[4] Enough is enough!

---

3. A German survey in 1979 showed that 30 percent of the potential clients were interested, the rest not. Cf. *Problèmes audiovisuels* (Paris: La Documentation française, June 1981).
    4. See *Le Monde de l'économie*, Feb. 1985.

2. The World of Gadgets

What do I mean by gadgets? In this context I am referring to mechanical or electrical objects which are amusing or entertaining and which we can take up or leave as we please. The electric knife to carve up the roast is a gadget; the gas-powered corkscrew that opens a bottle all by itself is a gadget. Always there is something of a game about them. They correspond to an older reality. I would say that the mechanical products of the 18th century were the first true gadgets.[5] These were marvels of invention, finesse, material skill, and mechanical knowledge. They might claim to be a scientific approach to the human mechanism. But they could serve only to astonish, to provoke admiration, to surprise court ladies, and to divert philosophers. They did not respond to any need, even a need for knowledge. Whether we like it or not, they were games, learned and noble games, but still games.

In its private dimension the gadget still has very largely the character of a game. But I will not study this aspect here because I will deal at length with the question of play in our society in part IV below. I will talk here only about useful gadgets. What then, are the features of a gadget? It is a technically very complex instrument which represents much intelligence, a combination of learned techniques, and considerable investment. It is now the main industrial product and an unlimited source of profit. It is an object which always involves "very advanced composition," and always (according to an absurd usage which is hallowed by tradition) very sophisticated. But a second feature is that the result of these efforts and skills does not correspond to any real need. By the very nature of a gadget, its utility is totally out of proportion to the considerable investment that it involves. Its services are completely out of step with the prodigious technical refinement of its conception. In other words, it entails an application of high tech for almost zero utility. This disproportion is what constitutes a gadget in the present sense. This being so, it will be apparent that for me the gadget is more than an odd little personal object. Yet I will begin with some examples from this area.

We have quartz watches, which will not break down, which tell us the exact time without varying more than a second in a year. What good are they? Will they get us to meetings exactly on time? Will they help us to get up in the morning more easily? Will meetings we attend end precisely at the time announced? Not at all! These watches are no

---

5. Cf. the learned work of J.-C. Beaune, *L'Automate et ses Mobiles* (Flammarion, 1980), though he tends to make over-profound philosophical inferences.

use whatever except perhaps for navigators (to plot their position). What is more, they also come with remarkable extra features which might waken us up to a charming melody or enable us to do astonishing calculations. An admiring friend told me one day that thanks to his watch, if I told him the day and year of my birth, he could tell me on what day of the week I was born and how many days I had lived thus far. I thanked him for his kindness but had no interest in such data. What is the point of so ingenious and learned an instrument when it obtains such absurd results?

Many engineers, each more skilled than the other, are working on a flat-screen television so as to prevent the slight distortion caused by the curve. Are we such lovers of art, such concerned aesthetes, that we cannot tolerate that slight distortion of the picture? Let us go ahead then! But if others are like myself, ordinary watchers, they are quite satisfied with what they see. What use, then, is a flat screen? None at all to anybody.[6] The same applies to the famous compact disc. It will give you an hour of music without stop, and its use of the laser produces no background noise, no static. What a marvel! And we had it presented to us on French television (March 21, 1986) by some idiot who raved about the beauty, the grandeur, the technical progress which has made possible the reduction of size and noise, and who finished his stupid address by telling us to throw away our old music and our outdated recordings. Advertising, of course; people had to be induced to buy the discs. But are we such informed musicologists or such lovers of music that the slightest speck of dust causes us to start? What good, then, is all this buffoonery, this creation of pure music, this deifying of the laser?

In reality, technique produces more technique whether it makes sense or not, whether it is needed or not. We are pressed to buy it. Thus there is the remarkable invention which enables us to see in a small corner of our television sets what is happening on other channels, so that we can better choose. Another fine invention for nothing! Then there is the remarkable oven which has a computer to tell us when the roast is cooked, or the microwave which cooks without heating up, as if there were the slightest need to invent a gadget of this kind except perhaps out of curiosity to see if it would sell! Are our meals better? Are our roasts better cooked? Is our gratin better? Obviously not! The end result, then, is nothing. There are also the household appliances furnished with "programs": electric ovens, washing machines, electric

6. I know that some say a flat screen is necessary to receive satellite transmissions. We will return to that.

irons, etc. Those who work them have to find the right button in
increasingly complex operations, the usefulness coming to light only
as they grow accustomed to them, that is, when the need has been
created. These things make life easier? Not at all! They go wrong, or we
press the wrong buttons. And what about the freezer, which one out of
three households in France now has? It enables us to do our shopping
(by car, of course) once a week. What a simplifying of life! How it
lightens our burdens! But how can we say that it really makes life better,
shutting people up at home and breaking off many social contacts? I
concede that in the case of big-game hunters the freezer makes it
possible to keep large pieces of meat for many months, or in the case
of those with many fruit trees, it enables them to keep the fruit (also
vegetables) "fresh" for a time. But is this worth the cost? (I could quote
from memory the millions of dollars worth of food that was lost with
the famous blackout in New York, not to speak of the lesser one in
Lyons, but these, of course, were accidents!)

We may refer also to the marvellous videophone, the first network of multiservice videocommunications (May 1985). The cost for
the Biarritz-Paris network was 600 million francs for messages from
200 Biarritz centers (1,500 in 1986). This basically useless gadget has
been called a new Concorde. The important thing according to M. Mex-
andeau is to give France state-of-the-art telecommunications technology. How crazy is technological discourse!

There is also the CB (citizens band), which provides entertainment on boring journeys. As in every case, an attempt has been made
to find some use for it. If I see an accident, for example, I can immediately notify the police and the ambulance service. But we know that it
is mostly used to keep in touch with other unknown drivers who have
the same gadget, and to carry on a conversation at a meaningless level,
or perhaps to make or receive a proposition of a dubious kind. Once
again, then, we have a fine invention that is perfectly useless and idiotic
and which might interfere with the useful communications of those
who drive taxis or ambulances. We might also refer to the use of
telecommunications for erotic purposes. The world of communication
has become in fact an erotico-communicational world of science fiction.[7]

Among other things that are socially reprehensible, such systems enable us to play with many lives. But in this world social disfavor
does not exist, since it has other terms of reference. This is still only a

7. M. Marchand and C. Ancelin, eds., *Télématique, Promenade dans les
usages* (Paris: La Documentation française, 1984).

game, and we will return to it when we speak of play later. What about the system which enables us from home to book theater seats, or to reserve a seat on trains or planes, or to know exactly when our train or plane leaves, or to find a telephone number that we need? We could do all these things quite easily before. The question, then, is whether such small gains are worth all that they involve in the way of laboratories, research, and capital. After the marvellous promises, there were also failures to live up to them. The system Claire de Grenoble went out of business in 1983. Others are used only when people have time to spare. There is also a big gap between the freedom that no other media can offer, the possibility of playing at will (since I do not know to whom I am talking and no one knows my name) in a game of masks and pseudonyms, and the miserable poverty of the messages and conversations in contrast to the communicational riches of the media.

Let us continue our slaughter of the innocents. What use is the telephone which will enable us to see who is talking? Is this so important (except in the case of two lovers)? It does, no doubt, enable us to catch fleeting expressions which might change the meaning of words. But all this gadgetry for a mere image! There is, however, a better use for the videophone. One simple call on it and we can be electronically guided through a town by the Tourist Office, arrange for a video, reserve a film for the desired hour, and make a doctor's appointment. All that! As Mitterrand has assured us, we have here the most advanced electronics and the true vehicle of economic recovery. No more than that!

Again, what use is the video recorder? It enables us to see films as we want and to record television programs that interest us. But do we not spend enough time in front of the television without adding to it, without doubling the brutalizing and dispossessing of the self which four hours of daily viewing produce? As regards television, what good is the ability to capture by satellite programs from all round the world?[8] Do we really need to see television from China, Pakistan, or Finland? Do we understand these languages? I have been in many countries where I did not know the language and the television sets in the hotels were totally useless. Yet attempts are made to justify channels that receive foreign programs even though they could be suppressed without loss. Is the price of a satellite worth the ability to listen to Dutch radio? Absurd! Pure gadgetry! Another worthless gadget is the all-terrain vehicle with four-wheel drive and big tires. I concede that it is

8. See C. Akrich, "Les satellites de télévision directe," *La Recherche* 140 (Jan. 1983).

useful for ethnologists, but the rest of the time all that it does is crush the undergrowth, destroy silence and nature, rush about beaches, bring down sand dunes, and poison the pure forest air—and all this to save walking a mile or two on foot. What use, except to wreck what is not already wrecked?

It is on vehicles that the imagination of (very serious) inventors of gadgets has had free play. One such gadget is the power window, which is useful only to spare us the exertion of turning the handle. A computer at least ought to be on hand for such a noble task! But that is nothing. Seats can be adjusted electronically with four different "memories" so that different drivers can find at once the right position. On-board computers can tell us how long we have been traveling, what our average speed is, what our gasoline consumption is, whether we are ahead of our schedule or behind it, and whether we are going faster than we planned. Then there are remote controls which enable us to lock and unlock the car doors or to control the heating and cooling, so that the car will be just the right temperature in the morning even though we leave it out overnight.

Nor is that all. There is a gadget to detect sleepiness and sound an alarm, another gadget to record the tire pressure, and another to tell us that a car is passing, not to speak of the many anti-theft gadgets. One of the most practical of these locks the brakes in such a way that a specialist has to be called in to unlock them. I am not making this up. I am simply selecting from the gadgets on display at the 1985 auto show. The automobile is obviously the gadget of gadgets. It is constantly being perfected so as to make it more expensive. The performance obviously has to be improved. Those who make cars have to provide work for studies on new engines. Power is their glory. They have to produce models that can cruise at 125 miles per hour (e.g., the Porsche 944 or the Ford Scorpio). This is crazy but inevitable. It might seem scandalous to call it crazy. Yet what use is a car that can cruise at 125 miles per hour? All highways have speed limits. There is also no doubt, in spite of all the debate, that fatal accidents are due to speed. The numbers prove the scandalous state of our lack of discipline. How long will the slaughter go on? Speed is active violence. Faster, faster, but where to, and why? We have full scientific certainty: A study by an organization for national safety on the highways shows that the number of victims seriously hurt grows 6 percent for every extra 6 miles per hour. An American study shows a 47 percent increase when the speed is raised from 39 to 70 miles per hour. A Swedish study of 28,000 accidents shows an exponential increase of risk with increased speed, with twice as many accidents for an added 17 miles per hour.[9] It is thus idiotic to make more powerful cars. But prestige is what counts. Also foreign sales! And above all, progress, technical progress! To complete our study, however, we must now move on from the automobile to the home.

What are we to say about that marvel, the electronic bathroom? In the bathroom you are now to have a mirror that will tell you about the weather, the state of the roads, and how long your journey will take. You can not only adjust the lighting but also start the radio and open and close various doors in the house. On the way home, in the car, you can adjust the bathroom temperature to your liking. Is not all this truly marvellous and so obviously useful?[10] But the whole house, not just one room, is now to be electronic. There are, it is said, four different uses: entertainment and information-education (audiovisual), domestic robots (household appliances, heating), security systems (for protection), and external communications (to control robots and a second home). The problem is that of interconnection. For example, you might be watching television and a message appears on the screen about the washing machine or to tell you that the roast is cooked. But central control is also possible. The Japanese are working especially on alarms, dividing houses into zones, each of which is visualized on a screen. A system of "simulation of presence" makes it seem to outside eyes that someone is home. Specific alarms are integrated with a central system so as to detect a burst pipe, smoke, or a failure in the heating or refrigeration. (The total system costs about $70,000!) We can easily see how indispensable such gadgetry is! We are verging on the absurd when we are told that here already is the house of tomorrow. You will pass it by like the rest!

I could extend indefinitely this list of expensive and ridiculous objects. I want to mention another one, the most common, the microphone. Its utility is obvious. It enables miserable singers who have neither voice nor talent to overwhelm rooms with their caterwaulings. Without this instrument they could only whisper little songs at wedding feasts. They do not amount to anything. But they have a mike which they clutch to themselves as a drowning person does a life jacket. This being so, they are needed. Today we have shows, mostly on television, for all hours of the day. The time has to be filled up, and to keep an audience, the shows have to be constantly renewed. There always have to be new games, singers, musicians, spectacles, etc. Television has many critics. For my part, I admire the creators of

9. Cf. "Automobile et vitesse," special number of *Le Monde*, June 23, 1985.
10. See *Le Monde*, March 1986.

television shows who are always producing something new, who give evidence of astonishing imagination and a vast knowledge of what is possible. But obviously they cannot offer us new singers of genius every day without end. They thus profit by the microphone.[11] Any imbecile with rhythm, a very noisy band, baroque costumes, and psychedelic lighting can put on a show that will enrapture the crowd of young people who want only to become the fans of a new gesticulating idol. But again we are entering the domain of games and shows that we shall have to study at greater length later. Let us continue with our list of gadgets.

If we pause for a moment to reflect, we realize that none of these objects is of the slightest use. Yet they are bought enthusiastically. Publicity undoubtedly plays a part here. Each gadget has its own advertising campaign. We shall examine this, too, further on. But advertising can succeed only on two conditions. First, there has to be a favorable mood created from "higher up." Every computer gadget will succeed because it is carried by the mood of our society. The computer is our salvation. Put a microcomputer in any gadget and it is sure to sell. The second condition is that the object be as sophisticated as possible. The more complex the mechanism and the greater the number of programs, the more people are inclined to buy. Rebates or incentives help. Soon a fad is created. Those who do not have pocket calculators or radio-controlled cars or video recorders are wretched folk whom one can neglect and despise. Fads are demanding. It is obligatory to have a touch phone, either apple green or sky blue. Consumer objects are just as much a matter of fashion today as clothing was a century ago. It is all useless, yet that is how it is. But a minister of state might reply indignantly on television that all this is very useful, for by buying these innumerable gadgets we keep the factories busy, give work to artisans and engineers, create capital, increase the gross national product, help exports, and act as good citizens. As we are regularly told, in an advanced society the problem of needs is a false problem. We have now moved up into the domain of freedom, culture, and well-being. It is useless to dwell on it. I will simply refer to the excellent chapter by J. Neirynck on "The third industrial revolution or the invention of what is useless," in which he shows clearly that the more we advance (especially in computers), the more we invent and create what is useless. But this is already leading us into the next section.

In the past pages readers might have smiled and shrugged,

---

11. Cf. Jézéquel, et al., *Le Gâchis audiovisuel* (Éditions Ouvrières, 1987), pp. 135ff., on the trickery of the false mike, false guitar, etc.

thinking I was making much ado about very little. But what we are now going to discuss should make us angry. I refer to social or collective gadgets, that is, objects of great collective importance which are still no more than gadgets. The two main ones in our society are those connected with space and the computer. Now, I am not saying that these are no use at all. My point is that their value is small compared to the investment of intelligence, skill, money, and labor that goes into their creation. Remember that it is this disproportion that makes a gadget. Naturally, I recognize the great space achievements: walking on the moon, etc.[12] But what do they really amount to? What real value is the enormous, gigantic, ruinous growth of booster rockets and satellites put into orbit?

If we reflect on the reality instead of being carried away with enthusiasm, we might mention meteorological satellites. But these do not allow us to make very accurate forecasts. Experience shows over 30 percent mistakes, though some of these may be due to faulty interpretation of the images transmitted. There are also telecommunication satellites, but these only allow quicker telephone links and the worldwide broadcasting of television shows, which we have already seen to be absurd. Then there are observation satellites, whether for civil or military use. As regards the former, we have referred already to crop forecasts, and the rest are similar. They also help to update maps, which is worth many millions of dollars. Above all, it is hoped that they will help us to discover new metals and hydrocarbons. A system launched in France in 1985 was not to be just a spectacle but was to be of scientific and industrial benefit, generating stock dividends. It was hoped to sell its pictures at $250 each. But the Americans, who have launched such satellites since 1972, are far less keen today and have gained from them much less than expected. In spite of promises, they do not give notice in time of earthquakes, avalanches, or floods.

The true usefulness of satellites is military. Observation satellites pass over enemy territory and make it possible to detect troop movements or the launching of nuclear missiles. Communications satellites keep the various military posts in a nation in constant touch. Combat satellites play a defensive role, as in the American SDI (popularly known as star wars). The idea here is that a laser beam from a satellite might destroy incoming missiles. At first the possibility was

12. Behind them, of course, is national prestige and competition. Cf. P. Langereux, "L'Europe spatiale à la croisée des chemins," *La Recherche* 138 (Nov. 1982).

regarded as almost certain, but numerous studies have shown that it is not. Furthermore, satellites armed with lasers could easily be used offensively. But the waste that it all involves is past imagining. Hundreds of satellites would be needed for an effective defense, and one satellite a week would be needed for replacements. At the moment France has only observation satellites (Helios) and communications satellites (Syracuse).

However that may be, we should not forget that three out of every four satellites launched thus far are military. I do not think that this use offers an adequate response to my question of their usefulness. Let us not forget that even if they are of no use, the cost is dreadfully high. One French observation satellite costs 800 million francs. Research on laser satellites had cost 600 million francs by 1986. If we look at another possible use, that of the economic exploring of mining resources on other planets, we recall that there were many dreams about this after the moon landings. There was talk of enormous reserves and new metals. But two or three years later this talk seems to have faded. The moon has been abandoned. We should not forget that the main aim of the Apollo program (there were four voyages) was not simply to go to the moon but to settle there and build factories. This aim has been abandoned by the Soviets as well as the Americans. In another century or so it might be possible to exploit other planets, but the expense is enormous and the usefulness of the fabulous exploits is nil.[13]

But does not space have scientific value? Space probes undoubtedly have great interest from the standpoint of scientific understanding of the galaxy and of some of the components of the universe. There are also space laboratories. But there is a great difference between purely scientific work and the laboratory experiments that are interesting from a technical point of view. As regards the latter, some chemical (pharmaceutical) products can be made which are possible only in a weightless environment. But there are only a few of them and their use is limited. Similarly, other products have been made that are useful in high-precision instruments, infrared detectors, and detectors of X-rays and gamma rays. The behavior of fluids in a weightless state has also been studied, as have also some polar phenomena, the components of the atmosphere, the behavior of rats and monkeys in a weightless state, and space materials. A wide-range astronomical camera has also been used, the conception of French scientists. In addition, there has been

13. Not to speak of the littering of space; cf. A. Repairoux, "L'encombrement de l'espace," *La Recherche* 159 (Sept. 1984).

research into the problems of blood circulation in a weightless state, along with other biological and psychological studies.

But scientific observations of the latter kind are of interest only if we decide to settle and live in space. We have seen, however, that this has been abandoned. The scientific observations are for the sake of science alone. We thus arrive at the vital conclusion that some vast technical enterprises serve only to advance scientific knowledge and nothing else. Note that! Some readers might think that science translates itself into useful techniques that pay off. Not at all! That is an outdated notion. Scientific discoveries do not necessarily give rise to practical techniques. These mostly come from technique itself. Science today often stays on the plane of knowledge. As I see it, that is not a bad thing. But we should not try to justify the billions spent on the conquest of space by inventing uses that do not exist.

Nevertheless, we cannot stop progress, and the French government decided in October 1985 to build a spacecraft (it is hoped by 1990). The National Center for Space Studies is in charge. But in view of the great cost it is hoped that Europeans will unite on the project and thus give Europe space autonomy. It has to be recognized that this project must be started from scratch. Ordinary satellites will not do. Computer programs are needed for such gadgets, and the programs for Hermes will be much bigger than that for Ariane or the observation system. In addition, materials that can withstand up to 20,000° C (36,000° F) are needed for the return to earth. For testing, such temperatures have to be created, and we can imagine how great will be the cost of providing them.

As a last example along these lines we might refer to the project of the biggest orbital station ever constructed (American, of course). All the plans and even the materials are ready and assembled. This is a station which is to be inhabited. It will be composed of four metal pylons (300 meters long), joined together and supporting all the rest, then of four big cabins (20 meters by 4), with many laboratories and solar panels for energy. President Reagan predicted that it would be launched in 1994. The interesting point is not the exploit but what is to be done with it. The director of the enterprise made the remarkable statement that they would find this out when they were in orbit. He could not have stated better that we launch out into vast technical undertakings without having the least idea what use they are. We need to ask why people are doing things like this. There is a mania for exploits, for technical success, which has remained the same ever since I formulated the principle that whatever technique can do, it has to do. There is also the need for prestige. If we are to remain among the bigger nations, we must do what the others are doing, and do it better, in the area into which the whole of the civilized world has now thrust itself. There is also the question of economic rivalry. We hope to sell our prototypes, our rockets, our satellites, our space labs, our orbital stations, our probes. But the true usefulness is almost nil. This is our first great collective gadget. We have to make people enthusiastic about something.

The second great collective gadget in our advanced world is the computer, which has the peculiarity of being able to link individual gadgets to the great universal gadget. Naturally, I know as I write this that it will be found scandalous. For all our hopes rest on this gigantic worldwide instrument. We must begin by taking note of all the propaganda and publicity which incessantly promotes the merits of the computer and especially the microcomputer (Lussato's "Little Kettle"), as though the macrocomputer no longer existed. Yet in reality it is the latter on which laboratories, banks, insurance companies, multinational corporations, and governments depend. We forget it because it is embarrassing, a centralizing organism which we cannot dress up, which we cannot come to grips with. That said, what is the usual line on the computer? First, the possibility of "decolonizing" it,[14] of bringing in a new world order of information, so that nondeveloped peoples will not be totally dependent for information on Western centers. Communication has been changed into a system of signals and commands which can increase the power of the large machines. Freedom of information was always accepted in theory, but was never possible for lack of means. Today the computer makes it possible. The modern technology of signs allows of pure and simple transfer. Nonaligned countries do not want a new order of closed information. What is needed is a pluralism of sources. This demands a truly universal expansion of means. Thanks to the free flow of data,[15] underdeveloped countries can finally get moving, there can be new activity and employment on a planetary scale, and the prices of raw materials can be prorated.

The computer smashes any monopoly in information or techniques. The computer can transform information and knowledge into useful operations. It is the tool of a twofold revolution, in both communication and economics. We are witnessing an astonishing growth

14. Cf. the three articles of J. Deornoey, "Empire des signes et signes de l'empire," Le Monde, Aug. 1983.

15. See Les Flux trans-frontières de données (Paris: La Documentation française, 1983).

in its applications: letters, accounts, forecasts, microprocessing. The computer figures in office automation, telematics, robotics, and factory automation. The computer can be used in scores of ways: for household accounts, for orders, for information, as office equipment for the execution and even the conception of work. Informational tasks can be automated so as to increase productivity, to speed up the rate of economic growth, to raise the level of education, to improve medical services, and to reduce the rate of pollution. Thanks to fantastic progress in perfecting the computer, everything seems to be possible. The creation of smart cards, computerized banking, electronic mail, data banks: all are available to help in the search for information.

It is estimated that by 1995 the equipment available for each inhabitant in France will have the power of dealing with 100 million instructions per second (100 mips) and of storing 20 million characters. In 1984 the total power was only 9 mips. But 200 mips are now expected, with a central memory capacity of 50 million characters. The sending of one page of 2,000 characters takes 20 seconds. The communication (by satellite) between two computers will be 1 billion characters in 20 minutes. A coaxial cable can carry 50 images or 1 million characters per second, and an optical fiber ten times as many. At the same time the price of these marvels is rapidly decreasing. Computers in charge of specific functions can cooperate with one another. Texts can be changed into vocal messages and vice versa. Simplified languages make it possible to get in touch directly with data banks. Telematics systems are also increasingly rapidly. In France in 1984 there were 800,000 videotex terminals in operation and more than 1,000 operational services with some 8 million calls a month.

Progress is incessant and seems to be unlimited. One might say that it involves an economic transformation, an intellectual transformation, and a communications transformation. All the services and institutions that are the framework of a nation are undergoing transformation. Everything has to be adapted. The computer will supposedly end all mindless and boring work. "Teletel [an electronic telephone directory] will open up our minds and spirits. It is said to be bringing a power to earth that will illumine everything with its light. We see a new omnipotence on earth. The ideal of socialized omnipotence is incarnate in modern technique and will bring great benefits to life."[16] Things are moving too fast to leave time for decisions. A debate on Teletel was suggested but technical progress moved so much faster than

16. Quotations in Marchand and Ancelin, eds., *Télématique* (Paris: La Documentation française, 1984).

discussion that videotex was being integrated before there was time to consider the related details or problems. The experiment was thus initiated without debate.[17] Confronted by this evidence of progress and utility, how do I have the audacity to talk of gadgets? I am not arguing that computers are not efficient or that they have not become indispensable for accounting or budgeting or juggling with fantastic figures or keeping track of sales or stocks. They play a useful part in economic and financial management.[18]

Computers are also indispensable for calculations in modern astronomy and mathematics. We simply note once again that technique makes possible a certain development in science without our really knowing whether the new knowledge makes sense. I will undoubtedly be told that the uses I have cited are minor and relate only to calculations when in fact the handling of information has far greater economic impact. The future of the computer is supposedly more oriented to the manipulating of information, of which only a small part has to do with figures. But it is precisely in this regard that I ask whether the computer has any *real* use and not merely a fictional, phantasmagoric, supererogatory use.

No one denies that data banks contain a vast amount of information. But who consults them? Do we really believe that 50 million French people are using these services? Only intellectuals, engineers, and journalists do so. In this way the gap between the upper level and the rest widens. We have only to consider the difficulty for houses equipped with videotex. A whole propaganda campaign was needed to get people to buy this equipment. Vitalis pinpointed the problem. We were invited to live as free, autonomous, rational subjects who could

17. For an example of the way in which all have to participate when a technical operation is launched, cf. a short article in *Le Monde*, March 1984. It deals with the Minitel project at Evry in which the mayor, inhabitants, and experts (who at first had no particular interest) all became progressively involved. Once such a project becomes the center of attention, no one thinks of anything other than to improve it, to find new goals for it, etc. General participation is supposedly a proof of democracy. Technicians orchestrate progress, and no one has the courage to say it is not wanted. There is a technological seduction that brings everyone into the bluff.

18. J.-P. Chamoux, *Menaces sur l'ordinateur* (Paris: Seuil, 1986), stresses that 90 percent of the applications of computers are in banking or commercial management and accounting. To arrest the computer in these areas would severely disrupt operations. Each day dependence on automated procedures grow. Of the remaining 10 percent, games take up three-fourths. This is the reality in spite of all the rhetoric. Data banks play only a weak role as compared with other forms of passing on information.

find the answer to our problems by asking the system, but also to live as submissive subjects conforming to the bureaucracy. In response to the latter demand we practiced diversion; in response to the former we resisted merely as objects with hyperconformism, dependence, and passivity.[19]

In fact, there has been much deception regarding Teletel, which was supposed to replace the telephone, television, etc. It was reported in 1985 that Minitel was making headway in business in spite of some problems with videotex. But few households were interested; only 1.5 percent had microcomputers. Teletel was supposed to make possible a socializing of technique and become the basis of administrative services and strategies. But no one seemed to put the question of its true usefulness. Can we really say that a microcomputer is needed for a household budget? Or that reserving seats for the theater, planes, or trains (which can be done by telephone) is a primary use? Or that it is indispensable to know telephone numbers or railroad or plane schedules, which one can easily obtain when needed?

I could continue with my list of the ridiculous services which will merely replace services that are already adequate. The only real service is in the area of messages, in which we supposedly have a new medium that will revolutionize psychosocial relations. But the new medium, replacing letters and the telephone, can so easily be abused that some restriction has been necessary. This reminds us yet again that except in management most of the use is for play, which we shall study later. The same is true of innumerable gadgets for automobiles and the home.

Gadgets to detect pollen so as to guard against allergies are useful, as are also those for surveys, or for assembling data in the case of oil slicks, but they simply do much faster what can be done already. Gadgets for constructing computer graphics, which are very costly, are mostly ridiculous. Some graphics, for example, those for engineers or architects which can represent in three dimensions what they have in mind, are also useful, but the same can hardly be said of advertising graphics, or especially of artistic graphics, like the dreadful portrait of Montaigne by M. Combes (of the Center for Contemporary Plastic Arts) at the opening of the Bordeaux Museum of Contemporary Art. This computer portrait of Montaigne is not in the least bit like him, makes him out to be ignoble and stupid, and forms a kind of antithesis to his *Essais*. It is an interesting illustration of the aesthetic capacity of the

19. A. Vitalis, *Les Enjeux sociopolitiques et culturels du système télématique* (Telem, 1983); idem, *Informatique, Pouvoir et Libertés* (Economica, 1981).

computer, and all its creations seem to be of the same order. Of its musical creations I have spoken elsewhere.[20] Here again we need to distinguish between an acceptable use by engineers and a totally superfluous, absurd, snobbish use in which the computer is merely a gadget.

In another area I will simply recall briefly a much debated issue that cannot be decided. In its political impact, is the computer an instrument of unequalled centralization or a means of remarkable decentralization? There are as many arguments for the one side as the other. But the reference is to mainframes in the one case, to microcomputers in the other. Furthermore, the only two orientations that seem to be serious are as follows. Thus far the only applications have been for the reinforcing of central power, and I do not know of a single example of decentralization by computer. But second, different conditions are needed for the two different results. As regards centralization, whether in Africa, Asia, or the USA, centralization already exists, and the computer has simply to go with the tide. As regards decentralization, it would have to swim against the tide, opposing usages, institutions, habits, and the demands of the social body, and doing a creative work of freedom. But to make this possible we would need thousands of people and associations that are autonomous, that are nonconformist (whether left or right), that think for themselves, that stand for something specific, and that do not share the current commonplaces. This would demand an enormous effort. We have mentioned already the lamentable failure of independent local radio stations in this regard. A hundred times more quality and will would be needed for decentralization by computer. Thus far we have been shown what individuals might create thanks to it, but when reference is made to the transmitting of information, we are not told *what* information. All that we really have is the invention of games.

The case is self-evident. The idea that the computer is a creator of freedom is a myth pure and simple.[21] The information that it handles is that which *this* society uses and can use. The computer can only confirm it. To be amortized, the heavy investment in computers demands that the system already installed be preserved. The methods of analysis and programming do not permit an evolution of services already in place. There is no computer revolution. There is simply a

---

20. See J. Ellul, *L'Empire du non-sens.*
21. See J. L. Leonhardt, "Informatique et société," in *Une société informatisée—Pourquoi? Pour qui? Comment?* (Presses universitaires de Namur, 1982) (forty-five essential studies).

computer shock that impels the socio-technical system to move faster in its own direction.

In general, we may say that we have had some experience of the choices that are available when a technique is at work. Always and in all circumstances technique has historically gone along with centralization and the concentration of power. "Without automobiles, planes, and loudspeakers," said Hitler in October 1935, "we could not have taken over Germany." "By the magic of the telephone and telex, centralization is even easier today. Orders come down smoothly from superiors to the lower echelons. Information circulates rapidly and discreetly, making possible both increased surveillance of citizens and total concentration of decision. Totalitarianism goes hand in hand with modern gadgetry."[22] "Totalitarian societies seem to be simple, logical exaggerations of the technological state of modern society."[23] "What would the new order be without the transmitting capacity of modern instruments . . . which is constantly growing. The Nazi government launched a big advertising campaign to get all Germans to buy radio sets. In May 1935 they could legitimately triumph, for in two years more than 800,000 sets had been sold."[24] These experiences seem to me to be conclusive, confirming a reasonable evaluation. Yet I cannot think that all the talk of the possibility of decentralization by such technical means as the microcomputer is a lie. There is no intention to deceive. It is a major example of technological bluff. It is a bluff that ensnares democrats and liberals, and that is all it is.

"The computer is not a means that can be used in the service of new social ends."[25] Once again, I am not doubting the marvels performed by computers and especially microcomputers. I am simply trying to show that these marvels do not really change existing society (except to speed it up and, as we shall see, to make it more fragile). Nor do they truly better the individual lot. The use of a touch screen instead of the traditional keyboard did not really change anything. The visual information terminal is a masterpiece, but it does not really change anything. In the politico-social realm it is simply an aid to decision and office automation.

As regards the aid to decision, we must differentiate between this and ideas which receive help from computers. The latter kind of

22. See D. Pelassy, *Le Règne nazi* (Paris: Fayard, 1982).

23. See C. Friedrich and Z. Brzezinski, *Totalitarian Dictatorship and Autocracy* (Cambridge: Harvard University Press, 1956).

24. D. Pelassy, *Le Règne nazi*.

25. Ohrenbuch in *Travaux de l'Institut d'Informatique*.

help is incontestable. An engineer has the idea of circuits and the computer designs them on the screen. If the engineer agrees he gives the green light and the computer, linked up with precision instruments, makes sure that the circuit is made properly. We have here the particular case of a graphic representation of an idea. The same process can be helpful in the routing of planes or cars. Technique aids technique. But the aiding of decisions is very different. We enter here the domain of economics and politics. According to an old idea, computers can register all the information concerning an issue. They can encompass all the parameters. The dream twenty years ago, then, was that computers would make useful, right, and wise decisions. Knowing all the data and all the rules, why should they not be able to decide definitively? They could also foresee various scenarios and combine factors in different ways. They could be told what the goal was. They would then point out the most appropriate scenario. Since they could not make a mistake, we ought to follow them.

That is a dream. First, our human models are always incomplete. Second, qualitative imponderables that the computer cannot know enter into all political and economic issues: How much bombardment can a population stand? What will be the level of courage of the last German troops? How apt are Japanese workers to mobilize? The next theory, then, is that of the aided decision. Politicians will make the decision but only after having received all the data that the computer can provide. In my view, we have here a great mistake, for people who are deluged by information become incapable of making decisions. An excess of information and parameters results in total paralysis of the process of decision.

We must return at this point to a distinction made by Jouvenel that I have often quoted. Some situations can be totally expressed in numbers and can thus be reduced to the formulation of a problem. There is then only one solution and no decision is needed. But political and social situations are not of this kind. They cannot be presented as a problem. There is a time, then, when we have not to solve but to simplify. In this decision to simplify all kinds of imponderables, inexpressibles, impressions, and intuitions will intervene. It is precisely the art of politicians (and perhaps also economists) to be able to sense what no computer can tell them, and then to simplify. The gathering of data is no doubt useful at a pre-pre-preparatory stage. Secretaries will work and rework them until they are reduced to broad schemas and hypotheses in brief reports that will be useful to those who decide. But no more!

As regards office automation, tertiary activities need to be automated and computerized so as to improve the conditions of functioning and the quality of the services rendered by the offices and administrations. It is "the ensemble of techniques and means aiming to computerize office work, especially in the handling and communications of speech, of writing, and of images" (cf. a circular issued in the *Journal officiel* of January 1982). The brakes on its expansion are the result of poor preparation and of beginning to use foreign materials, which consists of software roughly Frenchified and not adapted to the workers. To computerize administration there has to be vigorous training of all the personnel in a permanent process. A comprehensive governing schema has to be worked out in relation to the traditional functioning of the services. Counsellors have to be set up who will keep a permanent watch on the functioning. The public demand for the products and services has to be organized. But we are far from having fulfilled these four conditions. We are also far from making a proper use of the computer in the services provided, and we still have an abundance of overlapping, of wrong billings, and of useless paper.

From these various angles, and without insisting on the techniques, one might say that we are moving into a media society with no mastery over the constituents and no real knowledge of the effects or all the possibilities. We are advancing like the blind in the direction demanded by technique but also decided by politicians. It is the latter who have formed an image of the society to come (inevitably) and who are imposing the computer everywhere so as not to be left behind in the race. We shall return to this matter. For the moment let us simply state that telematics rest almost entirely on official decisions. Neither industrialists nor the providers of services have created a true market. If there are some successes, there are also failures. In some places people rejected cable television, but it was imposed by the government. The costs have been about 20 percent above estimates. Many instruments have not been used or used only for games. Many users say that they used them a little at first out of curiosity, then less and less.

The least that one can say is that for the general public none of these techniques is essential. I realize that it will be said that at first it is the same with all innovations, but finally people take to them. But this is simply to say that no gadget responds to any real expectation or need but that they all impose themselves by the power of the economic, political, or technical system. Technique always entails more technique.[26]

26. Vast sums have been expended by industry, government, and universities on computers. Government alone spent 17 billion francs in 1973, 70 billion in 1983, and almost 100 billion in 1986.

We shall stay in the realm of computers but may now engage in some reflections. An essential point that we have noted already relative to information is that an encyclopedic knowledge of all possible choices provides no criterion for choice. We may get information on every possible objective choice, but the logic of choice has nothing whatever to do with the examination of possibilities. Abundance creates disorientation. No place is left for the practical logic of exchange of information. No one can use this model. It rests on the substitution of technical rationality for social and moral rationality. The information offered is virtually credible but never true.[27]

The problem is the same as in the case of political decision. After a lengthy study of telematics at Nantes, Vitalis concludes that it did not change in any way the relations between the administrators and the administered. The public is simply a consumer.[28] The coming of new techniques of management represents an extraordinary windfall for the conservation and self-reproduction of an administrative system like ours. The perspectives offered reinforce the bureaucratic ideals of France. The vital point is the wave of belief in the omnipotence of things, in the mechanical enmeshing of startling findings, in the idea that we can only submit to what is happening. We can do nothing, for the power is outside us and we count for nothing.[29] The systems also allow anonymous conversations about trivialities, in which people can adopt false identities and engage in fictitious dialogues, in which anything is possible with no sanctions or social reprobation, in which they can express themselves freely (but express what?), and in which they can form groups, asking every member to pass on a daily message. These are mere games, but they are a substitute for human relations and in fact denote a basic loneliness and the boredom of having exhausted every experience.

All this might be regarded as a distortion of the real function of instruments such as Minitel. In a different way we find exactly the same abuse of freedom with Minitel as in the case of the independent local radio station. Most of the exchanges end up with pornography, prostitution, erotic dialogue, dirty graphics (anonymity guaranteed), lessons on new positions, the creation of a network of pedophiles. Mr. de Valence, director of the A-Jour group, who edits *Minitel magazine*,

---

27. See D. Boullier, "Télem à Nantes," in Marchand and Ancelin, eds., *Télématique* (Paris: La Documentation française, 1984).

28. Vitalis, "Les enjeux socioculturels de Télem," in ibid.

29. See P. Legendre, *Paroles poétiques échappées du texte. Leçons sur la communication industrielle* (Paris: Seuil, 1982).

notes regarding the pornographic messages that what was first said to be acne is now seen to be smallpox.[30] There are 20,000 calls of this type per day.

One of the great problems of data banks and other memory systems is that they guarantee the information contained in them. But simple manipulations can result in swindles. This is very common. Small swindles repeated millions of times can have serious effects. So can the tapping of information. As one of these swindlers noted, the more complex the systems, the more they help swindlers. Those who handle computers have to take serious steps to protect themselves: codes and passwords. But codes can be broken. There is also the great problem of protecting information, which must not be tampered with. Codes and passwords help, but they have to be changed frequently. Records must also be kept of all attempts to obtain information, but systems to keep track are costly and slow. Without pressing the point, I would emphasize that there is no sure protection of data. We have here a vast and uncharted territory of possibilities of fraud and data exploitation.[31]

Some pirating is an involuntary result of playing with the computer. The story is told of one operator who asked his screen to display lines of zero characters. No one had ever thought of this absurd possibility. The computer complied, and that generated a prohibited operation (division by zero), and the operator saw appear on his screen materials which enabled him to enter areas of the operating system to which he never should have had access.

J.-P. Chamoux has made a fine study of pirating and trickery.[32] He offers hundreds of minutely described cases in which experts use their knowledge for personal ends, pirating industrial secrets, plagiarizing works, and especially engaging in financial crime by means of computers. Electronic currency is now made possible by a vast system. Thus the automated exchanges of the bank of Lyons runs to millions of francs and can change and sell a billion dollars in fifteen seconds. This type of money opens up new horizons for fraud by means of small additions (usually of the order of 10 million francs) or by making it possible for funds to cross borders freely. The electronic bank is more vulnerable than the classical banks because there are fewer opportunities for control.

30. See *Le Monde*, Sept. 1986.
31. Cf. A. Grissonnache, "Des risques grandissants, mal connus, peu combattus," *Le Monde*, Sept. 1983.
32. J.-P. Chamoux, *Menaces sur l'ordinateur* (Paris: Seuil, 1986). Cf. also J.-C. Hazera, "La Sécurité informatique," *La Recherche* 113 (April 1980); and F. Bergantine, "La Sécurité informatique," *La Recherche* 143 (April 1983).

The vulnerability of payment systems is one of the great weaknesses of society as a whole. Frauds are often simple (e.g., erasing data, changing numbers, using lapsed identification numbers). There is also a great deal of insurance fraud, and in the case of programmers the possibility of programming frauds. But all this presupposes access to mainframes, which is now possible with systems like Minitel. All that is needed is a little patience. In this way confidential information can be pirated, as was done by a journalist of the *Canard enchaîné* (Nov. 1984).

Another important point is that fraud can result from children playing games on microcomputers and by chance finding a code which gives them access to a whole group of circuits.[33] In fact, it has been found that some children are "geniuses" in the field. Rawson Stovall at twelve years of age created hundreds of video games and published dozens of articles on the theme in a daily column in seventeen journals. Cori Grimm at thirteen became a graphics consultant in a computer firm. Jeff Gold at seventeen was taken on full time to protect against piracy. Musa Mustapha at fifteen works full time on special effects. Cyrille de Vignemont at twelve negotiated a first programming contract with Apple and at fourteen spends two hours on education, two on programming, and two on marketing (contracts, etc.). He has sold two software packages with hundreds of copies. Cyrille's programs are in increasing demand. I could cite other cases.

Are these children geniuses? Have they special gifts? No, there are too many of them. The proliferation of children who can reach the heights of computer creation brings to light a basic feature of the computer itself—it is infantile. When some simple facts have been mastered even children can understand it more easily than they can grammar or arithmetic, and once they are on this path, the complexities are no problem. They do not have to be geniuses. They do not need to be encumbered by hundreds of other branches of knowledge or questions. They do not have to know their own language, or history, or science. They have to know only the possibilities of their computers and the networks. They do not even have to waste time on human relations. Neither Stovall nor Vignemont has any friends. They are small examples of fascinated humanity. They obtain fantastic results without standing for any human or intellectual values. Vignemont

33. Marchand, *Télématique*, quotes the case of a ten year old who, playing with his father's Apple, connected by chance with Minitel and finally tapped out a whole series of passwords, which resulted in the abandoning of all the internal numbers of Minitel.

prefers correspondence courses and admitted on television (Nov. 1985) to not being a good student except in algebra. This shows that the microcomputer is above all a game and is infantile. But it is also very dangerous. We need to know whether it does not also "infantilize."

What we have written concerns microcomputers, but there is also a problem in the case of mainframes. This is their fragility. In contrast to what is often said, they are very fragile.[34] They have to be kept in a room at a constant temperature (20° C, 68° F) and humidity. Many factors can make them unusable, for example, a sharp rise in temperature, fires close by, lightning, a break in current, a power surge, a magnetic field in the vicinity (e.g., the installation of radar), the presence of an electric cable, excess humidity (through flooding!), or the penetration of dust or smoke into the chamber. Chamoux also points out that the computer is a type of machine, and that like all machines it can fail for no external reason. (The manager of an automated bank acknowledged that his machines broke down on average twice a month.)

Specialist critics wrote in *Terminal* (Oct. 1983) that computer science is a field in which error is the rule and programming mistakes take up most of the time of programmers. We thus need to resist firmly those who pontificate seriously about the infallibility of computers. When we consider how many important things on which our life depends (banks, police, etc.) are at the mercy of this fragility and vulnerability, we may well be very anxious about what might happen.[35]

We will conclude with some more general reflections. The first is that this whole field of computers is making society more vulnerable. The risks extend to the public and daily life. We are vulnerable

34. Cf. chapter IV above on the general fragility of the system. Another study of this matter may be found in J.-P. Chamoux, *Menaces de l'ordinateur* (Paris: Seuil, 1986), which is along the lines of P. Lagadec, *La Civilisation du risque* (Paris: Seuil, 1981).

35. In an article in *La Recherche* I also read of a possibility about which no one speaks, i.e., that of a nuclear explosion of moderate yield at a high altitude (300 miles) which would have no fallout or victims and would hardly be noticed but which would develop electromagnetic impulses that might cause a continental electrical blackout destroying all the broadcasting networks (telephone, radio, all that is electronic), and blowing all the components of computers, the circuits and cables. Every country would thus be reduced to impotence (military as well). There is no longer any need of war to conquer! This is especially important when we realize that every army is now computerized. Cf. J. Isnard, "L'armée et l'ordinateur," *Le Monde*, special number Sept. 1984. For a humorous novel on the theme, which shows great foresight, see J.-M. Barrault, *Et les bisons broutent à Manhattan* (Julliard, 1973). Cf. also G. Vezian, "Le retour de la grande informatique," *Le Monde*, special number, Sept. 1984.

financially. We are vulnerable to strikes by a very small number of people. We run the risk of greater social control and a levelling of behavior.[36] To be acceptable to electronic machines and to calculators, models of human behavior have to be reductionist. In relation to both things and people analyses and data have to be in simple classes for easy expression and comparison. A specialist tells us that; Chamoux is a specialist on network security and has founded a private research center. We have here the counterpart of the supposed "conviviality" of the computer, but no one, of course, talks about this.

In reality, we have to pay for the gain in efficiency with the risk of enslavement on the one side and actual vulnerability on the other. But let us pass on to what seem to be more sophisticated reflections of a different kind. Neirynck has noted that the world of computers fulfils all the conditions of a cult of initiates. It has every quality as well. It is clean. It uses little energy. It deals with immaterial things. We increasingly need its information in our society. To the public it is very mysterious. It corresponds to an ancient dream of humanity, that of constructing an automaton that will perfectly imitate humanity. Having shown why there is this passion for the computer, Neirynck then asks what use it really is. Does it better the lives of individuals? Does the spread of personal or domestic computers meet a real need? Or is the manufacturing of well-marketed microcomputers a solution in search of a problem? Do robots really make life easier for workers? In principle they do, but in practice they do not. (We shall look at the problem of productivity later.) Having noted the ineluctable march toward centralization, Neirynck concludes that the greatest danger of the computer is ideological. The less comprehensible it is, the more extravagant the statements to which it gives rise. In particular, it supposedly liberates us from the constraints of energy, reversing entropy! This idea rests on a confusion which Neirynck analyzes in detail. By giving pure computer information to a closed system, we do not diminish entropy. The real truth is simply that we can get spectacularly better energy performance by the computer. A computer linked to a network of electrical distribution can avoid the loss of power by better transit planning. This is valuable, and the computer is a remarkable tool in constructing a technical system at a low rate of entropy growth.[37]

To return to the human level, the computer can hardly fail to pose anew the question of time. It is a machine which greatly com-

36. See J.-P. Chamoux, *Menaces sur l'ordinateur,* p. 211.
37. J. Neirynck, *Le Huitième Jour de la Création. Introduction à l'entropologie* (Presses polytechniques romanes, 1986), pp. 208-12.

presses the time needed for planning, production, and management. As programming advances, a society is set up that is fully synchronous with a generalized synchronization of which the computer is the founding myth and omnipresent organizer. Real time is now looped in advance and works in an instant. The search for ever more synchronization weighs heavily on workers, as we shall see at length in relation to productivity. Chesneaux has given a detailed description of the impact upon human life at all levels of this instantaneity to which the computer accustoms us and which it progressively forces upon us. "The computer is the instrument of the absolute primacy of the present over the future and the past. It is the central benchmark of social duty, which must adjust its rhythm to generations of computers."[38]

I will close these general reflections with three references which are very interesting because they do not come from opponents of computerization. Quite the reverse!

First, I will take up the vital phrase of Neirynck: "the invention of the useless." The gigantic economic organization with billions involved; the whole political and ideological mobilization; the proliferation of gadgets; the inflamed rhetoric about the society of tomorrow— what is their point? Simply to convey information of which 9,999 parts out of 10,000 are totally useless. Neirynck describes this new technical revolution in the invention of the useless. It is not driven by any immediate necessity or express need but by the automatic process of technical growth and by the ideology which Neirynck calls the technical illusion, the climax of expectations which have not been met by promises of a society of abundance. Lacking abundance of food, we will have a superabundance of the empty nourishment of information.

The second reference is to B. Lussato, the apostle of the microcomputer, in which he finds every hope of salvation. Yet he is still uneasy about the vertiginous development of the computer. He does not know whether this is good or bad. He sees the risk of uncontrolled growth, for the more communications develop and computer terminals proliferate, the more people want to communicate, but the more they want to communicate, the more communication systems are put in place, and the less they can do without communicating. It is an addiction, as in the case of drugs. The marriage of communication and the computer (telematics) has a multiplier effect that we cannot evade.

The third reference is from a telematics magazine, *Terminal* (1984). Tomorrow everybody, equipped with a smart card, will be able to tap information, phone an aunt in New York, and use the television

38. Chesneaux, *De la modernité* (Maspero, 1983), pp. 35-49.

for forecasts and reservations. At home he can consult the telematics journal, and help the children to program their electronic games. If he has any remaining time, he will profit by linking up with his terminal, with the card as identification, in order to work some more hours at home. But the question is then raised what will be the world and the psychology of people who work, communicate, consume, play, and educate themselves from birth to death by means of a screen.

A good question! The computer is indeed a gadget whose usefulness is infinitely less than the bluff of technological discourse would have us think. Yet even though it is only a gadget, it can turn the world and humanity around and set us in the direction of nonsense.

CHAPTER XV

Waste

It has often been said that our Western society is a society of waste. To me this seems obvious. But it is often attributed to the excess products at our disposal, to poor economic management, and at times to administrative or political decisions. All these things, of course, play a part in waste. The root of the matter, however, is that waste is the ineluctable consequence of the technical system that is in constant development. Technique has to produce all that it can. All possible techniques have also to be applied (unless there are economic obstacles). These are two principles which inevitably lead to waste. We should begin by recognizing, however, that there has been waste among many peoples from the very first. The story of the famous potlatch is a good example, as is also what has been called the economy of ostentation. We may refer also to the great sacrifices to the gods or the dead. If we add up the number of animals regularly sacrificed according to the rules of the Torah we are amazed that poor peoples could "waste" so much. Yet this type of waste is not like our own. It is part of a sociological or religious structure that is necessary for social equilibrium or cohesion. It is precisely measured. A decision is made about conspicuous consumption and there is no variation. But very great uncertainty obtains in our society. We do not choose or decide; we are carried by the current, floating at the will of the forces that manifest themselves, wasting because we have too much, because the environment leads us to do so. In particular, we must distinguish between private waste and public waste, which in my eyes is much more serious.[1]

1. Cf. C. Gruson's report on waste (1970); P. d'Iribarne, *Le Gaspillage et le Désir* (Paris: Fayard, 1975); H. Guitton, *Entropie et Gaspillage* (Cujas, 1975).

1. Private Waste

Of course, we think first of the waste of food. When I go to restaurants I am always alarmed to see the number of plates on which the helpings have been only half eaten (enormous steaks, etc.), the remnants being thrown out and totally lost. In imitation of the American custom, the helpings served are much too large. Technique here is for nothing. All the world knows that we eat too much in the West. We are gluttons. Normal consumption would be 2,400 calories, but the average in France is 3,500 and in the USA 4,500. What we have here is double waste: too much eating of fat, sugar, and meat on the one hand, and the waste of caring for illnesses due to excess on the other hand (liver, arteriosclerosis, etc.). Raising the proportion of lipids by 10 percent translates into 15 percent more sickness and a 2.5 percent increase in mortality.

But this overeating is not due solely to greediness on the part of Westerners (and others!). It is due to the fact that more and more is produced and has to be consumed. For this reason we have advertising and constant campaigns by which alone the domestic market can be induced to mop up excess production. The latter, however, is the result of technical methods. As regards food, waste has another aspect. Modern methods, modern feeding, and modern implements speed things up. Thanks to hormones calves are fattened in three months. Each cow yields much more milk through scientific feeding. Trawling scrapes the seabeds. Plant varieties are created, including trees which produce two or three times as much as traditional ones. Pigs, calves, and chickens are battery raised, with execrable results. Why? To apply the most developed technical methods and in this way to make our products "competitive." We all know the result. Each year a great part of the harvest has to be sacrificed because it is too large for even swollen consumption on the domestic market. Similar results are obtained abroad due to the same methods, so that the products there are not more expensive. For the last twenty years, then, millions of tons of apricots, apples, and artichokes have been buried and millions of tons of fish have been thrown back in the sea. This is the direct result of techniques of high productivity.

The use of oil is a second area. It has now become the custom to use an automobile no matter how short a distance one is going. If one argues that this is the driver's fault, I reply that the driver is the victim of habit in a technical setting. There is waste, too, in traffic jams. Thousands of cars carry only a single person, but each car covers several square yards. How can we avoid traffic congestion, however, when we are incessantly exhorted to buy (advertising) so as to keep the factories busy, and every so often we are told in scandalized tones that no more that 75 percent of the people in France have automobiles? This is bad when having an automobile is the normal thing. It is unjust and unacceptable; 100 percent of the people must become owners. More congestion, then, and more waste of gasoline! The fact that wastage is inevitable may be seen from the failure of the campaign against it which was solemnly announced when the price of oil rose so drastically in 1975.

Is there waste that might be stopped immediately? We could begin with pleasure craft, which are nothing but harmful, providing neither sport nor true contact with the ocean. We could then turn to car races—another stupid amusement! It is no good arguing that we need them to improve performance and to test auto parts. What use are they when we ought to keep the speed limit at no higher than 65 miles per hour? Does it make sense to test engines and tires at three times that speed? Then we could reduce the number of costly, dangerous, polluting highways. Fourth, we could have fewer air force maneuvers—a frightful waste when even specialists agree that they serve no useful purpose. But nothing has been done in any of these four areas. We cannot close the pleasure craft industry (a sacrosanct argument used on behalf of the Concorde!). And what about national defense? (linked to the evolution of the outdated Mystère). As for highways, who can withstand the incredible power of this lobby and displease our modern mammoths?

Heating is another area of private waste. It has been rightly argued that older buildings are poorly insulated. With full insulation there could be a 50 percent energy gain. In this regard technique brings economy. But no attention is paid to the opposite fact that there is now a mania to have higher indoor temperature in winter (20-25° C, 68-77° F). As we have proved from fifty years of experience, without any discomfort we can live with temperatures of 15-18° C (59-66° F). The excess heating is pure waste. But we have to use the electricity which is produced. The oil companies have to keep going and sell their fuel. The enormous expenses incurred in opening new wells have to be recouped. How can a technical society stay alive if it does not find the most effective means to drill exploratory wells offshore and build ocean platforms? Progress of this kind must not be for nothing. Hence we have to use more fuel. And city dwellers have to live in an asphyxiating cocoon of heat.

Another form of waste relates to the rapid obsolescence of our machines. How can we use an older car, or an outdated heater (with no replacement parts available), or black and white television, or a record player that is not hi-fi? Where do we think we are? In Neanderthal caves, no less! We are told that the new products are more developed and perform better, so we must rush out and get them. It is not just a matter of fashion. The new engine does what the old could not do. But are the new gadgets necessary, or even useful? No one asks this question. Once they are produced and perform better, they are self-evidently useful and advantageous. Very quickly, then, we have to switch. The products we buy are made to last only for a limited time, and they cannot be repaired because the parts are not available. Hence we have to throw away things which could work for a long time were it not for a single part. It is essential that we replace them. This is how what has been called the throwaway society works. We have to change very quickly because the socio-technical machine works very quickly. Once we buy a thing we have to realize that it is ready to be thrown away. The unconscious process is at root what Iribarne has called the "cycle of the better": the more expensive it is, the better it is! The more recent it is, the better it is! In food, clothes, hygiene, or care, the infinite cycle which pushes up normal usage is almost solely responsible for the waste that we see. This cycle is the cycle of the better—a conviction created by advertising but which once created, like a conditioned reflex, finally functions on its own and without stimulus. The cycle of the better has of itself the power to make waste the first item in our budget. The only brake on its course is the limit of time and of individual ability. It will advance without end when progress makes us more efficient. To increase efficiency is thus to increase the potential for waste. Our daily experience confirms this. But it confirms even more collective waste.

2. Social and Collective Waste

It is hard to discern collective waste except when an audit brings it to light. Since it is caused by the various branches of government, no one can denounce it, and because of self-interest, opposition to it (e.g., that of politicians) is very superficial. In political debate attention might be drawn to a few individual instances but nothing is said about the fantastic techno-scientific waste. What do I mean by the branches of government? First politicians, then higher civil servants, then high level technicians, then research scientists, then experts. We might also include the most powerful business executives, some unions, and the media (newspapers, television, radio). These all advocate technical development and therefore none of them will denounce the failures, malfunctions, and general waste.

I will confine myself to waste deriving from technical operations and not deal with the waste relating to personnel, etc. Before turning to some monumental examples, I must first distinguish between expenditures on a human scale and those that are gigantic. The former, although they finally amount to important sums, are close to waste at the private level. For example, offices waste a great deal of paper. We must not forget that computers consume paper at a terrible rate. The least operation demands sheets and sheets of paper, 90 percent of them useless. An inquiry is needed into the paper consumption attributable to computers, attention being directed to the triple and quadruple sending of the same letter or brochure (and the postage incurred). It is said that when this happens the computer is badly programmed, but I would say that they must all be badly programmed, for every day from different sources I receive double and triple invoices and brochures and advertisements.

Everywhere, too, we find countless and useless photocopies. This is no doubt the fault of secretaries who photocopy everything. I could give many examples. But my present point is that they have at their disposal such simple and efficient copying machines that they run off ten copies when one would do. There is also an extraordinary multiplication of administrative forms that citizens have to fill out. Nor should we forget the need that business has felt to modernize offices. One advertisement admirably states that in order to humanize administration and combat bureaucracy, business should equip itself with X. I also think of the scandal of new government offices throughout France. There is here a waste of time, money, and materials which ought to be treated as private waste. My main concern, however, is with monumental waste: the waste of public projects, of badly planned projects, and of failed projects.

Among useless public projects I do not hesitate to classify those that have been seriously called artistic. Some of these have been successful, and the billions spent on them supposedly have an educative, commercial, and aesthetic impact that will give them a place in history! In other instances it was realized too late that they had only a kitschy charm, and some people wondered whether they ought not to be rebuilt. Others are simply scandalous, and in some cases it is not even known what to do with them. The trouble with such enterprises is that people have grandiose schemes in their heads, technique makes them possible, and the millions can always be found. We have vast schemes of urban renovation because our presidents want to leave behind them monuments comparable to those of Louis XIV and Napoleon. Thus building commences, and it will be seen later whether there will be any use for it. This is technical logic. And what about the headquarters for defense with its enormous deficit? The state has had to spend almost a billion francs to bail it out. I might refer finally to the magnificently useless and enormously costly Grand Louvre of Mitterrand with its glass pyramid, which has caused much discussion.

Let us press on! A marvellous example of waste is provided by the famous Parc de la Villette affair. Under Giscard in 1979, 800 million francs was set aside for a park and museum, and 200 million was spent on a museum of science and technology in the meat-market hall. The museum was so badly planned that it was decided in 1982 to restore this hall, but to do this 21,000 tons of iron and steel (three times the weight of the Eiffel Tower) had to be removed. The government decided instead to build a real museum of science and technology. In 1984, 4.5 billion had to be set aside for the park and museum, 1.7 for the museum of technology alone. It was argued that this was not expensive since the Pompidou Center had cost 3.2 billion. Furthermore, thanks to concessions, taxpayers would not have to pay more than 3.8 billion.

In all these cases, of course, the forecasts and estimates were much too low. Thus it was estimated that the Sports Palace (Palais Omnisports de Paris-Bercy) would cost 300 million (1979). But the costs had already run over one billion by 1986. Nor was that the end, for innumerable faults were found. The framework had to be reinforced, 3,000 mobile seats changed, aluminum supports substituted for steel supports, and the functioning of the mobile gangways improved. In addition, special arrangements have to be made for each event. It cost 1.2 million to prepare for the Six Day cyclists. Nothing but waste!

Let us take some other examples. We have stated already that France produces too much electricity. Yet work continues without a qualm on supergenerators. Super-Phoenix will cost 20 billion francs, an ordinary nuclear plant 6 or 7 billion. But in every area those who promote nuclear power have always displayed a desperate optimism. It seems that by 1990, according to the present plan, there will be ten or so more nuclear power stations than our maximum need requires.

Among other large-scale useless projects there is now the Channel tunnel. This will provide trains from Paris to London in three hours. But there will also be needed a motorway from Paris to Calais and a new port at Calais to compensate for the ferries, which the tunnel will replace, and to make it possible for larger ships to call there. There is also to be a high-speed train Paris-Calais-Brussels. The total estimated cost runs to 15 billion francs, though from previous experience we can double that amount. But the development of the area which is projected will never take place because the infrastructure does not exist for economic or industrial development. The only thing that is certain is that the many projects will demolish the human and social equilibrium of the district, as has always happened with motorways and high-speed trains (to which we shall return). As for the tunnel, it will no doubt be much easier to take one's seat at Paris and alight at London. Why should we not spare passengers the trouble of changing for ferries? But it will bring no development. Tourism? It will not bring any.

We might also refer to the obsession with bridges. In France there is the Havre bridge, which traffic has never justified, the Oléron bridge, the Île de Ré bridge.[2] We have a surfeit of bridges. The problem is always the same. We allow areas that owe their charm to silence, solitude, and secrecy to be overwhelmed by cars and tourists. But we have to have these great projects. Soon there is to be a Gironde bridge costing billions (including much speculation). I will be told that such things are really the result of speculation and ambition, not technique. This is true, and yet it is technique that makes everything possible. Without it these projects could never have been undertaken, for there is no more of a work force than under Napoleon or Louis XIV.

Another example of useless waste is the expansion at all costs of the telephone service. The French have to have 25 million phones. It is not acceptable that there should be a home without one. With fiber optics and satellites available, nothing less than 100 percent will do. Technique demands it, as with Minitel. The French are coming round. We have only to create the equipment and the need will slowly arise. But this leads me to my second class of waste, that of poorly planned projects.

The model here is obviously the Concorde. France and Britain decided to build this plane in November 1962. They would build a prestigious plane of great speed and with maximum comfort thanks to the application of advanced technique. It would also be a long project which would assure work for hundreds of people over many years. Though costly, it would also be profitable. But cost was not a major concern. The figures jumped around in an amazing way, but so what? Billions of francs could always be found. The result was the fastest and most modern plane in the world (though it is now outdated), but the costs and the time spent in construction went through the roof. The

2. This last bridge is remarkable because work on it began without an official inquiry or authorization. Condemned by a tribunal, the builders carried on as though nothing had happened. An appeal obviously does not stop such works; according to the doctrine of Biasini, when a public project has begun, no matter what legal obstacles may stand in its way, it must be completed.

prototype, begun in 1962, was not ready until 1967! The expenses were so great that it would have been better to abandon the project, but the usual argument in such situations prevailed, namely, that since so much had been spent to arrive at this point, we must not lose the money already disbursed. The 1.8 billion of 1962 had become 8.4 billion by 1969! The project was continued in the hope that the extra expenses would be covered by good international sales.

In fact, the USA and China seemed to be interested. But the USSR with its Tupolev 144 was also in the field. We recall the Concorde's arrival in New York. The noise was so loud that the plane had to fly at subsonic speeds, using supersonic only over the Atlantic. The time of the flight from Paris to New York was reduced by half (i.e., to four hours). But from 1985 a seat cost more than twice a seat on other planes (26,000 francs instead of 12,000). Passengers had to ask whether their time was worth 3,500 francs an hour. There were in fact few passengers, and each flight was a loss for Air France. The expected clients did not materialize. Four more planes were built and Air France was forced to buy them, with an equivalent number for Great Britain. In other words, Concorde was a financial disaster from every standpoint, whether construction, profitability, or marketing. Technique had made possible the building of an extraordinary machine that was beyond our social and economic capacity. Taxpayers had to make up the loss incurred by each flight.

The disaster was not quite so total with high-speed trains. Yet these are another example of waste. Once again prestige and high technology were the motivating factors. Since 1979 there has been little but praise and exultation regarding these trains. Two hours from Paris to Lyons! The technical performance once again gave France a leading place in the railway world. More than 40 percent of the population had been given a new form of travel. Nothing but that! It not only led people to prefer the train to the plane but also induced them to travel when they would not otherwise have done so. It seems that these trains made a profit, but that depends on the months and years. Other experts are more reserved, as we shall see. But what glory! Lyons has been liberated; "Lyons has lost its provinciality" *(Le Monde)*. It seems we must be within easy reach of Paris not to be poor, isolated provincials. By 1983 the high-speed train was carrying 6,600 first-class passengers a day compared to 3,000 in 1981. Hopes were high. After this success some thought that we might sell the train anywhere in the world.

Unfortunately, the British and Germans were building their own systems. France's first clients (Korea and Brazil) cancelled their contracts. The Japanese went one better by selling their trains in the USA, for they had long since built trains just as fast and comfortable as those in France and already had over 1,100 miles of track compared to only 350 miles in France. The question thus arose how long it would take to amortize the 15 billion francs spent in building this first high-speed train. It is not enough to say that the line has brought benefits. But after this first success, the system had now to be doubled. A line had to be built to the Atlantic with branches to Brittany and Bordeaux. The signal to go ahead was given in May 1984.

This leads us to some more general reflections on the project. It would certainly be good to improve communications with Brittany, though without all the work involved in high-speed trains. But the line to the Southwest is an absurdity. The journey from Bordeaux to Paris normally takes four hours and it would merely be reduced to three. Is the hour gained worth the 12 billion francs that we estimated for the project in 1984, and the additional billions that would actually be spent (there was already talk of 15 billion in 1985)? The decision rests on three mistakes. The first is commercial. It was announced that the train would be profitable, but that is not true. The fact that it is profitable in the Southeast (an economically strong region) does not mean that it will be so in the Southwest. The cart is being put before the horse. Rapid transport will not get the economy of the West and Southwest moving. On the contrary, it is economic expansion that justifies rapid transport. The opposite idea is absurd. Furthermore, it has been pointed out that a high-speed train is profitable only after ten years, seven for construction and three in use. During these ten years there has to be borrowing in dollars on the international market to ensure financial stability. But French railways were in no position to incur further debt in 1984.[3]

The second mistake is a very different one. It has to do with democracy. To build a new line it was proposed to proceed according to the nasty custom of government in all projects (motorways, high-tension lines, etc.): The large project would be divided and subdivided into smaller projects, and inquiries into public utility would be for each little project rather than the whole, that is, one segment of the railway line at a time. Of course, when the consent of the first communities is gained, construction will begin, and then it will be explained to opponents that the project is under way and there can be no more question of opposition. The interesting point in this case is that the first communities to be consulted on the new line turned it down. But that

3. The Paris-Brussels line (with branches) will cost 25 billion francs and will become profitable only in the year 2000.

made no difference.[4] The government stuck out its jaw and decided to
go ahead anyway. We are reminded of the time when the motorway to
the West and Southwest was built. Communities objected but the
decision was still made to go ahead, and once the first section was built
the remaining communities had no voice. But what does democracy
mean in these circumstances?

The third problem with the project is Jacobinism. There is talk
of decentralization! Mere talk! Behind high-speed trains is the disastrous notion that only Paris counts and that every "province" must be
linked to Paris. We must be able to travel faster from Brest, Bordeaux,
Marseilles, or Lyons to Paris. Paris is at the center of the network of new
roads and railways. But the real national and local need is not for links
to Paris but for cross-country routes (e.g., Bordeaux to Lyons, or Nice
to Rouen, or Toulouse to Strasbourg). This would be real decentralization and would be of real use. For on these routes the connections are
terrible. We should make lines that are of value to passengers and are
not just for prestige. This would be real escape from the tyranny of
state-of-the-art technology. But people and their needs count for little.
First we make the marvels and then come the projects. There is also
the Concorde argument. During construction it provided many jobs and
kept the wheels of industry turning. A budget for projects is thus voted.
What projects? No matter whether they are of any use or make any sense
so long as they are projects.

I also recall the great enthusiasm for the agreement regarding
the Siberian gas which was to heat France. This would put an end to
all our anxiety about oil. The pipeline of almost 3,000 miles represented
another striking technical achievement. The first deliveries came
without problem in 1983. In 1985, however, France tried to renegotiate
what had been extolled as the contract of the century (another one!),
as an unheard-of opportunity for France, etc. (January 1982). For this
contract had increasingly proved to be valueless. Only a billion cubic
meters of gas were received in 1984 (instead of the 4 billion forecast)
and 2 billion in 1985 (instead of the 6 billion forecast). What about the
8 billion for 1986? And all this from the great pipeline (a masterpiece
of which we were so proud) that was supposed to deliver 25 billion
cubic meters a year. Then suddenly everything came to a halt. As might
have been foreseen, the ground had shifted. There were problems even

4. The groups formed to oppose high-speed trains in 1980-81 claimed that
decisions were forced upon them. They objected to properties being divided,
villages cut off from their fields, electrical stations being set in open fields, and above
all the noise. But these human factors were of no importance compared to progress!

in the USSR. Many Soviet economists advised against any more contracts for the exporting of gas. Silence gradually descended. Siberian gas, however, had not ceased to pose serious problems after all the enormous sums expended on installation.

Examples abound of projects that swallow up hundreds of millions of francs with no other result than to provide contracts for certain industries and to meet the objectives of the technicians associated with them. This is the way of the "advanced" world. We recall the account given by J.-J. Salomon of the Mohole Project, which was launched by the National Science Foundation (1965) with just as much fanfare as the Apollo project. The aim was to bore through the earth's crust at the bottom of the Pacific so as to penetrate to the "mantle of the earth" at a depth of 20,000 leagues. Preliminary studies commenced with great publicity and to the tune of $125 million. But Congress refused to vote the budget (for once the argument of money already spent did not prevail), putting the simple question: What use is the project? Obviously none. Thus the project was stopped, as many other things ought to be stopped.

Another story relates to the commission for the development of the Aquitaine coast. A proposal was made to build a fine canal linking the great lakes of the Landes: Hourtin, Lacanau, Arcachon, Cazaux, Biscarosse, Leon, Soustons, etc. There would be nearly 200 miles of a fantastic tourist canal, and digging began. The first link from Hourtin to Lacanau was finished and opened with great ceremony. But some days later it was discovered that the lakes were not on the same level and Hourtin was beginning to empty into Lacanau. The canal was hastily blocked and the project abandoned. Tens of millions had been wasted by the commission, but so what? They had just forgotten to check the altitudes of the lakes.

A more serious problem arose with nuclear power stations. In May 1983 the minister in charge of energy presented a crucial report which concluded the labors of a group for long-term energy planning. He announced that the nuclear plants were overproducing, that the surplus would reach a climax by 1990, and that the program should be slowed down so as not to waste investment. Any jobs that might be lost could be replaced by work on finding new sources of energy. To prevent investment loss the nuclear program should be slowed down and other sources explored. But naturally there was no slowing down. Instead, the pace was accelerated, and the waste of investment continued (*Le Monde,* May 1983).

A final example of miscalculation in projects relates once again to the automobile. The automobile is today the great master of human life. It is the idol, the future, the economic solution, etc. Everything possible must be done to favor it. Among other things, we must provide roads for it, all kinds of roads from motorways to city roads. Nothing must stand in their way. A third fast road must be built on the narrow Cap-Ferret peninsula which will destroy the forest. Access roads must be built through the green belts around Paris, further reducing open spaces that ought to be left intact. An enormous plan of circular routes was drawn up to relieve the congestion of the traffic in Paris. The automobile is law. At all costs we must end bottlenecks, improve circulation, and satisfy motorists. Everything must yield to this economic, social, and psychological imperative. But a strange thing happens. The motorways are rapidly clogged and the circular routes are a hell of traffic jams and bottlenecks. We all know that.

The works of Ziv on traffic in the USA and those of J. Dupuy on urban life and traffic show that there is no mystery about this result.[5] The explanation is simple. A new road does not reduce congestion on older ones but simply increases automobile traffic. The mistake is as follows. Those who plan highways and bypasses begin with the idea that the number of cars passing a given point per minute will be the same in ten years, so that a new road will reduce it by half. But this is wrong. The new road will bring in new cars. Both statistics and experience prove this. Growth is induced. A new road does not respond to a demand but creates it. The logic is that of the development of the automobile market, which gives the appearance of being there to satisfy transport demands. The infrastructure of roads will ensure an important increase in traffic.

We might quote Ziv and Dupuy in full. But the government imperturbably proceeds to build its highways, following what was (we should not forget) a Nazi model. In the process it destroys the countryside, divides properties, tears up small towns (what matter!),[6] and shatters the life and human equilibrium of whole areas. These things must go and motorists must be satisfied. But according to the logic noted above they never will be. We have here a contest like that between the fortress and the cannon. The more roads, the more cars, and the more cars, the more roads. Waste pure and simple! Generalizing the rule, we come across the basic formula of Dumouchel and Dupuy: "The

5. J. C. Ziv, *Planning Model for Private Goals: A History of Urban Transportation Planning in the U.S.* (Ithaca: Cornell University Press, 1977); J. Dupuy, *Urbanisme et Technique* (Centre de recherche d'urbanisme, 1978).

6. See J. Hussonnois, *Les technocrates, les élus et les autres* (Paris: Éditions Entente, 1978).

attempt to overcome scarcity (by increasing the quantity of goods and disposable resources) increases it. This is the problem of scarity: it is totally independent of the quantity of goods and disposable resources."[7] We should regard this as a veritable law. It is the law of the relation between the technical system and the technical society.

We look finally at failed projects. There are countless numbers of these as well. I can hardly make a list. This is more a matter for research, for we find such projects locally. They are projects on which preliminary studies have cost millions, the project has been started, but then it has been abandoned. Whenever I go to Paris, on leaving Orleans, I have bitter thoughts as I see the route of the aerotrain. Thirty miles of viaduct support a rail on which this train was to have speeded from Orleans to Paris at 250 miles per hour. The train itself had been built. But the trials were a disaster and the whole project was abandoned. When we think of the expense of buying land and the initial work, it might be worth calculating the cost. We might also think of the giant windmill of Ouessant which was supposed to provide electricity for the whole island but which simply broke in pieces in July 1980. The wind was too strong for it. In general such absurd failures are hushed up. I will mention two or three from my own area.

Among the big projects of the Aquitaine coastal commission was (1) a plan to develop the beaches of Capbreton-Hossegor. A sea wall was constructed to "enrich" the beach. It had a little lighthouse at the end. But the first big equinoctial storm smashed it, leaving only some concrete blocks. All the local sailors had foreseen this result. (2) Around Arcachon, because of the terrible pollution caused by the tourists attracted there by the commission, a sewage system was built. This was a sizable work, but unfortunately the population increased so much that the system was overloaded and a few years later it burst. While it was being repaired, millions of gallons of polluted water escaped. But what matter? (3) The next idea was to connect the sewer to an outlet far out in the ocean on the edge of the Atlantic trench. This was a grandiose scheme, but the first equinoctial storm smashed the outlet. The German firm which built it stated simply that they had never realized that Atlantic storms were so violent. The sewage was thrown back upon the coast, so that at high tide one could see floating in the Arcachon basin the effluents that had been taken from the area. I repeat, we need a list region by region of these abandoned projects that we planned poorly.

7. P. Dumouchel and J.-P. Dupuy, *L'Enfer des choses* (Paris: Seuil, 1979).

3. Responsibility

I now want to raise an unanswerable question. We are in an idyllic situation. On the one hand billions and billions are wasted for nothing; we have to apply better techniques. On the other hand we have resounding failures. But in the middle there is no one—no one is responsible for anything. The contractors were not responsible for those homes of children that were swept away by an avalanche or for the dams that break, etc. Whom would we hold responsible? The scientists who are there at the beginning? But they do only theoretical studies. The upper-level technicians who do the practical studies and planning? But they simply make proposals. The experts who examine the plans? But they only give advice. The politicians who decide to carry them out? But they know nothing about technical questions and simply rely very reasonably on the labors of the technicians. The civil servants who see to the execution of the plans? But they only obey the politicians. The other technicians and supervisors who do the work? But they simply carry out the plans of others. No one at all is responsible for anything. We are in the same situation of not being able to fix blame as at the Nuremberg trials. No one was responsible for the massacres at the concentration camps.

My own view, however, is that we ought to establish a very strict rule of responsibility, even if only by adopting the reports of auditors. This is the only way in which to save billions for taxpayers and to restrain the technical madness. First, politicians are responsible. Politicians and administrators must be made personally responsible for what they decide and do. In the 19th century there was a good reason for personal freedom from responsibility (except in cases of crime or malfeasance). It would ensure the independence of political decision and the anonymity of public functioning. The only sanction lay in the hands of the electors, who could refuse to reelect an unsatisfactory representative. But we now have totally new conditions. We can no longer accept a freedom from responsibility which is a cover for waste, disorder, and contempt of the public. We must make politicians, administrators, and technicians personally responsible for useless, unjust, and unsuccessful projects that are shown to be such.

Sometimes, accidentally, technicians are called to account in this way. But it is always at the level of execution: the engineer whose dam gives way, or the captain of the Amoco-Cadiz who was unreasonably held to be totally responsible for the disastrous oil spill. The responsibility must be placed at the top, where the decisions are made. I know that the objection will be raised that decisions are very complex, that no single individual can make them, that decision is a process which involves many people. In such important matters, then, we must be very strict. In various ways all those who help to make the decision must be held responsible for it.

As regards politicians, those who decide in favor of useless and wasteful projects must lose their mandate and be refused the possibility of reelection. As regards the higher civil servants who prepare the papers and who often make the decisions themselves, and the upper-level technicians who draw up the plans (often full of mistakes, as in the case of nuclear power stations), we should apply strict financial sanctions as was customary under the Roman Republic and at some periods under the monarchy. Those who make such mistakes must pay for them. If this were done, it would perhaps cool the ardor of civil engineers who take a percentage on such works.

I believe that the rule of responsibility is a basic one today. A good means of bringing it into play is the ancient Roman institution of popular action. If we take seriously the fact that we are all citizens and that we can control the decisions made by our representatives, unjust and absurd decisions can be rightly attacked by any of us. Being citizens is enough to give us an interest. Everyone laments the widespread irresponsibility of society today. To fight against it we must begin at the top, where the rot starts.

CHAPTER XVI

The Bluff of Productivity

1. The State and Science

Before going to the heart of this subject (i.e., the use of productivity as a kind of magic, the promotion of an ideology of productivity, and finally productivity as a bluff), I must say something about the relation between the state and science, since everything else depends on this. The matter is simple enough. The state, directly if it is socialist, indirectly if not, controls the economy. It wants good production, a balanced budget, good exports, adequate domestic consumption, and continued growth. To achieve these goals the state needs a technical efficiency above that of other countries. Technique is closely dependent on science. The state, then, has to promote science and orient it to a high level of technical production and to progress. But science can develop only with the help of the intricate technical devices that are beyond the means of even the strongest corporations. It can do its work, therefore, only if the state focuses all available national resources on the primary goal of scientific and technical research. This gives us the expression which first appeared in the USA in the 1950s: research and development. The result of research and development is productivity.

As may be seen with angelic simplicity, the problem arises at this point. (It is no longer the problem that I sketched in 1950, when I showed that state intervention is always disastrous for science and technique.) Productivity is the final goal. Productivity justifies the costs, the apparently unreasonable investment. Productivity is the hope of technical culture, of its rationality, of the scientific and technical pursuit. But this is the bluff, as we shall try to show. We have here a typical example of the "scientization of politics" and the "politicizing of science," both of which are linked to the same phenomenon of the domination of technique that is the reason for new social conflicts and individual initiatives.[1]

The tremendous initiative that the state has at its disposal in regulating technique brings increasingly to light the tension between the complexity of its operations and the rationality of its decisions. Industry and the scientific community have such influence in directing technical research that the impartiality of the government has little credibility.[2] We have seen many examples. Technical decisions are also made within administrative structures and by procedures over which individuals, as citizens, have no control. The issue in the debate between Habermas and Luhman (as recalled by Salomon) is whether democracy is possible in the most decisive questions and their economic, technical, and social implications (e.g., the elimination of peasants).[3]

For Luhman it is a peripheral and provincial idea to think that individuals can influence the state in areas which are beyond their competence and in which the process is autonomous and contingent. Luhman rightly refers to the growing autonomy of the apparatus of government. Habermas talks about the dependence of the government on the interests of the better-organized groups, the weightiest of these being the technicians and scientists. Thus the combination of political administration and the technostructure results in the total elimination of individuals. But the character of the combination and of the power of the state is strange, for the state less and less directs the economy. Its planning is for itself alone. Unemployment brings to light its total powerlessness (in spite of the omnipotence of its decisions!) and the multiplying of the victims of modernization (peasants, workers, etc.).

The modern dynamic has basically gambled: on the one side, the priorities of profitability, productivity, efficiency, and competitiveness, and on the other, as effects, the turning of the countryside into a desert, unemployment, the manufacturing of useless goods, and the reducing of consumption to banality, all with an appearance of ease, comfort, and health! But we must not ignore the fact that the system seems to rest on a consensus. In spite of the repeated failures of the politico-technical conception—as Chesneaux put it, "the calculated

1. See J. Habermas, *Technik und Wissenschaft als 'Ideologie'* (Frankfurt am Main: Suhrkamp, 1968), partially translated in *Toward a Rational Society*, tr. Jeremy J. Shapiro (Boston: Beacon, 1971), pp. 50-127; idem, *Legitimation Crisis*, tr. Thomas McCarthy (Boston: Beacon, 1975).
2. See J.-J. Salomon, *Prométhée empêtré* (Paris: Pergamon, 1981), pp. 92ff.
3. Cf. the works of J. Habermas cited in n. 1 above, and Luhman, *Theorie der Gesellschaft oder 'Sozial-technologie'* (Frankfurt, 1971).

plan opened up an abyss before being a failure"—the same policy has been triumphantly pursued by both left and right. The state and big business have an uncontested monopoly in the field of great techno-logical advances, the only qualification being that a large part of French production (about 20 percent) is now controlled by foreign capital. The left can do nothing about it. In spite of protestations, it has kept up the nuclear power program and the expansion of high-speed trains. The state is the prisoner of the technique that it thinks it directs.

This is strikingly shown by the great Eureka plan which takes up the old theme of research and development and which is so grandiose that even though worked out by France it requires European cooperation. Financed by the state, it implies that large industrial corporations must follow it, engaging in programs of research along the lines of the plan, which is to be the powerful engine of technical development. The experts chose six sectors as a basis on which a technological Europe might be quickly established. (1) Optronics (the systems which enable the light of photons to be transformed into electricity, pressure captors, discharge it, amplifiers of light, linked by fiber optics). (2) New materials (composite materials with a base of glass fibers, of carbon, ceramic, alloys of titanium, etc., for engines, the automobile, the space industry). (3) Large computers (the fifth genera-tion, where we are overtaken by the Japanese). (4) More powerful lasers and particle beams. (5) Artificial intelligence (to improve the "dia-logue" [?] among humans, machines, and systems experts, to recognize forms, etc.). (6) Very fast microelectronics (the USA has devoted $676 million of its military budget to this).

These six sectors do not exclude others that are less urgent (e.g., biotechnology). When the state commits itself on this scale, we have to consider that it is a reasonable venture in view of the stakes. The main stake is productivity, which will make us militarily, technically, and economically independent, which will ensure foreign trade, and which will develop industry, thus reducing unemployment. We shall see concretely whether this is true. For the moment we may simply state that what is at issue is staying in the race and upholding national prestige by multiplying communications and increasing the efficiency of the technical apparatus (for no particular purpose!). More and more, technique is at stake. All other stakes are false. There is no real political, economic, or scientific stake. The pseudo-stakes that are supposed to decide the fate of France are all shams.

When we look more deeply at performance in technique, we see that it is not really possible to think of any other stakes. E. Morin has shown this very well. As he puts it, scientific knowledge is less and less produced as an object of thought or reflection for the human mind and more and more accumulated for recording by computers, that is, for use by supra-individual entities, and supremely by that which is supercompetent and omnipresent—the state. At the same time, and correlatively, this science blinds us. The face of our world and society and destiny is divided up into pieces by a scientific knowledge which is incapable today of conceiving of individuals, subjects, or the nature of society, or of elaborating any thought that cannot be put in simple, formal, mathematical terms, but which is very capable of giving to the authorities new technical powers of control, manipulation, oppression, terror, and destruction.[4]

The relation between the state and science is reciprocal, and it has become essential. Neither can live without the other. Talk of productivity and the economy is a pretext. The real reason is power on both sides. Less than ever (in spite of scientific innocents who believe the contrary) is there any such thing as pure science!

We could stop there. But I thought it might be interesting to add the main points from J.-J. Salomon's report *La politique française de la technologie*, which was requested in May 1984 by Laurent Fabius, which is document No. 61 of the Centre d'évaluation et de prospective, which was sent to the prime minister in June 1985, but which was at once pigeonholed, never presented or discussed, and published clandestinely. This report is terribly honest. It denounces the policy of the arsenal. It discusses commercial fiascoes, the real plan, the real plans for the computer and telematics, and the Concorde. It challenges the three priorities of defense, the atom, and space when the most competitive countries are those with the smallest programs of military research. It shows how ridiculous is an economic system in which the state is both client and investor, both banker and entrepreneur, both administrator and executive, in enterprises that have no guarantee of technical or scientific success. All the great technological projects upon which France has embarked since 1966 have failed except for one, which had an unexpected result—the development of the software industry. These strategies supposedly put public services in the service of an industrial policy but in fact put industry in the service of public services!

Will things be any different in an industry that has not yet been

---

4. E. Morin, *La Méthode* (Paris: Seuil, 1980), 2:299ff. Cf. P. Feyerabend, *Against Method: Outline of an Anarchistic Theory of Knowledge* (New York: Schocken, 1978); Bryan Wynne, "Sociology of Science," *Science, Technology and Society* (three special numbers devoted to the sociology of science, 1984).

developed, biotechnology? Here, too, state intervention is indispensable, and heavy, departmentalized structures are being set up without the necessary specialists being available. It would be better for the state to stay in its own domain: education, regulation, patents, no bureaucracy (though Salomon has confidence in experts!), consultation without the power to decide, and association with industrialists. Yet Salomon still believes in research and development and in the need to make technical and professional training a priority. Aid must be given to research and development and the economic results of science improved. But being very critical of the omnipresent state, Salomon was ignored. This shows how sensitive is the relation among the state, science, and technique and how important is self-justifying discourse.[5]

In this complex of science, the state, technique, and the economy, we should finally recall the extraordinary difficulty of technological transfers.[6] These correspond to the universalizing of technique but are dictated by other than purely technical imperatives. On the one side are economic imperatives (e.g., improving the balance of trade) and political imperatives (favoring allies and discriminating against potential enemies, though these imperatives will often coincide

5. A fashionable slogan in 1984 was: "Less of bureaucracy, more of the state." (I am thinking especially of an article by Strauss-Kahn in *Le Monde,* Nov. 1984.) The state, we read, guarantees national unity, supports firms in trouble, and inspires public projects (with boasting about nuclear development, space, the airbus, and Ariane). The state has pumped a good deal of money into industry. The only countries that are achieving the modernization that technological changes demand are those in which there is most state intervention. The state offers pragmatic motivation and thus increases efficiency. It can neutralize the risks of modernization, lay down the rules, and fight against social and economic rigidity. Remarkable! It might seem that we are reading Mussolini. And all this, as in fascism, in a battle against bureaucracy, the root of all evil. Nowhere do we find the simple question: With what tools is the state to perform all these miracles? The state needs arms and hands. It needs the tools to analyze situations, evaluate problems, propose solutions, make concrete contacts, and distribute funds. This is all administrative work. There is no state without administration. But administration, it is argued, is not bureaucracy. Of course it is. Administration today has many tasks to perform and it has to apply increasingly detailed and complex rules. It proliferates as the functions of the state proliferate. It has also to engage in public relations. It is equipped with an increasingly efficient, modern, and fast apparatus. This is why bureaucracy is omnipresent. If there is no state without administration, there is no modern administration without bureaucracy. To say that administration is not bureaucracy is infantile.

6. See Angela Stent Yergin, *East-West Technology Transfers: European Perspectives,* Washington Papers no. 75 (Sage, 1980); Stephen Sternheimer, *East-West Technology Transfers: Japan and the Communist Bloc,* Washington Papers no. 76 (Sage, 1980).

with the economic). These transfers of technique may be between East (the USSR) and West, between the USA and Europe, or between the USA and Japan. But a problem here is that the USA does not want technique transferred to the USSR, and yet it cannot be sure that France or Japan will not sign contracts with the USSR. On the other side are transfers between North and South to help developing countries. But here transnationals interfere, on the one hand following their own interests, on the other imposing the concept of a new international order.[7]

One gets the impression that official arrangements in these areas are idealistic and illusory. For it is not very likely that transnationals will obey them. It is estimated that the transnationals are responsible for 90 percent of the transfers! Again, as regards the Third World, the problem is not to increase the flow of techniques but to let the Third World develop its own techniques, its autonomous supply. Finally, we have to take subcontracting into account and the fact that the Third World countries have very little negotiating power. In all these cases technique is viewed solely as merchandise. Today, however, I think that we must see it as the merchandise par excellence, that is, as that which grants it an independent place in political calculation.

2. Productivity

Our theme is productivity, which is a very simple idea. Thanks to the development of technical means of production, it is a matter of obtaining a higher production of goods for the same quantity of work. In more learned terms, let us call productivity a ratio or measure of efficiency in relation to production and the factors of production (labor, fixed capital, circulating capital, etc.). Productivity rises when output increases faster than input. But the interpretation of these changes depends on the specific concepts of production and factors of production, and on the quality and content of the standards used. The usual standard today is that of production per hour per worker. Nevertheless, it is hard to have a single production index for the multitude of different products and especially for services. In general, it is granted that there can be an increase of production per hour of work for four reasons: (1) using other factors (equipment, new raw materials—the problem of the global productivity of factors); (2) the structure of the workforce, which can be modified, for example, by bringing in better qualified

7. See J. Touscoz, et al., *Transferts de technologie* (Paris: PUF, 1978).

workers; (3) economizing so as to increase efficiency without altering work procedures; (4) bringing in new methods and infinitely more efficient machinery as a result of technical change (much the most common reason today). In fact, the great search in our world is for new methods and equipment which will enable us to transform our resources into the desired products. This technical change sees for itself a twofold mission. It must restore long-term economic dynamism by new techniques that can increase market demand and build up investment capital. It must also loosen economic constraints that are an obstacle to policies of regulation.

Thanks to gains in productivity, a rapid technical change can reduce inflationist tendencies and the external constraints of the international market. It can also work against restrictive economic policies and improve the employment situation through new ventures. The slowing of economic growth, the rise in unemployment, and the toughening of the world economic war have made more evident the need for the development of applied science and of techniques relating to production, for it is on these that productivity and competitiveness depend. Productivity is all the more necessary due to the world economic recession. So long as competition was only domestic, productivity was relative. But the speed of transport and the development of information have now exacerbated the problem of competition, of profitability, of economies of scale, and of mass effects. No business can justify itself except by increasing its productivity.

Between 1963 and 1983 international trade increased twelve-fold. That is what we are told by both left and right. During the same period both Thatcher and Mitterrand stated that research and development are the primary means to solve our economic problems and to stimulate growth. On April 10, 1986, Chirac stated that all economic growth depends on developing our export capacity, which implies advance in productivity, technological research, and the promoting of traditional sectors. This is also the central idea of political economists (e.g., Barre) and of most economists and technicians.[8] It is so ingrained that there is research into productivity for productivity's sake and no other basic reason. Productivity justifies itself in any area in which it appears. It is a good thing in itself. It is a standard of judgment. Reasonable researches from an economic standpoint are now neglected if they are not affected by a coefficient of productivity.[9]

8. Cf. G. Schmeder in Salomon and Schmeder, *Les Enjeux du changement technologique* (Economica, 1986).
9. See J.-M. Ferry, "La robotisation," *Esprit,* Jan. 1985. Ferry also raises

Naturally, this doctrine and its accelerated application have already aroused fears. I will mention four. The first is, of course, the fear of unemployment. Rapid development of communications and automation inevitably put more people out of work than are demanded in the new jobs created by computer research, production, and servicing. The computerization of the Bank of France led to the removal of 40 percent of the work force, and the same was true in insurance. A detailed study shows that the result of robotics is more unemployment than new employment, a free hand being given merely to the fourth sector (communications).[10] In effect, if every worker produces more in the same time, there are only two solutions; either reduce personnel or reduce the hours of work. This is why there is a strong movement for the reduction of working hours with no reduction of pay. We cannot indefinitely increase the amount of goods produced. The domestic market is quickly saturated and competition makes the foreign market very risky, since it is dependent on many other factors apart from productivity (e.g., financial fluctuations, dollar quotations, etc.).

The second fear is that the constant appearance of new and different technical instruments will eliminate qualified workers who cannot readapt and acquire new professional skills. The levels of qualification, according to the unions, tend to become lower. It is true that proportionately there is more unskilled than skilled labor where there is automation.

A third fear is that fundamentally productivity by means of technical improvement is simply an extreme development of Taylorism in the form of computers, automation, robotics, and industrial automation. Tasks are parceled out, the rate increased, and work not interrupted. This finds confirmation in the total elimination of "dead time" by the application of computers.[11] That is to say, results are immediate. It is stressed that this elimination can take place not merely in industry but in society as a whole thanks to the speeding up of social, economic, and, of course, work rhythms. This total elimination of dead time is

---

another vital problem. The age of information and robotics is causing us to lose all sense of the traditional principle of social justice: "To each according to his or her work." This principle no longer has any meaning!

10. Cf. examples in the 1986 Bordeaux thesis of Gellibert, "Le Choix de la biomasse comme énergie. Révélateur des mythes et des conflits de la société technicienne": the possibility of using the nitrogen from the air and the nitrogen from legumes (with the practice of letting the land lie fallow) in order to avoid the misuse of nitrous fertilizers. This would economize both land and energy but it is neglected because it does not have enough technique.

11. See Bressand and Distler, *Le Prochain Monde* (Paris: Seuil, 1986).

plainly a great source of productivity in the world of work. By constant observation of workers and procedures it is possible to rethink the whole system of production around internal communication networks. Ultra-fast video makes possible fine tuning. The traditional rhythms corresponding to the speed of perception by the hand and the eye are no longer the obligatory terms of reference but an obstacle.

The fourth and final fear found among workers and unions is that if workers adapt to all this they will finally internalize an ethic of productivity.[12] That is, they will themselves try to improve production on their own initiative and rationalize both their work and their daily lives. They will come more and more under the pressure of attitudes both at work and outside work. The pressure of work will shape their whole personality.

When they come up against these fears (especially the last two), specialists on the subject dismiss them. There will not really be any acceleration or compression of time. In automatized and computerized work, time is split. There is a human time in the physiological rhythm and a machine time measured in nanoseconds (billionths of a second). This is the time in which machines command machines. We are no longer in the era of work in atoms (G. Friedmann) but in that of work in "elementary particles" (Bressand and Distler) which the machine gathers together and joins. Everything is in networks, and human beings are simply directors of networks.

None of this has anything in common with the older industrial society. We need a total renewal of thinking, organization, and conceptions in terms of network structures. There are no longer any divided and repetitive activities. The human role is that of the intelligent coordinator who controls good functioning. But how many directors of networks, or coordinators, are needed? And what about the rest? Furthermore, in spite of the solid scientific basis, is not all this a purely theoretical idealization, since none of it exists as yet? However that may be, the fears that are felt correspond exactly to the hopes that are set in productivity.

We must now look a little more closely at the actual situation. A stricter, nonideological examination shows many gaps and illusions in all the talk and projects. I would argue briefly that it is false to say that productivity necessarily brings growth and that growth will eliminate unemployment. It is false that productivity is an unfailing result of scientific and technical research. The formula "research and develop-

12. See Mercier, *Vie quotidienne et Nouvelles Technologies: La Société digitale* (1984).

ment" is inaccurate. Research cannot engender productivity. Productivity may result from other things besides technique. It is false, finally, that productivity responds to human needs and carries within itself the solution to the crisis. Let us look first at the complex combination of technique, research and development, and productivity. It is important to note that contrary to the general idea and uncontested dogma, it is not countries that have devoted most money to scientific and technical research that have the highest rate of productivity. For some years this has been the great problem that has occupied American researchers, economists, and sociologists.[13]

It is surprising to be able to state that among developed countries France is second in the world in the absolute level of productivity, with a growth rate five times higher than that of the USA. Contrary to the usual belief, productivity in France is higher than it is in Japan. And in general Japan is behind France in techniques. We shall see later that the Japanese challenge is through better organization rather than technical superiority. Between 1977 and 1983 the average growth rates per year were USA 0.6, Japan 3.4, Germany 2.1, and France 3.[14] The surprising point is that the countries with the most research and development are low in productivity. The rate of growth in the USA fell from 2.4 per year in 1968-73 to 1.9 per year in 1974-79. There was a deceleration of growth of 1.5 percent between 1960 and 1982. Japan is even worse, with a deceleration of 4.5 percent between 1970 and 1982. In France the rate was 1.4 percent. In the USA there was a recovery in 1982 with an extraordinary boom in investment, so that there was talk of the crisis being at an end. The growth rate went up 2.9, but it was a flash in the pan, as some had foreseen, given the enormous deficits in the budget and trade. The year 1987 set a record for business failures in the USA (some 200 important companies failed) and the various recoveries that were announced were fragile. The budget deficit has continued to grow since 1950 and reached 6 percent of the gross national product in 1985. The sector that was affected worst was agrobusiness. There the decline in revenue has been 30 percent since 1983, and it is now the lowest since 1932. Farm exports have become increasingly difficult, so much so that in 1986 Reagan had to subsidize the growers

13. See J.-J. Salomon and G. Schmeder, *Les Enjeux du changement technologique* (Economica, 1986).
14. See Lester C. Thurow, *Organisation sociale et productivité* (Economica, 1986). Bourguinat, *Les Vertiges de la finance internationale* (Economica, 1987), poses the difficult question how a country that normally exports capital can suddenly become a debtor nation, as the USA has done, with a debt exceeding that of Brazil.

of corn so that they could sell at a competitive price abroad, especially in the USSR. A new deal for farmers was launched in 1984 but it does not seem to have been very successful. There has been a continuing exodus from rural areas, and only 2 percent of the population of the USA now lives by agriculture. In 1985 it was recognized that the global growth of the American economy was still modest (*Le Monde*, Nov. 1985). Expansion in 1985 was much weaker than in 1984 (around 3 percent). In 1986 there was economic stagnation and increasing debt (*Le Monde*, Aug. 1986). The annual deficit rose to $212 billion in 1985 and the total debt to $2.1 trillion. The foreign trade deficit was $148 billion in 1985 and over $170 billion in 1986. Many big corporations ran into difficulties (oil and automobiles). Federal debt doubled in five years.[15] Vergara stresses that the creation of jobs in the USA is a "myth."

Nevertheless, in raw figures the USA was investing ten times as much as other advanced countries in research and development. If the relation of research and development to the gross national product varied slightly (2.9 percent in 1960, 2 percent in 1980, 2.2 percent in 1985), it was still an enormous sum, as much as $56 billion in 1984.

Thus productivity and development are not linked directly to research. Sometimes the very opposite is the case. Being ahead in innovation is not necessarily an advantage. The American computer industry had this experience. It developed much faster than its counterpart in Europe. Too fast: There was a serious crisis in the sale of computers in 1985, which shows that the talk about computer production being the salvation of France is false. Was the market saturated? Perhaps not, but obviously a threshold was reached, and on this threshold it became difficult to continue adapting firms, unions, and people to new forms of employment.

As is well known, we may also adduce the example of Japan as that of a country whose economic advance was not on the basis of research and development. The growth of Japan after 1955 had nothing whatever to do with investment in research. Japan simply exploited discoveries made elsewhere, especially in the USA. It secured the licenses for what it regarded as important patents. After 1980 Japan then began to invest heavily in research and development, with many

15. Bourguinat (op. cit.) emphasizes that in this situation the "new finance" has brought the whole American economy into a speculative bull market. After five years of an overvalued dollar, the economy seemed to be shattered, with imports rising 19 percent and exports only 5 percent, and a particularly drastic reduction in high-tech exports (from $27 billion in 1980 to only $6 billion in 1984).

patents and technical innovations. But at the same time we also note a decline in productivity. This decline has gone hand in hand with research and development, though naturally I am not contending for any causal relation. Japan is falling slowly into an economic crisis with lower productivity. An interesting point is that to stave off this crisis it is now engaging in the manufacture of armaments. In 1985 the countries with the most important innovations were the USA with 64 percent and Great Britain with 17 percent, but these were precisely the countries in which the rate of productivity did not increase and economic development was very weak.

It is thus pure ideology to associate technical progress and economic development by way of production. But if the relation between research and development and productivity is weak, American economists and sociologists have been trying to understand why there has been this decline in the USA. If it is not technical progress that brings increased productivity, what is it that brings decline? They have come up with some interesting findings.[16] There has been especially a change in the labor force and the view of work. Many poorly qualified young people joined the labor force between 1960 and 1980. There was a general decline in the competence, qualifications, and experience of workers. There was a serious decline in discipline. There was a great rise in turnover. Much less effort was put into work; there was much less interest in it. On the whole one might say that ethical and disciplinary problems, greatly aggravated during the 1970s, essentially explain the decline in productivity in spite of new materials.[17] With the same machinery and methods Ford obtains 20 percent higher productivity in its German factories than in its American factories. In other words, what we have here is a human problem. The vital element in productivity is the qualification and motivation of the labor force.

In all the evaluations of innovations and discoveries, however, what is completely ignored is a factor to which we have referred already, that of feasibility. An invention means nothing unless all the factors are present to give it a place in actual industrial expansion. Research and development have to be mediated. Japan proves this point. The prodigious success (or miracle) of Japan is due primarily to the Japanese conception and organization of work. People work with a company for life and are not afraid, therefore, of losing their jobs.

16. See Martin Bailey, *Capital, innovation et croissance de productivité*, p. 53.
17. See Lester C. Thurow, *Organisation sociale et productivité*, p. 75.

Salaries are poor but there are good bonuses. Self-management is the rule. Workers maintain quality control over their own products and participate in investment decisions. In other words, Japan has succeeded in interesting workers in increasing productivity. The greatest amount of flexibility is sought (to the benefit of workers, unlike that in France), as are also the best means to evaluate the technological know-how and expertise for a given enterprise. The great problem is setting up labor-supervisor-management relations, and the participation of the workers in the total life of the company. If there can be "hard productivity" by research, technical innovation, and capital, there can also be "soft productivity" by motivation, cooperation, and equipment. To this, perhaps, corresponds the concern (quite new to the Minister of Research) to bring the social and human sciences into research, along with their psychologists, sociologists, jurists, etc., who play a part in programs of technical research by studying the social and economic impact and the changes that will be demanded in corporations, etc. (*Le Monde,* May 1985).

Finally, as regards the decline in American productivity, there has been an emphasis on the excessive growth of the third sector (including all the jobs relating to the computer), since productivity in the service sector is very low. Thus it has been admitted that in the USA a transfer from the second to the third sector (health, law, police, commerce, administration, etc.) means a loss of 37 percent in productivity. Between 1977 and 1983 the productivity of laborers increased 6 percent but that of white-collar workers only 0.8 percent. Industry, however, has become increasingly bureaucratic. There are 57 million white-collar workers for only 30 million blue-collar workers. We thus slip once again into the problem mentioned above. The growth of information services is well and good, but we cannot eat computer paper or wear what the service sector has to offer.

We are not to think, then, that productivity always denotes the production of usable goods. Jean Voge has estimated that each time productivity increases 10 percent, the part corresponding to "information" (i.e., essentially the cost of organizing work) rises 20 percent (the cost of information has a growth quadruple that of producing usable goods). Naturally, there are efforts to improve the productivity of services (electronics, computers, rationalization), but these efforts do not solve the problem. In the consumption of knowledge, it is hard to give knowledge use value when it functions as exchange value. This explains the growing complaints of consumers about the artificial obsolescence of goods. Many techniques are abandoned as soon as they are refined. They are relegated to marginal zones because they are no longer profitable. The logic of production is that information exists on its own, while that of demand is partly to impose its use.[18]

We must now ask whether productivity creates employment or unemployment. We come across some surprising facts. Martin Bailey argues that developments unfavorable to the growth of productivity in the USA have not had serious effects on the labor force. Even during the period of stagnation civil employment increased by 15 percent.[19] C. Freeman makes the generalization that the level of employment goes up in tandem with a growth of productivity so long as this is weak.[20] In other words, a dip in productivity does not prevent the hiring of workers and a rise does not necessarily mean either a drop or a rise of employment. Even if it were true that growth would lead secondarily to a growth of employment, this would be limited and would not end unemployment. Rising productivity does not accelerate the rotation of capital—quite the reverse. Nor does it necessarily mean economic growth, which in any case would not overcome unemployment.[21]

Expansion creates jobs but also eliminates them in sectors under reconstruction. Economic growth has never created many jobs: 125,000 a year in 1959-71, 60,000 a year in 1971-84. Thiot has an interesting table showing a growth of the gross national product of 162 percent from 1959 to 1983 and only 11 percent growth of employment. In France and Japan the figures are 475 percent and 32 percent. But we have to take into account many other factors. Thus growth, by offering more consumer goods, demands higher wages. Standard wages in 1950 would now be at the poverty level. We have also to take into account demographic progress and the difference between the demographic curve and the employment curve. Here there is only a weak relation between growth and employment. In sum, it is wrong to believe that unemployment will be solved by higher productivity and the resultant economic growth. But we must now look at the opposite.

We may take the typical case of the robot. One ordinary robot replaces two workers but has different capacities according to the purpose for which it is intended. A robot is very expensive (e.g., a robot for soldering cost $25,000 in 1981). Moreover, the savings in labor is compensated by the operating costs: maintenance and the extra energy the robot consumes (the robot is an enormous consumer of energy).

18. See J. Beillerot, *La Société pédagogique* (Paris: PUF, 1982), p. 125.

19. Martin Bailey, *Capital, innovation et croissance de productivité*, p. 53.

20. C. Freeman, *Technologies nouvelles et Avenir de l'emploi*, p. 91.

21. See F. Thiot, *Le Monde*, May 1986; Couria, *Informatique et Emploi* (Conseil économique et social, 1984).

Again, a robot begins to be profitable only after producing so much. The amount varies according to the industry. Finally, models of robots soon become obsolete; often a model has to be replaced before it is paid off. Paul David cites a German study of five branches of industry and twelve manufacturing operations that concludes that between 1990 and 1995 robots will save 10 percent of the labor force (though some journalistic studies give a figure of 25 percent for the USA between 1985 and 1990). David thinks that the most likely figure is 5 percent when we take every factor into account.[22] It seems that in the choice between building a new factory equipped with robots and gradually automating an older one, robots are at a disadvantage. In fact, the installation of robots in an existing factory is hampered by the interdependence of techniques. A robot requires a whole series of related equipment, that is, another conception of the business. Its place is really in a new factory with automation and computerization. This new factory relates machines to machines without interference. The networks are the links, and they fully integrate production. Computerization becomes a process that affects every sector of automation. Information automation replaces mechanical automation. Thus a new financial and business strategy is needed. We have here a wholly new conception of business and economic life (cf. the "networks" of Bressand and Distler). This will perhaps result in a new dynamism of growth.[23]

We must now deal with the assumptions. Research and development rests on the premise that technique is in the service of people to enhance their well-being. But O. Giarini along with many others has

22. Paul A. David, *La Moissonneuse et le Robot,* op. cit., p. 109. The Bureau international du Travail observes that the use of robots is also not making as much headway as expected: 44 percent of the English firms that use robots have run into serious difficulties and 22 percent have abandoned them. Germany is investing only 5 percent of its capital in robotics. Yet forecasts are optimistic. By 1990, there are to be 70,000 robots in Japan, 60,000 in the USA, and 20,000 in France. I do not know the basis for these estimates. Though robots will not solve all problems, they are described as a viable option that will not create unemployment. Europe has the most unemployment and uses fewest robots, while Japan has little unemployment. But there can be no question of enforcing the use of robots; workers must agree (*Revue internationale due travail* 1 [1986, Bureau international du Travail, Geneva]).

23. Cf. the interesting article in *Le Monde* on robots in the clothing industry (R. Clavaud, Oct. 1983). Although computerization and robotics will supposedly be the salvation of the industry, new factories will have to be built, robots will do only simple tasks, and the range of products will be reduced (one model for T-shirts, three for underpants). A rise of 5 percent in productivity may be expected. Fewer textile specialists will be needed but more laborers and engineers. Costs should be reduced, but there is no mention of the price of the robots or the new factories!

shown that in reality there is no relation between economic growth and enhanced well-being.[24] He takes as an example the common notion that all paid work produces an added value that generates well-being. But this is not accurate. Technical work often means reduced value. Thus industries for eliminating pollution simply recover an existing value; they do not enhance well-being. Yet they increase the gross national product and thus contribute to productivity! This investment represents a supplementary cost due to production that reduces well-being.

At times the reduction of value is greater than the addition. Thus technique, by its own development, can increase productivity and yet at the same time diminish the positive economic benefit. Either way, however, there is growth! Our technical society, says Gellibert, is like a top which can stand on its point only in virtue of its speed. The least slowing down and social disorder appears at once (farmers threatened with ruin, and fertilizer, pesticide, and insecticide factories over which the specter of unemployment looms). We run into the problem of useless work, but no matter what the consequences, we must keep the top spinning and increase productivity, even for destruction.[25]

Concluding this sketch of productivity, I will recall an older thesis that is often forgotten and needs to be brought to mind constantly. It is the thesis that our industrial or postindustrial organization and technical or computerized society are not for the purpose of creating consumer goods or enhancing human life and well-being but solely for the producing of profit. Solely! All else is pretext, means, and justification. Marx demonstrated this in the case of capitalism but it is equally true of socialist society as we actually know it.[26] Real productivity is solely for the purpose of profit. All the rest is accessory. The calculations are controlled, then, not by science or technique, but by the strategy of profit.

The supporting market is first selected. Thus in the case of rockets it is said that within two or three years there will be a demand to put satellites in orbit. Then there is investment only where control is assured. Then the greatest possible development is attempted. Finally, with no shame, what is least profitable can be discarded (cf. the textile and metallurgical industries, though with subsequent regrets), and investment made in a new supporting market (which will probably

24. O. Giarini, L'Europe devant l'âge post-industriel (Paris: Futuribles, 1977).
25. J. Gellibert, "Le Choix de la biomasse comme énergie" (dissertation, University of Bordeaux, 1986).
26. Cf. the essential book by Michael Voslensky, Nomenklatura: The Soviet Ruling Class, tr. Eric Mosbacher (New York: Doubleday, 1984).

soon be saturated, since others are doing the same).[27] This clearly explains the reversal to which we referred in the relation between technique and economics, but it has become necessary only to the degree that the sectors of technique have multiplied inconceivably (so that there are many choices and we cannot exploit all of them) and economic resources are limited.

A strange phenomenon thus confronts us. Efficient modern techniques may exist which have useful qualities (e.g., not polluting), but they never see the light of day because the big economic and financial groups choose other techniques which, they think, will bring more profits. Thus the oil companies do all they can to dismiss the biomass, solar energy, wind energy, and geothermal developments. In other areas, too, we see gentle, appropriate techniques that have no chance of success because they will not yield maximum profits. In the last resort there is no relation between profitability and knowledge. That is our last word on the famous expression: research and development.

3. Entropy

According to Ingmar Granstedt, since 1964 the growth of capital per head in industry has not been accompanied by an equally strong growth of technical efficiency. The total productivity of factors has not kept pace. Production will henceforth grow less quickly than the amount of capital put to work.[28] This judgment leads us to a brief reminder of the theory of entropy, which is the exact counterpart of research and development, and which, if verified, will give further evidence that politico-technological discourse is all bluff. Entropy has successively been discerned in physics, astronomy, biology, and philosophy (Lalande, 1899). In the 1960s it was realized that the second law of thermodynamics might apply to other fields, and especially to economics and technical progress.[29]

Entropy is a state of absolute disorder, corresponding to disin-

27. For an excellent account see Gellibert's thesis, "Le Choix de la biomasse comme énergie."

28. Ingmar Granstedt, *L'Impasse industrielle* (Paris: Seuil, 1980).

29. Especially following Shannon's theory of communication (1948). It might be of interest to recall two implications of Shannon's formula, namely, that the entropy of a source relates directly to the number of elements that it contains, and correlatively that it diminishes in direct relation to the dissymmetry of the probabilities of each of these elements. This has important consequences for the interpretation of the evolution of the technical system.

tegration and a complete absence of "information." The law of entropy simply restates mathematically a law which has been known for at least a century, especially in farming—that of diminishing returns. Georgescu-Roegen put it in theoretical form when he criticized the general idea that technique is of unlimited capacity.[30] It is supposed to develop exponentially. The superficial justification for this idea is that one technical advance brings another. This is true, but the advance is not cumulative, as in the case of demographic growth. Even if technique continues to progress, this does not mean that it has no limits. Growth has an upper limit, which in the case of technique is fixed by the coefficient of efficiency.

Entropy applies also to economic systems. These can continue to grow so as to keep going, but with no greater usefulness or betterment of human well-being. As Georgescu-Roegen proves with detailed examples, technique, too, is subject to the law of diminishing returns. Furthermore, by its own development it contributes to the diminishing of economic returns. This is simply to show in another way what we have demonstrated in the preceding sections. The costs of research mount, the time it takes grows longer, and the costs predicted are exceeded, as in the case of nuclear power stations. The appearance of new techniques means inevitably a faster obsolescence both of the technical tools themselves and also of the products, so that they have to be replaced before they have paid for themselves (cf. computers). This idea of entropy (and of information as "neg-entropy") has become widespread,[31] but since 1984 the scientific world has reacted vigorously. It has shown that sociologists and economists have misinterpreted the second law of thermodynamics, that entropy is not a notion that can be applied to everything, and that it does not apply ineluctably. We thus need to reason judiciously.

I accept many of these reservations. It seems that the enthusiasm for the idea of entropy was extreme, and I distrust universal generalizations. I prefer to stay with the simple but very powerful idea of diminishing returns in both economics and technique. This seems to be so firmly demonstrated that it is hard to contest. Yet I think it is correct that a new technique might arrest the decline and renew the movement of productivity. This is simply an application of the idea that new information checks the possibility of entropy.

Nevertheless, the growth of waste, of technical aberrations, of

30. Nicholas Georgescu-Roegen, *The Entropy Law and the Economic Process* (Cambridge: Harvard University Press, 1971).
31. Cf. Henri Guitton's excellent book, *Entropie et Gaspillage* (Cujas, 1975).

adverse effects, of slowness in the transition from innovation to application, and of subtracted values reflects the global reach of technique and leads inevitably to a lowering of productivity. Economic phenomena tend to be more and more unstable (flow, networks), and when there are no stable phenomena (technique in its globalness may be a stable unity, but only individual sectors are economically significant, and these technical subsystems are most unstable), no theory can be formulated. This is why economic science seems to be restricted to the short term, to the period when the variable factors can serve as data and a mechanical and quantitative analysis can have some meaning.

A further point is that the more technique develops, the more, as we have seen, it specializes. Every new solution is better, in general, in a narrower field of application than the technique that it replaces. It may be more efficient but it applies in a narrower economic sector, so that the innovations bring diminishing economic returns from a larger standpoint. According to Giarini, producing a narrower selection of goods more quickly simply increases quantitatively and qualitatively the demand for supplies at the various stages of production, places heavier demands of input and output storage, and increases the number of intermediaries. The obvious result is a considerable reduction in technical productivity.

This trend is accentuated by the slide to the third sector which technical progress not only facilitates but induces. Scandalous though it may be to say this, the computer, which results in fabulous third-sector growth, is a decisive factor in economic entropy and therefore in technical entropy, even though innovations and technical marvels do not cease to multiply. However that may be, in spite of modern critics J. Neirynck has taken up and generalized the concept of entropy in a systematic study of all the places where it may be found.[32] Obviously, we do not have here a mathematical or statistical demonstration but a rigorous investigation which seems to lead us to the incontestable conclusion that the greater the increase of technical power, the greater the entropy of the complex system that we might for a moment regard as a closed system. It is certainly a total illusion to think that the computer, as information, can play the role of neg-entropy and reverse the process.

It is at this point that we need to look at a crucial article by a scientist, Claude Riveline.[33] Riveline runs through the many uses of

---

32. J. Neirynck, *Le Huitième Jour de la Création. Introduction à l'entropologie* (Presses polytechniques romanes, 1986).
33. Claude Riveline, "Manifeste pour la désinformatisation de la société," *Pandore* (Feb. 1982).

computers but he then asks whether it is not all a gigantic mistake. Computers have never increased the returns of administrative work, given the number of operators. As regards the drawbacks, he gives many examples of the damage they do. But are they not essential to science? Few scientists work with them. Their main use is for calculation. Confronted with "a mystery of nature," scientists used to reflect. Today the questions are translated into logical systems which computers can handle. But they are very selective. What masks the poverty is the quantity of elementary facts that they can absorb and their speed of operation. But intelligence is efficient preoccupation with essentials. Why search for essentials, however, when we can deal with everything at once?! The appearance of a new tool or a new application is hailed as a triumph even though its virtues do not support this. What produces enthusiasm for computers is not that they are useful and efficient but that they give the illusion of being intelligent.

PART IV

Fascinated People

We must begin by recalling a question that I dealt with a long time ago. When someone says that people can and should master and handle technique as they wish, my simple question is: What people? People in themselves? They do not exist. You and I and average citizens? I can refuse to have a telephone or a videorecorder, but what does that change? Politicians? They know nothing and can do nothing. The upper class? This has power in its own domain but not over technique. Technicians? They, too, are limited to their own sphere, and their interest is to apply and perfect their technique as best they can. The technostructure comes into play here. The application of technique enhances the social (and financial) status of technicians, and this leads them to further applications of their technique. Finally, scientists? But often scientists do not know what will be the technical consequences of their discoveries (remember Einstein), and their passionate interest in scientific research prevents them from exercising self-restraint. Therefore, no one!

The idea that people master and use technique as they wish is meaningless and absurd, as we have shown already (see chapter VII above). In recent years it has been modified. We are now told that people in the West are fascinated by modern technique. Fascination means exclusive fixation on an object, passionate interest, the impossibility of turning away, a hypnotic obedience, a total lack of awareness, and finally exteriorization of self (either possession or dispossession, according to where one is situated). But again we must distinguish. I am not saying that everybody in the West is fascinated. Contrary to what a simplistic judgment might suppose, it is the most educated people with the most developed personalities who are the most fascinated.

323

At this point we come across a phenomenon that we also found in our study of propaganda and which might seem to be surprising. Those who are most susceptible to propaganda (and advertising) are the intellectuals, while the hardest to reach and to budge are those who are rooted in traditions, whose ideas are fixed, who live in a relatively stable environment (like farmers up to the 1950s), or who are in structured relations (like members of unions). The last group is not exactly fascinated by technical modernity. A very interesting recent work has shown that ordinary people or those who belong to the middle class may watch television, but if they are asked about it they are very guarded and have not been much influenced by it. This creates serious problems of interpretation for the investigator.[1]

In fact, those who are fascinated by technique are the intellectuals, the technicians, the scientists, the upper classes, the journalists, the various shapers of public opinion, the artists, the priests and pastors (when they want the church to change and to adjust to modern tastes), the responsible economists (bankers, etc.), the professors (who have suffered enough from being told that their teaching is worthless!), and the high-level administrators. These are the ones who are fascinated and who show no critical spirit, or who, when they believe (like many artists) that they are engaging in violent criticism of our society, fail to see that they are simply reproducing in a kind of parody the technical world itself with all its perversity, thus strengthening the perverse effects and in so doing reinforcing the myth. I demonstrated this for most of the arts in *Empire du non-sens*. It is to this essential group of responsible people that we refer when we talk about the fascinated.

A first dominant feature of these people is the presence among them of a number of images that we might call myths if it were not that the word is given a different use—modern but fundamental images which at one and the same time engender, validate, and render incontestable certain judgments, attitudes, and choices. We have already dealt with some of these, for example, productivity, science, and rationality. Some are constant, others spring up, are grafted on to the former, and are both new and incontestable. There is, for example, the myth of Japan. The West has been invaded by motorbikes, cars, watches, computers, and videorecorders that are Japanese, and all of them (in spite of duties) at prices lower than Western prices. The commercial and economic stakes are high.

At the same time we learn that this is not dumping. Nor is the

1. See D. Brethenoux, "Étude de la réception télévisuelle: Sémiologie T.V., réception T.V." (dissertation, University of Bordeaux, 1985).

Japanese labor force exploited. There is practically no unemployment in Japan and the standard of living is high. Even the boldest experiments are the rule in Japan. The Japanese have an incredible number of robots and the fastest trains in the world. Their territory is limited but in spite of the high density of population Japan is not overpopulated. It is thus the absolute model. It may be noted that the groups of people listed above are always on the watch for the absolute model. For many among them (though not all) it used to be the USSR, then China, but in truth there was too much disagreement between these (intellectually seductive) countries, and then the technical reality! Now the ideal is Japan because Japan also seems to have solved social problems thanks to fabulous technical progress.

Japan is thus being studied with a view to imitation. For the only way to compete is to imitate the leader (as in 1930 when everyone imitated the USA and gambled everything on the automobile). We see at this point the difference between average people and fascinated people. The former are content to buy Japanese products without another thought. The latter form an idealistic image of progress incarnate. They feverishly study Japan. They learn of the striking role of cooperation between the state and private industry, of the complex elaboration of the economic plan (though the work is done smoothly, the ministry does not impose policy), of the creation of links between the ministry and industry, of the success of a kind of general mobilization for productivity, of a deliberate readiness to make collective decisions in concert, and of the progressive emergence of a consensus which is also a process of taking cognisance, analysis, agreement, and negotiation, in which the unions are on good terms with the owners and the workers are strictly supervised.

But above all Japan is the ideal because it was the first to find the vacuum in which to provide new goods. Hence it is less the (difficult) Japanese structure that we try to imitate. Instead, we are trying to play in the same field (robotics, etc.), which is an obvious blunder, since we will never catch up with the Japanese. Or else we are trying to find our own vacuum which no one has yet filled and in which we will have no competitors: more refined, sophisticated, extreme high tech, which can still be marketed. However that may be, the Japanese model fills our minds, fascinating us and pushing us in this direction.[2]

2. In general, a dubious feature about Japan is ignored: the very low value of the yen. When the yen rose in value, Japan's international commercial advantage was reduced. The truth is that 50 percent of the workers are employed by the innumerable subcontractors, are severely exploited without defense, and accept

It is these obsessional images which seize the intelligence and characterize what I have described as fascinated people.

---

their lot. The works of Satoshi Kamata may be biased, but we cannot ignore them (*Japan in the Passing Lane: An Insider's Shocking Account of Life in a Japanese Auto Factory* [New York: Pantheon, 1983]). The idea that Japan has no unemployment (or a rate of only 2.6 percent) is also an illusion. In fact, the organization of the job market is so different that comparison is not possible. There are different forms of unemployment and a crude instrument like the "rate" cannot measure them; cf. P. Saucier, "Le chômage au Japon," *Travail et Emploi* (Ministry of Social Affairs, May 1986).

CHAPTER XVII

From Information to Telematics

1. Information

The world of information can be a terrible one. Information has enjoyed a triumphal onward march. It is viewed as the condition of all intellectual, social, and economic development. A surplus of information is a surplus of culture and personality. In the theory of entropy, we recall, it is information which will renew the system and stop the ineluctable slide into entropy. But information is seen in many different ways. Costa de Beauregard has taught us to distinguish between knowledge-information and organization-information.[1] I receive information which remains with me as knowledge but which I quickly forget. Etymologically, however, information *(in-formare)* has the sense of giving form. It shapes conduct. If the same information is given to many people, by being led to adopt this conduct, they form a coherent group. Thus the information given by bees tells other bees where there is nourishment, in what direction, and how far away, so that they all know where to fly.

This is precisely the role of information in a "primitive" society. It is useful. It tells about hunting possibilities, dangers, and relations to the spirits. All the information that members of tribes receive is classified by them as useful or not. Lévi-Strauss has shown that Indians receive a phenomenal amount of information about nature that would be completely strange to us. When useful, the information is passed on to others, and the group acts accordingly. This is the model of information in all traditional societies. The information of pure knowledge was much less extensive, was reserved for special groups (the Greek philos-

1. Oliver Costa de Beauregard, *Science et Conscience* (Stock, 1980).

ophers), and was little appreciated (cf. what travellers reported about strangers in the Middle Ages, which was received with skepticism on the ground that those who came from afar found it easy to lie).

There is also the information which seems to be knowledge but which in reality is not. On the one hand, we have information in the spheres of art and music (singing, dancing, theater, clowning, story-telling, minstrels, and troubadours). This type of information is indis-pensable for the coherence and fellowship of the group. It is not just a matter of aesthetics or entertainment. As has been said again and again for the last fifty years, festivals have a basic role and the information relative to them has the same role. On the other hand, there is also religious information. This plays a part in organization: popular pro-cessions, the actions of religious orders, the preaching of crusades or missions, etc. Information of this kind seeks to galvanize and shape the group.

After this brief survey of information in the past, we must now assess the complete change in our own day. Confronted with what now passes for information, we note at once the intellectual and conceptual gulf that separates us from the computer. What is information for the computer? Information is defined as data. Facts and ideas are formalized in such a way that they can be communicated or manipulated by different procedures. But the data have first to be represented. This representation is used throughout. The process consists of handling the data, which may or may not be memorized. It is interesting to note that in analyses of the information handled by the computer, we find again the ideas of knowledge-information and service-information, but the words have now changed their meaning. The knowledge at issue here is comparable merely to the predigested knowledge of an encyclopedia, which gives a certain picture of the world but bears no reference to reality. The new status of knowledge-information makes of the world and culture a superficial reality and a language of artifacts.[2] Service-information replaces organization-information. Services are useful bits of information to guide us in the jungle of the modern world (classified announcements, administrative notices, timetables, etc.), which in re-ality integrate us into this world rather than stirring us to action on the basis of our own judgment and assessment of the situation. But let us stop for the time being at information as it is generally circulated.

We are deluged today by a flood of data, by an uninterrupted flow of mixed material about everything and nothing. Now we have to

2. See M. Mirabail, et al., *Les Cinquante mots clés de la télématique* (Privat, 1981).

distinguish between the information which comes to us from outside by way of the senses (millions of data a minute, though we spontaneously filter out most of it, or else we would go mad), and the data that are directed at us by the procreators of information who want it to reach us and want to make us receive it. On the one side the data come from the natural world; on the other they are fabricated by other people with a view to making us do something.

Now 999 of every 1,000 data do not concern me at all, but they still smite my eyes and ears, they still assault me, for they are designed to make me feel concerned, to control by feelings and ideas and likes and dislikes, and finally to commit me to action, modifying my opinion, attitude, and behavior. These data invade my imagination and subconscious. They constitute a mental panorama in which I have to put myself. It is extraordinary that as regards data, no less than energy and merchandise, no one asks about the value of their distribution. When we want to evaluate the culture of a people, we look at the number of newspapers and broadcasts it has. More data necessarily mean progress, as may be seen from the urge to create satellites which can distribute more information, without asking whether the human brain is not already satiated.

These data are neither useful knowledge nor organization. They are incoherent and useless. They produce disorganization. It is in this regard that the theoreticians of communication-information, the philosophers, and the scientists show the weakness of their basic paradigms. The important thing for them is that communication be perfect, without loss or addition, that the information transmitted be perfectly received by the recipients.

No effort is made to find out whether the information serves any useful purpose, whether it makes any sense, whether it is worth broadcasting. The important thing is that the information is there and that it is transmitted well. The millions of artificial data received each day are completely incoherent. They transport me into different worlds. They do not hang together. They help to shape a split personality. Happily, this does not happen all the time, but this invasion by empty and useless data, this circle of information, has still singularly modified our personalities. Four traits, as I see it, proceed from this invasion by information.

We note first the result of a process that I have already studied: disinformation through excess of information.[3] J.-C. Simon writes:

3. See J. Ellul, "L'Information aliénante," *Économie et Humanisme* 192 (March 1970): 43-52.

"The multiplying of information is not without negative effects. A study in Japan (1975) shows that 90 percent of the information produced is not used at all even though very efficient means of distributing it are available. Our society has reached saturation point. There is a tendency to reject information in general. Similarly, most of the advertising mail is tossed in the wastepaper basket without being opened."[4] Those who receive, receive nothing at all. Their brains are stuffed with incoherent, uncoordinated, purposeless data that they cannot register, master, classify, or memorize. The spontaneous reaction of their organism, then, is total rejection, pure and simple. They throw it all out, including the data that might be of interest, which they miss because they are inundated by the rest. Disinformation results because the information received is neither knowledge nor organization. It goes in one ear and out the other. Or else it constitutes a kind of confused mush that I have often seen in students. Nothing is correctly linked either to a coherent whole or to a network. Those who have a precise ideology (e.g., communism), escape this confusion. They receive information that conforms to the ideology and reject the rest. But this is disinformation of another kind and it leads to the creation of jargon.

The second trait follows on from the first. It is a broken vision of the world. Everything is accidental (a direct and inevitable result of event-by-event information). There is Chernobyl, a famine in Ethiopia, war in Nicaragua, war in Lebanon. Nothing is correlated or thought of in a coherent way. Biased judgments are constantly passed.[5] Each event is considered in isolation. Chernobyl does not cause us to question our nuclear program. The only problem is knowing where the "cloud" is going. There is a similar refusal to consider the probable consequences of such a group of events. When the same events happened ten or fifteen years ago and had specific consequences, one might at least draw out possible consequences.[6] But no. Excess of information goes hand in hand with a culture of forgetting. The mass of information produces a blind life with no possible roots or continuity.

The third trait that results from this obsessional information is that we become exclusive consumers. The consumer society has been condemned enough, and we have heard enough criticisms of the consumer attitude and enough exhortations to be active and re-

4. J.-C. Simon, *L'Éducation et l'Informatisation de la société* (Paris: Fayard, 1981). An interesting point is that Simon is himself an ardent advocate of the information society.

5. For an example see J. Ellul, *Un chrétien pour Israël* (Monaco: Rocher, 1986).

6. As I tried to do, with no success, in relation to the crisis in South Africa.

sponsible. This is all good. But we need to look at the causes, and the first and decisive cause is excess of information. I do not say advertising. What I have in mind is the information spread by radio, television, and newspapers. We cannot absorb this information. We do not inform ourselves. We are fed information. The result is a general attitude that we do not think we are capable of caring for our own needs. The commercializing of life is not an end. It is experienced as a means of gaining information and acting upon the world (an illusion). We become consumers because what ought to produce our initiative is inhibited by excess of information. "Consumers are not the decision makers. The decision makers are not those who pay. Those who pay are not consumers." Information leads to obligatory consumption in the same way as suburban living leads to the obligatory use of the automobile. The lack of consumer power has often been stressed. Consumers take what the producers give them. But this lack begins with a failure to find out what is useful. The mass of unimportant information which is circulated produces the passive consumer attitude. Educated as they are, consumers constantly want more. It is only thus that they feel they are alive. They swallow the information even though they later disgorge it and seek something new, as they also seek every technical innovation that is presented as indispensable and redemptive.

All the previous traits converge on a fourth, namely, a confused sense of impotence. What do you want me to do in face of the disasters of which I am incessantly informed? There are two sides to this. The first is the one that I have just indicated. What do you want me to do about the war in Lebanon or the famine in Ethiopia? I will no doubt be asked to sign petitions and support relief agencies and Amnesty International. But I have so often been misled; I have signed petitions with deceptive aims and been cheated by associations that were supposedly giving aid to the Third World. A feature of this information is that I cannot verify personally that what is being done corresponds to what I am told. Because of the confusion of causes and arguments, I abstain.

The second aspect takes us deeper. The infinite multiplicity of facts that I am given about each situation makes it impossible for me to choose or decide. I thus adopt the general attitude of letting things take their course. This is one of the most essential orientations of Western society. But the course that things take is essentially that of the process of technical development, as may be seen in numerous political and technical texts.[7] We thus arrive at a formula which seems

7. An exceptionally clear example is the report *Prospective 2005*. Sept

to be a veritable law of our society: The more the number and power of means of intervention increase, the more the aptitude and ability and will to intervene diminishes. We think that we live in a society that is open to "conception," but in reality I think that an apter way of viewing our society would be as a society of "contraception."[8] Information is the main carrier of contraception.

2. Television

We can hardly engage in detailed study of television.[9] We shall look at it only from the specific angle of this chapter. Television is one of the chief forces that exercises fascination in our society. We have only to watch little children in front of a television set. Its power to fascinate is much greater than that of the cinema.[10] We may also quote the hours spent in watching it (4 hours a day on average in France, 7 hours in the USA).[11] These figures give us some idea of its influence on ideas, opinions, and political orientation. On this level television has much more power than any other medium. It affects the psyche and the personality. It is the great agent of transition from a society of writing to a society of pictures.[12] But this can be taken in two ways, or rather one can see two kinds of consequences.

There is the orientation of Marshall McLuhan and that of Guy Debord. We are thus told that the Gutenberg era is now over and outdated, so why not accept it? We are entering the society of the spectacle.[13] As regards the first orientation, intellectuals who accept progress argue almost unanimously that printing led us out of the oral society to the society of writing, and that this produced tremendous

---

Explorations de l'avenir (C.N.R.S., 1986). We shall discuss this matter in the chapter below on technological terrorism.

8. Cf. the pill, a technical tool that is the product of a conception but that results in contraception.

9. See the fine passage in M. Henry, *La Barbarie* (Paris: Grasset, 1987), in which he criticizes television and shows that it is destructive of culture.

10. Cf. D. Brethenoux, "Étude de la réception télévisuelle" (dissertation, University of Bordeaux, 1985).

11. For children under fifteen the weekly average is 18 hours, so that some children are watching as much as 25 hours.

12. Cf. J. Ellul, *The Humiliation of the Word*, tr. Joyce Hanks (Grand Rapids: Eerdmans, 1985); Abraham Moles, *L'Image communication fonctionelle* (Casterman, 1981); *La Recherche* 144 (May 1983) (a special number on "La Révolution des Images").

13. See G. de Broglie, *Une image vaut dix mille mots* (Plon, 1982).

intellectual and cultural progress (a point that might well be contested!). But now we are taking a new step. The new tool is calling us to move on from the society of writing to that of the image. We have to adapt, and we may expect equally important new progress. The door is opened for a new and no less vital and expansive culture. It is all waiting to be invented.

I would like to temper this enthusiasm by taking seriously McLuhan's own formula: "The medium is the message," or "Massage-Message." What this really means is that television has no message apart from itself. It does not transmit anything, whether information, thought, or artistic creation. It is itself the message. What it implants in us as message is itself. The pictures that it presents have no meaning. This is why they must be short and striking. Dancing is more televisual than yoga, a papal visit than meditation, war than peace, violence than nonviolence, the shouting of a charismatic leader than reflection that expresses ideas, conflict and competition than cooperation. Ecology does not go over well on television. Non-messages go over best. All that remains is a general haze out of which only the screen itself emerges. We are given no information about reality.

There is in fact no information on television, only television itself. An event is not news unless television carries it. One moment, some weeks, it is excited about Biafra, the next about Cambodia and Pol Pot, the next about the Boat People, the next about Israel, the next about South Africa. We are shown the same pictures again and again (a process that is becoming more common all the time). Suddenly millions of viewers are worked up about injustice in Israel or South Africa. But then it is gone in a flash. The situation, of course, is still the same. But television cannot follow it any longer. On television everything has to be very simple (pictures!). There are the good guys and the bad guys. Again, viewers want something new. Only what is new is interesting. Things must not go on too long, even tragic things, or they become boring. There is a total confounding of the important and the new. The taking of hostages is very important. The progressive seizure of Cambodia by Vietnam is not important because it is not new. In effect, viewers are watching a show. There has to be action and it must not go on too long.

When television stops dealing with a question, the question no longer exists. This is what shows us that television is itself the message. Television does not communicate information. Information communicates television. We are merely consumers of information, that is, of that which television has dramatized. This is why the televised message is really a massage of the brain, of knowledge, of memory. This massage causes all that we have seen yesterday to disappear.

An expression that caused great excitement was that television is changing our world into a small village. I do not accept that. But I agree with Debord's concept of the society of the spectacle.[14] This has given rise to many misconceptions. It has been taken in a very simplistic and not very interesting way, as though it meant that we are living in a society in which there are many shows (television, video, film, advertising). That is not the issue. Debord has something else in view. His point is that the media are transforming real life (politics, wars, economic problems) into mere shows; for us as viewers the real is a picture and a show. Our own lives become shows. J. Piveteau saw where it all begins.[15] Television is a screen between us and reality. Viewers think of television as the screen on which reality projects itself. The sense of immediacy, of being present, means that we are there at a bombardment or an accident. Television plays with reality. Between us and life it sets up a screen on which shadows act. But we take the shadows for the reality. This leads us to equate all reality with these shadows on a screen.

We thus have a detachment from reality which we have to take note of if we are to understand the actions and opinions of people in the West (at least in the class denoted—not all people). This phenomenon is the key to what we said earlier about disinformation. "Disinformation results from confusion between reality and image" (Piveteau). We are detached from reality because time is eliminated. Television is instantaneous. This is extolled because it means that we are present everywhere, though it also means that the pictures grow old after a week or two.

There is a change in our relation to time. Delays are eliminated. So, too, is duration. As we said above, an event that goes on too long is not interesting. It is explained that if something has to be said on television, it must not last more than a minute and a half, or the viewers will no longer follow it. Thanks to television, the instant has become tyrannical.[16] When election results are given, the event is not the outcome itself but the triumphant fact that it can be known at once. Speed of communication is the valuable thing. It shows how capable

14. Guy Debord, *Society of the Spectacle* (Detroit: Black & Red, 1977); cf. J.-C. Missika and D. Wolton, *La Folle du logis, la télévision dans les sociétés démocratiques* (Paris: Gallimard, 1983).

15. J. Piveteau, *L'Extase de la télévision* (INSEP, 1984).

16. Piveteau uses the happy comparison of quartz watches, which give us only the instant and remind us that the instant that follows is different.

one network or another really is. Television suspends time. It does not make me attentive to any lasting reality.

Television also abolishes the relation to space. I see what is happening everywhere. My interest is in what is on the screen. A little experiment is very significant in this regard. Enable the inhabitants of a large complex to televise programs that their neighbors can see, not special programs, but scenes from everyday life, and immediately people who are completely indifferent to their neighbors will become passionately interested in what appears on the screen. Interest is aroused only when something is on television. If something has been filmed and is put on the screen, then it is important and interesting. What counts is not what is seen spontaneously, but what we are given to see.

In these conditions there is naturally no reason for human relations to be formed. Thus the idea of the global village is a snare and delusion. It is not really true that no matter who may talk, only that which is put on the screen exists. But where there are no human relations, there is no longer any participation (in spite of the supposed participation of viewers in televised games). Television sells illusions. It has to try to give an illusion of participation, though this is strictly impossible (J. Cazeneuve).

I watch, but necessarily, because of the screen, I remain at a distance. This becomes my general attitude. What I see on the street has the same reality as what I see on the screen. When I meet a beggar or one of the unemployed, I look at this person in the same superficial and disembodied way as I do at the living skeletons in the Third World that television shows me from time to time. We have here an extreme detachment from reality. The living world is confused with the tele-vised world. In other words, television exploits a profound human tendency, especially in intellectuals, which Kierkegaard analyzed under the "category of the interesting." But the attitude of life which was acquired by an intellectual exercise has now become an automatic product of the external world. It expresses disengagement from reality and is obviously balanced by an engagement in unreality that television induces.

It is not by chance that the idea of engagement in politics came to light at the same time as television. It is a direct product of television. Politics is given the status of reality, whereas daily life and interhuman relations are scorned and ridiculed. Along with the ridiculing of "char-ity," sexual liberation and the abandoning of all morality are signs of detachment from reality. I realize that cultural and scientific broadcasts will be adduced in response. But in fact these programs obey the essential law of television: They must go quickly, not give long explanations, contain dramatic turns of events, dramatize; the setting and encounters are the important thing. A kind of continuum is established between reality and fiction. A simple public can gather less from them (at least when it is allowed to voice its opinion) than a public which is already living on images and the production of rhetorical and metaphysical images. However that may be, the complex detachment from reality is a decisive factor in the fascination of people today.

I need only add two or three remarks on this central point. I stated that there is no real message. I ought to have been more precise and said that for me it is a question of clear and constructive messages, messages which can be expressed and discussed. Undoubtedly, there are messages. This is what is so dangerous. For these are not messages that can be conceptualized. They are subliminal (according to the thesis of Bretonnoux). Television acts less by the creation of clear notions and precise opinions and more by enveloping us in a haze. We must certainly not overrate the importance of the subliminal. We must not think that a film or clip can have a kind of magical effect, that it can introduce into the subconscious an image which will have determinative effects. That is clearly not the point. As Jézéquel says, there is less and less talk and more and more manipulation. Electronics can greatly alter the picture and take away any message. Television puts us in a world of falsehood, trickery, and deception (e.g., singers with their wooden microphones!). (We should read the pages of Jézéquel's remarkable work *Le Gâchis audiovisuel.*) But given the power of pictures to shock and to create impressions (and this is the only truth in the saying that one picture is worth a thousand words!), when thousands of pictures are forced upon us, upon our subconscious, and all with the same basic message, this power finally becomes a determinative component of our attitudes and opinions.

It is thus that television and television advertising are, as I see it, determining factors in the eroticizing of society and the growth of violence. No film or magazine ever had the same influence. For years now we have had erotic scenes and acts on television every day. The same is true of scenes of violence. Every television show must have one or the other. It seems that no show can be put on without a couple making love or a fight or a murder or a science fiction monster.

Again I am not referring to merely one effect or one show or one bit of information. It is the constant repetition evening after evening which forms the existential mental setting day by day for viewers. I am not making any moral accusation. What we need is awareness of the reality. I know the reply, namely, that we must cast off taboos and keep up with the times. Unfortunately, this means that television controls our times, shatters the taboos, and creates the public need for shows of this kind. When people argue that the television industry is simply responding to the public taste and demand, they forget to add that this demand is the creation of television. Television is responsible for the kind of general mental climate of which violence and eroticism are a part.

We should not be surprised, then, at the mounting violence and exaggerated eroticism throughout society. But no one worries about it; it is not interesting. What counts is the technical improvement of the equipment. This is the only question. As one specialist among hundreds put it, when we are dreaming about multiplying channels and intensifying the consumption of pictures, it is time to think about improving the channels and pictures. We thus have the progress toward high-definition television, flat screens, the spread of compact discs, direct television satellite, etc. Jézéquel is very explicit. The more we spend on increasing the number of channels, the less we spend on basic research and artistic creation. The statement to which I just referred is quite plain. We have to make the public consume more images. There is no real public demand. This is the work of technicians on the one side and the organizers of society on the other. We constantly find them talking about "having to." This kind of talk is by no means innocent. The reference to a need of images is a falsehood. People are being plunged into an artificial world which will cause them to lose their sense of reality and to abandon their search for truth.

There is no need for the heavy apparatus of Orwell's *1984* or the subtle biological manipulations of Huxley's *Brave New World*. It is enough to put people in a fictitious world. What we find in television is only the beginning of what we shall find throughout the rest of this study. It is quite wrong to say in reply that television will be personalized with cable and the possibility of creating private systems. The debate about ownership is grotesque, for so far as society and individuals are concerned, it is all one whether television is in the hands of the state, the networks, or large financial groups.[17]

Nor does the argument carry much weight that videorecorders confer much more freedom by liberating from bondage to scheduled programs. An inquiry by Piveteau seems to show that this is true only for films and scientific broadcasts but not for sporting events or political events. But video in any case does not free us from television.

17. Experience with cable simply gives evidence of unimaginable deterioration. Cf. J.-P. Jézéquel, et al., *Le Gâchis audiovisuel* (Éditions Ouvrières, 1987).

It simply adds to the hours of viewing. We have our normal hours and then an extra hour or two for shows that we could not see when we wanted to do so. This gives us so much less time to live our own lives and so much more time to live vicariously. That is the result of video.

A final question is whether average people really want this or like it. I have already quoted Brethenoux, who shows that the television viewers whom he studied almost all found in television an amenity, habit, and dream, but did not think it important. They were not criticizing the programs or content, but the thing itself. Yet they recognized that they could hardly do without it! An experiment conducted by Télérama seems to confirm this detailed study. In 1986 Télérama asked twenty families to do without television for a month. The inquiry was strict; Télérama even took away the sets. Naturally, the effect was a shock. Suddenly, there was so much empty time. When there is time to spare, it is so easy to turn on the set and take what is available. The experience of empty time which we have to fill on our own by conversation, by relations with other people, by reflection, or by reading, has become a traumatic one for our generation. These people were suddenly confronted by an inner void. They had nothing to say to other people. The details of daily life were not interesting. They themselves were empty. It is this existential void that has been, throughout human history, the driving force behind all cultural and social creating.

The results of the experiment were conclusive. The families became "marginal" in their neighborhood, but within the month they had a "feeling of freedom and holiday."[18] They came to see that television insidiously enmeshes us. A few could not stand being deprived and visibly pined. But most found, for example, things in their children that they had not known before, having almost completely ignored them. One couple that had no television and was given one quickly found that the result was less reading and less listening to records. But those who did without television found that there was much more talking between spouses and between parents and children, and that they now had the time to see friends and to read that they had not had previously.[19] At the end of the month some families took back their sets with no pleasure and put them in a corner. The most

18. Cf. the accounts by J.-C. Raspiengeas, *Télérama*, April 1986; and C. Humbolt, *Le Monde*, April 5, 1986.

19. One extraordinary statement from a wife was that after ten days she realized she had been living for ten years with a man whose qualities she had forgotten.

remarkable thing to me, however, is that when the couples were asked whether they would let their sets be taken away for six months, 19 out of the 20 agreed. In a broader poll 37 percent in France said that they could live without television, and 51 percent thought there should be no broadcasts one evening a week. But the majority was ready to accept television for the sake of the children. To me this is the decisive point. Children are in fact fascinated by television. They live in it and are shaped by it. They demand this drug. One of the subjects of the experiment stated honestly: "Without television, I do not exist." An English sociologist quoted in the inquiry showed that five years are needed to suppress the reflexes caused by television and to restore earlier cultural habits.

In conclusion I will quote F. Fellini's remarkable analysis. "Television has mutilated our capacity for solitude. It has violated our most intimate, private, and secret dimension. Enslaved by an invading ritual, we fix our gaze on a bright screen which casts up billions of things that annul one another in a dizzying spiral. Peace comes only when we turn it off. At eleven o'clock or midnight great fatigue descends upon us. We go to bed with an uneasy conscience, and in the night, with closed eyes, we try to renew, like a broken thread, the inner silence that was ours."[20]

3. Telematics

A third carrier of fascinating information is the most modern, telematics.[21] We have to remember that this is only one of the new media (television satellites, independent local radio stations, cable television, video, etc.) which are in competition with the old media of state radio and television and the press. We have to ask the following questions: Are these new media wanted by the public? Or by those who want to express themselves? (and are these representatives?) Or are they imposed by technicians, financiers, or the state? If they are imposed, for what purpose? For information? For pluralism? For money? To serve the

20. F. Fellini, *Le Monde*, Jan. 1986.

21. Among the many books cf. especially the report of Simon Nora and Alain Minc, *L'Informatisation de la société* (Paris: La Documentation française, 1978); M. Ader, *Le Choc informatique* (Denoel, 1984); M. Marchand and C. Ancelin, eds., *Télématique, promenade dans les usages* (Paris: La Documentation française, 1984); F. Poswick, in *Une société informatisée—Pourquoi?* (Presses universitaires de Namur, 1982); M. Mirabail, et al., *Les Cinquante mots clés de la télématique* (Privat, 1981); F. Holz-Bonneau, *L'Image et l'Ordinateur* (Aubier, 1986); also, of course, the government report on the plan for technology, especially the computer.

public or to serve political or financial groups? In no study have I found even an outline of answers to these questions. Telematics has been developed because the means were there to do it, and this arouses a passion to do it. That is all.

But first we must make clear what we mean by "telematics," which is not at all obvious. The reference is to the totality of services (apart from the telephone and telegraph) which can be secured by those who use a network of communications that enables them to get information or to do certain things (e.g., consult data files, commercial and banking transactions, teletex, video, or telecopying). The commission on the vocabulary of computers refers to services which, having their nature or origin in computers, can be furnished by means of telecommunication networks. If the computer is the sum of means of handling information (acquisition, checking, arranging, storing, calculating, and transmitting), the tele-computer consists of the techniques and related applications for the linking of telecommunications to the computer. The reference, then, is to all the new services linked to telecommunications that are in turn linked to computers. It is a subsidiary of the tele-computer, at the junction of telecommunications and the computer, but the information sent along the networks is of interest not only to computer specialists but to average people. Its influence is daily, and it touches many different subjects that can inform different social strata.

This equipment has made possible a veritable explosion of services. France by 1985 had 800,000 videotex terminals. There are more than a thousand services in operation. Some 8 million calls are registered a month. The services are accessible to the public twenty-four hours a day. The number of daily requests for information amounts to 1,700. Most of them come from banks, insurance companies, municipalities, chambers of commerce, the press, information services industry, and transport. What is transmitted is not an inert current but information, that is, power.[22]

The stake, then, is not just the adoption of new technical tools. The very notion of information is called into question. The result is that economic, cultural, and political choices have to be made. This leads us to the real question: What is at stake when information is handled by the computer and made generally available? Our interest here is not in the different systems themselves, nor in the sectors in which they are applied. The question is much more basic. To startle readers, I will refer to a headline in *Le Monde* which spoke of telematics

22. See Nora and Minc, *L'Informatisation de la société*.

being an explosive force in education; that is splendid and we ought to rejoice (Le Monde, April 17, 1986). We must seriously ask what is really at stake. Mirabail thinks it is right to talk about stakes in this context, but what is at stake, he says, lies neither in the field of hopes and fears nor in that of profit or loss. At stake here is no less than the world, humanity, society, the different experiences that we can have. At stake is the profit of a civilization preceding its transformation, which is viewed as inevitable, as we shall see later.

Telematics follows the computer in its radicalizing of everything that went before. It systematizes the reorganization of learning and revolutionizes conceptions and methods of work. It makes a break between two worlds and prefigures the second of these. But we can say nothing about the latter point apart from what the infrastructures of new technologies foreshadow: data banks, communication networks, telecommunication services. The risk is that of an indefiniteness of thought which is unable to define the meaning or nature of new social and cultural objects (Mirabail).

Handling information raises immense problems. Political problems: Who releases information and to gain what power? Economic problems: What are to be the new materials, industries, jobs (at the expense of what others, and for how long is profit to be discounted)? Sociological: Who is to profit from the new potentialities? Psychological: What will be the changes in behavior? Cultural: What cultural changes will there be?[23]

But first we must analyze the new way of handling information. At this point we rely on Françoise Holtz-Bonneau.[24] The processes move on from the very small (the reduction to minimal units) to the very big (the exhaustive). The main added value is strictly material. Access is possible at a distance. There is first a necessary process of reduction to minimal units of information. Prior to handling there has to be separation and classification with the strictness, exactness, and logic without which computers cannot function. Everything relates to computers. (Although they do not constitute new information, they demand certain processes if they are to be operational.) The result is reduction and taxonomy, but this raises the question what is to be left out. Condensation is required. Each page in data processing must contain the maximum number of data. This entails the risks of linguistic impoverishment, of a loss of the riches of vocabulary and the suppleness of thought (but is it a matter of thought in information of

23. See Poswick, in Une société informatisée—Pourquoi? pp. 291-95.
24. F. Holtz-Bonneau, L'Image et l'Ordinateur (Aubier, 1986).

this kind?), of schematism and rigidity. But does not the very fact of condensation enable new types of information to emerge? In fact, condensation works best for factual information. But the constraints of condensation can result in the distortion of information (effects of evaluation that can hardly be changed).

If, however, there is condensation in the shaping of information, one of the features of the telematics system is that it is so exhaustive, that its ability to inform is so capable of extension. Being exhaustive, it offers two new types of possibility: data banks and means of access to data that exist elsewhere on other support systems. But the risk is increased of being drowned in a flood of information that we cannot assimilate (as noted already). Furthermore, the possibility of automatically having the proper answer to my question means that I do not look for the answer in the mass of documents. Things go too fast. In true research, we all know that three things happen: We come across "lateral" information which modifies our view of the question; we find a document which by association suggests another problem; we have a new idea by intuition. None of these can happen, however, if, having formulated the problem, I receive a solution. In other words, the whole contingency of research disappears. Data banks produce blindness, making sense only in the short term. Power rests on the quantitative force of these data banks, namely, whether this or that country possesses 80 percent of the global economic data, etc.

But what about the qualitative side? Does not a country risk arriving at a time when the quantitative empire breaks up because it has no sure qualitative foundations? Holtz-Bonneau asks such questions, but I am not sure that we can follow her suggested answers. In fact, "telematics involves a reorganization of all that can be expressed and therefore of all discourse that can deal with meaning. The telematics revolution carries with it a linguistic revolution. Thought is trapped in the net of new language that gives form to the system of an inner emptiness with no existential significance or claim or history."[25]

It is an effect of the boasted "transparence" of the computer that it achieves convenience of usage but only by weakening research into ends and processes, that is, thinking which is not yet reduced to logic. The telematics system is significant in itself but has no significance beyond what it contains (services and sociocultural integration). The interference of symbolic language, of systems of signs, actually establishes a time and place for us, a world of meaning (even though we cannot attribute any meaning to it!), the opening of a history (not just

25. M. Mirabail, et al., *Les Cinquante mots clés de la télématique.*

the descriptive passing of time), and the possibility of true symbolizing (though we cannot admit that technique covers all anthropological questions or that videotex terminals cover all human imagination).

With its exclusive power telematics constitutes for us a world that is characterized by the construction of its data and by a language whose transparence closes our eyes to contingency and to the occasional significance of relation to reality. It robs history and morality of their weight and merges time and distance into the instantaneity of documentation. It results in great inequality between those who make full use of its services and the rest (90 percent) who remain outside except to gain a little information, for it has to be adapted to the state of knowledge of different groups and thus reinforces social distinctions.[26] I am astonished that all those who use this global instrument of telematics act as if they have understood it so long as they can manipulate it skillfully and no more.

I will leave aside the familiar issues of freedom,[27] identification, identity cards, the danger of a unique signifier, the risks of centralization, social immobilization ("a place for each but each in place"), and the attempts (unsuccessful in my view) to safeguard privacy and to control and limit the use of telematics. I will simply quote Mirabail to the effect that the widespread use of the computer raises the problem of our very identity. In this new mode of culture and communication our only real identity is administrative. Linking society by telematics results in a levelling out of differences. We are far from the convivial society, the informational agora[28] in which difference is the basis and point of exchange.

P. Lemoine finds in the identity crisis one of the main cultural effects of computerization.[29] A transformation of identity results from the modification of administrative identity brought about by computerized technologies. The facility of communication has this result. Satellites bring better access to networks and these (ideally!) make possible interaction and exchange. This is why there is all the rhetoric about the "relational society" or the "friendly computer." But we always come back to our simple questions: What relations? Between whom?

26. This confirms what we have said about interested persons, i.e., only special groups.

27. Cf. the basic work of A. Vitalis, *Informatique, Pouvoir et Libertés* (Economica, 1981).

28. This recalls our analysis of McLuhan's "communicational village" and television.

29. P. Lemoine, "L'Identité informatisée," in *Les Enjeux culturels de l'informatisation* (Paris: La Documentation française, 1981).

Human exchanges and relations are not isolated, independent, transportable commodities. Two corporations can exchange financial or commercial information, but this has nothing whatever to do with human relations. In fact, there is a temptation to "gloss over the difference in nature between human communication and communication through these artifacts."[30]

There are indeed few points of contact between ideas of identity, relation, and communication in the sense of human communication and the same phenomena at the level of machines.

As for Thélème, the first "convivial telematics" which is supposed to overcome the limitations of time and distance by messaging and teleconferences, it pretends to be telematics for the people, but this is a bluff. For one thing the people at large do not need conferences. The millions of workers (apart perhaps from union leaders), farmers, small-business people, employees, etc., do not have any use for this tool. As for messaging, it is useful for those who do a great deal of business by messenger or phone, but not for others. The system is also very costly. Even though neither the monthly rent of a Minitel (100 francs, or about $20) nor the subscription to Thélème (100 francs for individuals, 1,000 francs for groups) is expensive, the charge for connections (and this is what counts) is very high (230 francs, almost $50, an hour, which in practice might well be each day).

A further point is that words are typically misused when there is said to be "community" between people who do not meet and who communicate only by teleconference with the help of computers. It is sheer bluff to talk about fellowship or community in such circumstances. It is audacious to say that teleconferences are the same as clubs or societies or bistros or salons. We have here an inhuman outlook which can abstract from clubs, etc., all that is specific and simply retain the fact that one can communicate. There may well be conviviality in them, but there is also selection and the exclusion of those who do not keep the rules, as R. Klatzmann points out.[31] Klatzmann is right to talk about "electronic nomads." These living contacts between people far removed can become more real and significant for many of them than the real communities in which they live, and they thus change into electronic nomads with no roots in a place or a human setting.

With analysis of the stakes of telematics, there has been much enthusiastic technological talk to the effect that telematics will

30. A. Giraud, J.-C. Missika, and D. Wolton, *Les Réseaux pensants* (Masson, 1980).

31. R. Klatzmann, "Thélème," *Autrement* 1982 (special number on computers).

guarantee our recovery from the economic crisis by producing networks of communication and information that are typical of modern
humanity and that will necessitate a new and universal culture. When
France adopts a new planned strategy to keep pace with the USA and
Japan, this will lead to a concentration of national resources on certain
objectives and will thus socialize risks and protect weak industries
against market uncertainties.

It is indeed very hard to judge and to decide, to weigh up the
pros and cons. Technique moves much too fast for reflection, which is
complex. In part, reflection must be sociological in the broader sense,
studying the effects on social structures, on relations, on groups, but
also on language, and consequently studying intellectual, cultural, and
psychological effects. But reflection has also to be political, for nothing
can be done without financial support from the state, and the political
effects of telematics also need study. Finally, reflection must be
economic, as is clear. But while specialists, politicians, and intellectuals grope their way with only inadequate data at their disposal (for these
techniques have not yet been applied on a large scale), technique itself
forges ahead with prodigious speed. It thus seems that in this situation
the reversal of attitude took place in 1982 when Gérard Thery said
about Teletel that the time was past for hashing and rehashing the
question and the time had come for systematic distribution of videotex.[32] In other words, enough discussion—it is time for action! Let us
go ahead and see what happens. We could apply to telematics (and
computers in general) what Piveteau said about television: we have
here a technical device which has been nationally developed and
deployed without the least prior study or reflection.[33]

A particularly fundamental point seems to be that these techniques come equipped with some irreversible effects. People, society,
language, and thought processes inevitably model themselves on it. If
we ever perceive that we are on the wrong road, it is impossible to turn
back. What is done cannot be wiped out, for we are now accustomed
to living in a world of pictures and trivia, and we cannot be taken out

32. Cf. Charon in Marchand and Ancelin, eds., *Télématique, promenade
dans les usages*.
33. I had myself a small experience of how this can happen. The government
offered the Reformed Church a half-hour broadcast on Sunday mornings. Everyone
was full of enthusiasm. I alone suggested that it might be as well to do a preliminary
study of the congruity of this medium and propagation of the gospel. I was violently
criticized as one who rejects progress, modern methods, openness to the world, etc.
Not half an hour of reflection was given to whether we should offer a gospel show
on television.

of this world. As Roqueplo says, our heads are full of codes already. According to C. Durieux (*Le Monde*, Feb. 1981), "the only real problem is to codify democratically the use of these means of expression." This is a revealing statement; the media are not said to be the problem but their "democratic codifying." The problem is the more serious because this expression has no meaning. For how can we codify (i.e., regulate in stable fashion) something that is always changing, that takes unexpected forms and has unexpected applications and that is, therefore, always alterable?

We have seen with Distler and Bressand that law cannot regulate technique. This applies more than ever at this point. No court will ever sanction any departures! And when Durieux talks of codifying "democratically," what does that mean? Asking the people for advice? That is unthinkable when the situation is so complex and the equipment so difficult. Asking the elected representatives, the politicians? But they, too, are incompetent. The important thing, however, is calling this the "only" problem. It is stupefying that other things can be dismissed so cavalierly.

In fact, we have to realize that we are in a process of universal transformation without really knowing what is happening. This confirms our older judgment regarding the autonomy and collective supremacy of technique. By way of compensation we might consider, as Piveteau did in the case of television, that in order not to seem dispossessed, people deify the technical device. It is universal and spectacular; it defies my attempts to master it; it performs what would usually be called miracles; to a large extent it is incomprehensible. It is thus God.[34] We are justified if we give up any attempts to control it and simply ask for its services. The true gods today are not economic but technical, and they take their most direct form in television and telematics.[35] This is in keeping with absolute fascination.

34. As Piveteau puts it, television has itself become a religion, a kind of God. Like God it is everywhere, it is watched everywhere, it speaks everywhere. It is a practical god in a human dimension, much less mysterious than our older God. By way of advertising it has a revelation for humanity. Christianity should see that it has lost the two messianic messages that are at its heart, the one regarding the Savior, the other regarding the world to come. As a religion, television also has rites, grand liturgical ceremonies (*L'Extase de la télévision*).

35. For a fine analysis of telematics and divinity see Mirabail, et al., *Les Cinquante mots clés de la télématique*. Telematics reminds us of Babel, Golem, and Hermes. Its divinity lies in the rapidity of its progress and the multiplicity of its spheres. Thus computers can print three pages a second (R. Myers, "Les Imprimantes d'ordinateurs," *La Recherche* 123 [June 1981]).

CHAPTER XVIII

Advertising

We cannot pretend to offer here an exhaustive study of the phenomenon of advertising. Since the studies of Dichter and Packard there have been hundreds of books on the question.[1] But fundamentally there has been much more concern about the scientific character of advertising and its falsity than about its socioeconomic character. Inasmuch as the important thing for me is the relation between individuals and the technical system, my emphasis will be on the last aspect. Problems of motivation, studies of veracity, and inquiries into the source of needs are not vital in this context. Everyone seems to be agreed that we must dismiss any idea that advertising is not needed. But the need, like the style of arguments in its favor, is acquired.

Advertising is a psychological act which claims to be based on a science. At present it is there and I need not concern myself with its reality (except to thwart it, like Holtz-Bonneau). It is itself a technique. All economists agree that it is indispensable to promote sales. Mass production means mass sales, and these are not possible without advertising. An interesting point, however, is that studies from only ten years ago are all out of date in comparison with advertising today. Something has happened that has changed everything. For one thing, there is the intervention of the computer, and for another thing, there is the fact that what has to be sold today is always technique. What was said earlier is all true, of course, but there has been a complete change of scale.

1. See, e.g., Vance Packard, *The People Shapers* (Boston: Little, Brown and Co., 1977); E. Morin, *Sociologie* (Paris: Fayard, 1984) (a chapter on advertising); Françoise Holtz-Bonneau, *Déjouer la publicité* (Éditions Ouvrières, 1976); G. Durand, *Les Mensonges en propagande et en publicité* (Paris: PUF, 1982); idem, *La Publicité* (Paris: PUF, 1984) (with bibliography); O. Reboul, *Le Slogan* (Paris: PUF, 1975); idem, *L'Endoctrinement* (Paris: PUF, 1977).

Advertising budgets illustrate this point. In 1975 the advertising budget in France was estimated at 7 billion francs, but in 1986 it had passed 70 billion. This is understandable when we recall that making advertising images by computer costs one million francs per minute. "Advertising keeps television alive. Government television derives money from the temptations that it sets before me so as to have the means to warn me not to succumb to the same temptations. It is a rather sophisticated system!" (Piveteau).

Advertising plays an indispensable role in financing the audio-visual system. It provides 25 percent of the resources of the public networks in France in spite of restrictions. In 1984 it contributed 3 billion francs (about $600 million).[2] In countries with no restraints it contributes 40 percent. We recall the great debate about private channels, but there had to be development at all costs, the conflict between culture and technique being nowhere revealed more clearly. Those responsible for public channels reckon that restrictions impede the growth of television (a crime against technique) and that the pressure will become so strong that there will inevitably be deregulation and a full development of the economic market.[3] But since we must not lose sight of humanistic and cultural interests, it is usually explained that "audiovisual advertising has become a means of cultural expression with its own code, language, and rites. It is the emergence of the consumer society."[4]

Finally, the press and television must be freed at all costs, rigidity must be ended, and the greatest flexibility must be granted. This is a general requirement of all technique. We will not touch on the problem of the rivalry of the press and television but simply give some numbers. Advertising in the daily press has fallen by 12 percent, whereas that on television has risen by 68 percent and cable television is making inroads of 20 to 30 percent. There will be a corresponding rise in receipts. All this shows that advertising is increasing inexorably and that any impediment seems scandalous, not merely financially, but because it will retard the whole economic-technical system.

Inquiries into motivation have also changed their character

2. The advertising budget on French television in 1987 was 8 billion francs (about $1.6 billion), and the cost of producing one hour of advertising was 42 million francs (about $8 million).

3. After these pages were written, deregulation came and advertising has achieved total domination over private television. Private television serves those who pay for it. Frequent reiteration of messages eliminates critical reaction. Advertising can also decide whether a show is put on or not.

4. M. Le Menestrel, *Le Monde*, March 1984.

since people have to be made to buy more and more useless gadgets, as we have seen. What might have been regarded as rational or Freudian motivations will no longer do (in spite of the importance of eros and freedom). The public inclines increasingly toward the extraordinary. We shall return to this matter. Everyone can see that the style of advertising has changed. It is not only the use of new technical means, of computer images, of new cinematographic processes, nor the changing of subjects and the ways of treating subjects. There is an altogether different matter.

Advertisers are clearly aware that they must present new and unexpected products in different ways and that astonishingly new methods are at their disposal. They need to introduce into their goods libidinal qualities that are not intrinsic to them (E. Morin). In this way they will give rise to unconscious urges appealing to the Freudian id. They also need to follow the route of individuality, of the cultural construction that is called personality, appealing to the Freudian ego. But there is more than that to the change in advertising and its current role.

The style of advertising has changed because its function has changed and at the same time its status has changed. Advertisers are carried along by a change in the object of their work which has outdated all former inquiries into motivation. They are thus forging a new type of advertising which is adapted to its new status but which is in some sense outside their control.

Thus far advertising has been an indispensable aid to distribution and business, inducing people to buy products. Today, one of its aims is still to induce people to buy products but with the difference that these are not ordinary goods but technical products, sophisticated goods of high technical quality (even though the product be only Coca-Cola!). As technique has changed its status, so has the aid to selling it. Advertising is now the driving force of the whole system. It exercises an invisible dictatorship over our society.

We have passed through the intermediary stage when viewers were progressively transformed into consumers. The transition was not easy. Reflexes had to be conditioned if it was to be made. But today the situation is different. On the one side we have mass production. New high-tech objects are made available to the public. People do not easily see the use of such an object, but they are ready to react as obedient consumers. There has to be mass consumption of high-tech products. Indeed, these are the key to all economic development. The economic success of the businesses which produce this engine of development is necessary for technique itself. As we have seen, technique can continue its triumphant march only if the public follows, buying the maximum number of computers, tape recorders, videos, photocopiers, high-definition televisions, microwaves, and compact discs. For every new technical advance there has to be a public that is ready to buy the latest product. Before winning foreign markets one has to have a secure base in the domestic market. The massive buying that ensures technical development also makes possible the continuation of scientific research.

Alongside state investment there is also investment by big corporations which do not want to risk their money in research unless there is a chance of this opening into a technical production, and this will continue only if there is an assured market. Advertising guarantees this market. This is why I can say that advertising is now (not alone, for the state, too, is indispensable) the driving force of the system: science-technique-merchandise. Previously advertising had the task of making people buy in order to provide profits for capitalists. That is still partly true. In a capitalist system profits are essential. But advertising now has a new status.

This change has entailed a modification of methods.[5] It is not so much a matter of motivating people to buy products, or creating new needs, or making people into consumers. These are still objectives, but it is not by such means that advertising now functions. If one considers technique a system or milieu or "nature," then to make people buy what is offered they must be integrated into the system and made parts of the whole. There must not be the technical world on the one side and on the other the individuals who are to buy its products. As individuals are already part of the system as creators and producers, they must also be integrated into it as consumers.

It is advertising which even in the farming community makes people buy computers, electric milking machines, huge tractors, and other complicated farming equipment (with the result, as we have seen, of disastrous overproduction). Farmers resist propaganda and advertising, so they had first to be integrated into the technical system and then they could be convinced that they no longer wanted to be backward but must buy for their farms the most modern equipment.

In the technical system, then, advertising is a central and dy-

5. Today, e.g., there is advertising for advertising. A need is felt to convince and convert (cf. the full page in *Le Monde*, Dec. 1982). This is hardly the best way to go, since average readers will hardly read it and it is better to approach those who handle advertising for big firms directly. But the aim is to accustom people to advertising and to convince them of its excellence and necessity.

namic factor. Its role is threefold. First, it has the older, classical role of selling products. We have seen the importance of this. Second, it must make new products known. This implies a new orientation. For the new products are not just new brand names of cookies or appetizers. The public has to be initiated into the secrets of new automobile gadgets or irons with fifteen settings, and a simple presentation cannot adequately do this. The public has to understand. They have not only to be shown how to work washing machines, etc., but must also be acclimated to the necessity of technical progress and of the exchanging of an older model for a new one. But the demonstration or explanation must not be boring. It must not clutter the brains of buyers or lose their attention. Advertising that explains too much misses the mark. An example was a union advertisement on behalf of a factory which explained in great detail how efficient it was, how qualified were the workers, how it was a factory of the future, and why it was ridiculous to want to close it. This kind of arguing might be useful around a table but it does not influence the public. It has been found long ago that advertising must not be argumentative.

The same might be said of Marcel Dassault's pieces in *Le Monde*. They might be very amusing but they could not change anything. A scholarly approach is no good. Interested people have to be introduced into a world that is to be theirs and in which the object presented (often furtively) becomes indispensable. A simple example, though not pertinent to my project, might be that of youth, the joy of water sports, the beauty of a young couple. Here is a desirable world, and what completes it is the tonic drink which symbolizes youth. The more serious and profound result, however, is that by innumerable sequences there is integration into the technical world either by direct pictures of technical situations or by an appeal to the advantages of some technical product. The technical world itself is often related to nature; the great aspiration of people today is to be on a beach and to watch television by the ocean.

Advertising on every television channel plunges us into this world (which we know concretely) of an idealized and stylized technique, of technique transformed into a work of art, in which the object for sale takes its place spontaneously. In this regard advertising continues an orientation that it has had for a long time: one of the instruments of social control. Its messages are meant to model lifestyles and attitudes, to adapt them to their setting, which is made up of objects like that which is being sold. The difference today is that this world is now wholly technical and that what were once large-scale techniques have now become marketable micro-techniques. At the same time, as advertising seeks to integrate us into the total system, it also wants to fascinate us. Modern advertising plunges us into a strange, weird, surprising world. It makes use of unexpected statements which seem to be irrelevant and simply stir up curiosity.

The weird element is important for two reasons. First, it compensates for the more stern and austere aspect of technique, and second, it separates the viewers and readers from their ordinary, daily, familiar world, which although already informed by technique is not yet fully technical. Faced with an excess of technique, viewers and readers might retreat defensively into the everyday, the familiar. Advertising, with an explosion of astonishing, seductive, amusing, and interrogative images, fascinates future consumers, who by means of it can enter a world of dreams that might be a little crazy but is still desirable and appealing. Advertising has been greatly helped by computer graphics.[6] Thus in addition to the traditional role of presenting a model of people and of the life they ought to live (which is still a role today), and in addition to the more recent but already traditional role of inserting viewers into the society that is presented, and in which the object to be bought will guarantee a place, advertising now has the decisive role of introducing them into the technical world. An interesting point in this regard is that the French, unlike the Italians, Americans, and British, are not hostile to television commercials. In other countries people do other things during commercials, but the French watch commercials: 75 percent of viewers watch 85 percent of commercials, and 50 percent watch as many as 90 percent. Thus the commercials have their full effect. This is a remarkable indication of the lack of a critical spirit in France regarding television (Le Monde, June 1986).

We have now to bring into the domain of advertising a different factor. I refer to the media's general support of techniques and of technique. There has previously been discussion of indirect advertising when in the course of a movie an actor would hold up a bottle of whisky with the label showing or when leading competitors in various sports would wear clothing or equipment advertising various products, but I have in mind something different. Since 90 percent of the products we are asked to buy are the products of elaborate techniques (even our

---

6. Cf. Jézéquel, Le Gâchis audiovisuel (Éditions Ouvrières, 1987), who points out that advertising has become an art for a whole current of thought, that the production of music videos is very sophisticated and expensive, that the investment is recouped by the sale of videocassettes, and that Shakespeare now counts for less than a box of noodles or dog-food.

foods), there is indirect publicity for these products when there is publicity for the technique itself even though there is no label or allusion to the products. I have in mind the nearly permanent advertising for technique, high technology, etc.[7]

For a whole year I have never seen a news broadcast on any channel without something being said to the glory of technique. I maintain that this is advertising for technique, whether it be the person who does such marvelous things that is glorified, or the nation. What we have here is direct advertising of technique itself, usually by showing the most spectacular innovations, without labels or invitations to consumers, but as a way of enabling viewers or readers to enter the world of daily technical miracles. Open-heart surgery might be portrayed, or the implanting of an artificial heart, or the many uses of lasers in surgery or in star wars, or the various aspects of space exploration, for example, spectacular launchings, the placing of satellites in orbit, linking up with existing satellites, and operations within spaceships or shuttles.

In general robotics is less important. We have daily demonstrations of computers but the robot is not a favorite topic. Perhaps it arouses some fear. Perhaps there is nothing very extraordinary to add, for we are used to automated factories with functioning robots. Perhaps science fiction movies have used robots so much that viewers are blasé. But automobiles always cause a stir. We have to see Formula 1 races and rallies and all-terrain events, culminating with the infamous race from Paris to Dakar.

In general genetic engineering has less success. It is perhaps too repellent, and there is nothing much to "show" apart from the test-tube baby, which arouses no great enthusiasm. I will not go into the pictures transmitted by satellite or new manufacturing procedures. My point is that every day we are plunged into high tech. Along with the news of great political events (both national and international) and important economic and cultural developments, there must be equal time, and perhaps more time, for what is often a long sequence on technique. This shows us better than anything what integration is.

Newspaper advertising for technique does not lag behind. It has here a commercial character, but I would say that it is even more interesting. Thus we find a full page with the decisive heading: "Tech-

---

7. Cf. the remarkable article by A. and M. Mattelart, "La technique est l'évènement," *Le Monde diplomatique*, Oct. 1979, which argues that the technical event is at the center of advertising talk. In the USA there are 20 to 30 technical events per minute in commercials.

nology Is a Global Resource," and we then see ears of grain or giant peas (there must always be a reference to nature). A subheading tells us: "Human development is by way of technology." It could not be stated better. Here is a veritable credo.

Commercials are a favorite field in television, not only for financial reasons but also because advertising is the one type of broadcasting that really allows action and stirs the viewers to action, thus reconciling the watching viewer and the active viewer. For we are no longer divided here into spectators and actors. As advertising sees us, our integrity is restored. We can act tomorrow and buy the product that is extolled this evening. The complete person promoted by television is thus solely the consumer. Consumers are presented with marvelous devices (photoelectric devices, the communicator that helps those who have difficulty with speech to express themselves, photocopiers with cartridges, etc.). The basic idea is that technique is the key to everything. Thus we might read that the steel of tomorrow will come from the computer, for it alone can solve the problems posed in forging new steel that can be put to totally different and difficult uses. We have here a demonstration of the integration of different techniques into one another.

Before taking a look at computer advertising, we should consider a thoughtful weekly which contains an advertisement: "Live in Europe 1." What is the point? We see a small boy eating breakfast, very joyful, and the projection of images: a sinister Arab emir, air raids, and a raging military dictator who recalls Hitler. What a capital broadcast! What charming ideas to put in the head of a small boy! But everyone has breakfast, and it is a matter of proving that this brand name gives energy. It is not enough, however, to depict a dynamic person. We have to have a great deal more, and our symbol is a space rocket and a satellite in orbit. Only technical images are evocative enough to tempt buyers whose minds are filled with space concepts and images. Then there is a radiant couple with electricity sparking from their bodies and the exhortation to go electronic, with an invitation to visit high-fi centers. Finally, there is a double page. On the one side is a handsome young man in traditional dress, stiff, lifeless, with a stupid expression. On the other side is the same young man in casual dress, intelligent, concentrated, and enthusiastic. What has brought about the change? He is now "on" instead of "off" because he listens to a Slim-Line (a stereo radio-cassette). That is enough to bring about the change. The terrible thing is that I believe this is accurate—but it is not a change for the better!

I will conclude with a look at computer advertising in magazines. This is worth a look because the presentations are so significant.

There is the "computer friend"; with such a computer you have four friends in one: one for play, one for study, one for an introduction to computers, and one for family management. How remarkable! It reminds me of the story ten years ago of a Japanese airport with a robot that solitary passengers who were sorry to leave could go up to and have someone take them by the hand and wish them a pleasant journey. We are reducing friendship and affection to a machine that functions.

Here we have one of the very significant aspect of advertising for the computer: it makes the computer familiar and friendly, thus dispelling the mystery and the magic. The computer is your good servant, it will do your pleasure, it is a loyal comrade. The advertising plays on all the emotions of future consumers. Computers are also instruments of pleasure. Forms are drawn and come to life accompanied by a strange music. Thanks to computer X you can be a poet and inventor and will have pleasure. But three advertisements of this kind seem the most remarkable. First, there is the appeal to the emotions, to pure sentimentality. Thus at the time of the success of Spielberg's *E.T.* there was the moving appeal: "E.T. needs you, he is lost without you." We could save the day by using computer Y. But we had to hurry, for the powers of E.T. were failing. On the occasion of a sentimental success a false sentimentalism was thus put to work. Here we are totally in a world of falsehood, dreaming, and illusion. But it works. The second method is to call the computer a treasure. It can solve all our problems, family problems, problems in human relations, health problems. Computer Z will solve them all. I can bring them all to it. The computer, then, is an antidote to doubt. We need have no more doubts. We may have doubts about the existence of God or the truthfulness of our spouse or the morality of our children, but if we submit to the computer, thanks to it we will be cleared of doubts. (I have taken this from public statements.) The one who made this assertion did not consider all the implications of what he or she was saying. Among other things, if there is no longer any doubt, there is no longer any intellectual life.

Between this mythical and sentimental appeal and what follows there is some interesting advertising for Apple computers: "Take a bite and you will be convinced." The computer is not just a machine that helps us to calculate. It also gives us new personal powers so that we can convince others. There is an interesting allusion here to Eve and the forbidden fruit which opened her eyes and mind to all that knowledge makes possible.

We now pass to the second great strand of advertising, the more banal claim that the computer gives us great power and has endless applications. Power: a fine cat's head with fascinating eyes. The engineers of telecommunications see what others do not see. They are endowed with a sixth sense, that of larger and more forward-looking vision. The most powerful center of vector analysis is being put in the service of French industry, we are told: unlimited power which will enable us to deal with billions of data. This power is available not only for industry and big business but also for all who want to enhance their abilities. The computer will make us 100 percent surgeons, insurers, or bankers. Without it we were only half-surgeons, etc. Now we are 100 percent.

But again what is being said has not been fully weighed. We *are* 100 percent bankers. That is to say, we are nothing else. Our whole being is included in our profession, thanks to this technique. We are wholly integrated into the system, thanks to the combination of the computer and advertising. The stress is always on power. Thus we are shown encyclopedias and piles of books. We cannot read and know them all but our computer can do it for us. We can make it our memory, our power of comprehension, our knowledge. One of the great problems in education as I see it is that we are producing people with empty minds who can no longer work out data for themselves but can only manipulate computers.

As for the endless number of uses, the point here is the flexibility of the computer. It has the suppleness of a gymnast. This image relates to the concern to show that computers are human and are not mere machines. They have the flexibility of the human brain, and this means that we can apply them to any use. The most elementary advertising simply lists some of these uses, for example, in ships, cars, or planes. Computers can also test various possibilities by simulation and do away with the need for models in architecture, etc. But there is again an appeal to the average person. The computer will simplify daily life! A double page of cartoons might depict average people in France such as tradespeople, accountants, typographers, and secretaries, all of whom need computers. Another double page then contains cartoons showing the user in a hundred different situations, always resorting, whatever the difficulty, to the computer.

I have listed some of the forms of advertising because the computer is indeed the great agent not merely for the selling of a product but for integrating us into the world of this product. Whether it be by television or the press, the process is the same. This one gadget will irresistibly bring us into the movement of technique. By incorporating us into the technical system it will then lead us to buy the object in question. This advertising is possible only in virtue of technical equipment. We are thus set in an interesting circle. The technical object influences television (for example) which in turn influences viewers in favor of the technical object. This is true even when we do not have commercials, for example, when there are programs on space or on Formula 1 races. These are all hymns of praise to technique. Technique gives the televised program a dynamics, a power that fascinates us. I recall a striking formula on the occasion of a space launching: "Space is business but it is also a spectacle."

CHAPTER XIX

Diversions

With diversions we take a giant stride along the path of abstraction and addiction by means of the technical society and fascination. We are referring to diversions not just in the sense of amusement but in the sense of Pascal: being diverted from thinking about ourselves and our human condition, and also from our high aspirations, from the meaning of life, and from loftier goals. People "have a secret instinct driving them to seek external diversion [which includes war and algebra for Pascal, so that it is very different from amusement] and occupation, and this is the result of their constant sense of wretchedness."[1]

"Man is so . . . vain that, though he has a thousand and one basic reasons for being bored, the slightest thing, like pushing a ball with a billiard cue, will be enough to divert him. . . . However sad a man may be, if you can persuade him to take up some diversion he will be happy while it lasts, and however happy a man may be, if he lacks diversion . . . he will soon be depressed and unhappy" (136). "Leave a king entirely alone, with . . . no diversion, with complete leisure to think about himself, and you will see that a king without diversion is a very wretched man" (137). "All the major forms of diversion are dangerous for the Christian life" (764). "Man is obviously made for thinking. Therein lies all his dignity and his merit; and his whole duty is to think as he ought" (620). "Anyone who does not see the vanity of the world is very vain himself. So who does not see it, apart from young people whose lives are all noise, diversions, and thoughts for the future? But take away their diversion and you will see them bored to extinction. Then they feel their nullity without recognizing it" (36). "Being unable

---

1. Pascal, *Pensées*, tr. A. J. Krailsheimer (New York: Penguin, 1966), § 136 (p. 69). Hereafter references in the text will be to the sections of this edition.

to cure death, wretchedness and ignorance, men have decided, in order to be happy, not to think about such things" (133). "It is easier to bear death when one is not thinking about it than the idea of death when there is no danger" (138). "We run heedlessly into the abyss after putting something in front of us to stop us seeing it" (166). Such distractions, then, divert us from existential and essential truths and realities.

But also involved is an element of dissipation. Pascal notes plainly that one diversion quickly has to replace another. We jump endlessly from diversion to diversion without stopping, without stepping aside, without realizing what we are doing. We have to fly off in all directions. And thanks to technique our society has now made this possible for the first time in history. In Pascal's day the diversions to which he referred were those of the rich and powerful. The middle class and the peasantry had no part in them except at special festivals. For the most part, too, they were individual diversions. War as a diversion was not the war of foot soldiers but that of kings, nobles, and great generals. These diversions were for the few. But modern diversions are universal and collective (even when we are alone, each before his or her own screen).

We have already seen how the computer, telematics, and television are means of diversion. We will not go into that again here. We may remark, however, that self-evidently we would not accept the permanence of our diversions if we knew them for what they really are. This is why, as in the case of all our base, vile, and dangerous pursuits, we have to cast a large veil of idealism, grandeur, and seriousness over them. Things that are only diversions are declared by the authorities and the media to be an enhancement of freedom. We now have freedom to walk on the moon, to choose among many channels on television, to save time by plane and train, to drive at 150 miles per hour on roads (a motorcyclist was arrested for driving at this speed on a busy road in August 1986); we have the freedom not to conceive, or to make test-tube babies. Do we not see that we are overwhelmed by freedom? Conversely, we have a poor, foolish, mediocre idea of freedom if we call all these exploits freedom! For diversions are always against freedom inasmuch as they are against conscience and reflection. We shall now take a look at four types of general diversion: games, sports, the automobile, and some forms of art.[2]

2. I will not go into the associated question of medicine, which might be regarded as a diversion insofar as it prevents us from asking what sickness really is. Instead of seeing sickness as an integral part of human life, we now regard it as an accident, an unjust inconvenience, an unbearable affliction. Normal life is incom-

1. Games

Since Huizinga's *Homo Ludens* we all know that play is one of our essential human characteristics.[3] But I think that there is a great difference between the games that people have always been able to invent and those that have invaded our own society. In traditional societies we find collective festivals which gave society an opportunity for release and renewal (cf. the studies of Mircea Eliade and Dumézil), but these were exceptional events. There were also the great Olympic and Isthmian games of Greece, which were truly constitutive of Greek society. At the other extreme are children's games, which seem to have existed since Neolithic times. Then we find the games of aristocrats and nobles (races, hunting, tournaments, also cards, chess, etc.). Finally, there are village games on special days with jousting, dancing, etc.

Games are indispensable, but they are rare and they are played with others in a relation that has a metaphysical and social dimension. Rome seems to have been the only society which after the 1st century B.C. had permanent collective games. Juvenal's expression "bread and circuses," which has become famous, related to circus games and plays in the stadium or amphitheater. These performances were free, paid for by the magistrates out of their own fortunes. It was a common means to secure their election. But they were reserved for free Roman citizens, who in principle did not work and passed their time with these entertainments. This is all very significant, for games then began to change their character. They became pure spectacles, whereas previously they had involved participation as joint play.

Games as they have developed in our own society during the last thirty years bear no relation to traditional games. They are a new phenomenon. Between traditional games and the new games we find

---

patible with it. Time spent in the hospital is bracketed time. We should come out as we went in. Sickness is not a time for reflection or change; it is something to throw off as fast as possible. We cannot bear to suffer and to prove ourselves by resisting suffering. No need to suffer—all the anesthetics one could want are available to relieve us. No need to worry—all the tranquillizers one could want are available to help us. A serious study of the role of sickness is needed. What would Pascal, Mozart, Beethoven, Lautréamont, or Baudelaire have been without their maladies? But medicine is part of our enormous technical arsenal of collective modern diversion away from the self.

3. Cf. Johan Huizinga, *Homo Ludens: A Study of the Play Element in Culture* (Boston: Beacon, 1955); R. Caillois, *Jeux et Sports* (La Pléiade, 1967); M. Griaule, *Jeux* (1938); Anatole Rappoport, *Fights, Games, and Debates* (Ann Arbor: University of Michigan Press, 1974); Eric Berne, *Games People Play* (New York: Grove, 1964).

money games, set up by the middle class (e.g., in casinos or on racetracks). In principle these were meant for the wealthy, but they have gradually become popular. In France the setting up of the national lottery in 1934 made money games popular. Many young people found this morally scandalous. They thought it immoral to win so much money without working for it. This reaction was rooted in the work ethic. But the really new thing was that the state was now organizing games, and this was in keeping with the state's universal growth. The national lottery was a lure to make people pay painlessly, for it brought in big revenues.

Among modern games we must distinguish two main types. First there are national money games, all deriving from the national lottery. Then there are television and computer games, which have been all the rage for the last ten years. As regards the various forms of lotteries, including those that seem to involve some form of sporting participation, the true issue is their exceptional popular development, the passion of all the people, one might say, for these mirages. We do not intend to treat this as a matter of morality. We will simply recall that the state sets up these deceptions primarily to make money.

Nevertheless, I think that there is a more profound if unconscious reason for them. People's attention must be fixed on winning $10,000 or $100,000 or more either by pure chance or by guessing the qualities of a horse or vehicle. When they are excited about this they will not think about anything else. Thanks to the multiplying of the number of games, they are excited about it every day. Every day there are bets to make and results to await on television. Why should they be interested in anything else if they might become millionaires this very evening? These games (thanks to television) have become an obsession.[4] When we see people lining up to buy tickets, it is obvious that nothing else really interests them, that they are blind to what really constitutes life in depth or decisive politics or culture. A whole people is fascinated by this mirage, by what is an illusion in 999 cases out of 1,000. This is the first aspect of this by no means innocent diversion, which the state sets up to divert by the lure of easy and decisive winnings.

Yet it is not the most vital aspect of the invasion of games. For we are now immersed in a world of games by television and the computer. Among the many uses of telematics and video it is admitted quite openly that one of the four major uses is play. In the Hall of

4. Piveteau was not wrong to speak about the "ecstasy of television"; see his *L'Extase de la télévision* (INSEP, 1984).

Videocommunications in 1983 it was stated that electronic games were one of the most important items in the market for microcomputers and computers, though naturally these games were advertised as a positive activity inasmuch as they belong to videocommunications, help in education, create numerical images, etc. According to a questionnaire of June 1983, microcomputers are used 75 percent for games and only 25 percent for other activities. The games are in fact extraordinary and their fascination has been an essential factor in development. In 1980 there were only 20,000 consoles for games in France, but by 1983 there were 500,000 and by 1984, 8,600,000. In addition to these games, which we shall take up again later, we must also include the time spent in watching television and television games, which attract the general public.

Now none of these games has anything in common with ancestral games. There is in them no element of play or dreaming or amusement. They are primarily commercial. They are the products of massive technical equipment. They present factitious and coded situations and not an expression of the instinct of play. They amount to no more than graphic abstractions or simple operations on a screen. There is no place for improvisation. The computer does not "reflect" on a situation with x possibilities but unfolds an infinite series of situations with two possibilities. They are role games or simulation games or society games derived from Monopoly or Wargames. There are hundreds and thousands of them. The market for them changes rapidly. People pass their time to no purpose.[5]

Constantly gripped by the possibility of having something new, for these games are soon worn out, the players quickly lose interest. Yet the importance of the games is so great that a noble newspaper like *Le Monde* thought it good to devote twelve articles to a series on "video games in twelve lessons." This was an elegant intellectual way of participating in the general mindlessness. But can this really be presented as an aspect of intellectual development? In effect, if one is competent, one can create one's own programmed games, set up the rules, invent new situations, and program the computer for it.

The essential thing, then, is the possibility that the computer offers for creating games that are extraordinarily rich in possibilities (P. Berloquin). There is free play with images, texts, and melodies. Anyone can create, as anyone can paint! Games have been invented with computer materials that would not be possible with traditional

5. See Chesneaux, *De la modernité* (Maspero, 1983), p. 116. A single line of computers offers 1,500 video games that "let loose the passions."

materials. This is obvious, since we have here only fleeting images! Furthermore, the rules are included in the program. They may not be known, but they cannot be broken. Players may discover them with various attempts and groupings, and that is part of the game! The rules are not given. Players are in a closed situation. If the creator is not careful, they become not so much players as spectators, and the game is less a living game and more a playing script.[6]

Furthermore, it is possible to create a real football game or tennis or boxing match in which you participate from your armchair. Play is made with the elements of indeterminacy as the ingenious computer projects the necessary images on the screen. There is also the passion for videos. What a scandal it was when the French government tried to block Japanese imports of videos, thus depriving us of more games! We have here a true collective whim on the part of French adults. They are like four-year-olds stamping impatiently in front of toy stores. The cultural aspect is that we can now choose our own films. Unfortunately, the overwhelming favorites seem to be pornographic and science fiction films. We have here yet another manifestation of the infantile mentality of the public.

We must mention the jewel in the crown of computerized intelligence—chess. Does not this aspect justify all the others? To have constantly at one's disposal an indefatigable partner that can make different moves at our pace, is not that ideal? Is it not a sign of advanced culture? Not at all! True chess players know the importance of the psychological factor, of the implicit relation with the opponent even when no words are spoken, of sensing the reactions and moves of the opponent. Chess is played against and with someone. One of the main points of playing games is that they create a social bond. But if you play with a computer you are alone again. The essential vice of all electronic games (including Wargames) is that you are alone with the machine. These games are games for solitude. It is not the same thing to play chess with a friend or another enthusiast and to play chess with a computer. The computer can never be your neighbor. There is no social bond with it. This has its effect on individuals and has global consequences for society. The multiplying of these sterile games means that the little time left over from television and mass spectacles is snapped up by them.

Games, then, cease to be social cement and become a factor of dispersion, of enclosure within the solitude of fascination with ma-

6. See P. Berloquin, "Jeu et informatique: l'effet Golem," *Le Monde*, Aug. 1985.

chines. They become spectacles instead of being exercise for everybody and participation in a common creation. They become pastimes that take hold of us like drugs. We are fascinated by the screen, by images, by possibilities, by the unknown that comes out of the box, by lights and flashes. How many people have I seen in a trance, unaware of what is happening around them, unable to tolerate a human presence that breaks the concentration of their relation to things! The computer is not a companion; it is a vampire. It is false that the games are cultural or educational. They teach us only how to use a device. They are no more than an agreeable means of adapting us to the computerized society. Above all, they are fascinating.

It is well known that the average person (in France) spends at least two hours a day before the screen, watching television. We can probably double this time once there are computer games. In a curious way it is said that the obsessed and impassioned mentality of the players, hypnotized by the games, is that of only the tiny minority of those who frequent places with video games. But the enthusiasm has obviously spread everywhere as the games have become available in great numbers. We have to ask what free time is left for those who are thus obsessed. What time do they have to acquire the knowledge that is needed to live in our world, to act as good citizens? What time do they have for friendships or for social relations that are true relations and not just those of fascinated people in front of a little screen watching the same game or show, which are no relations at all? It has been realized for a long time that the family is being destroyed by the placing of the dinner table in such a way that during the meal it will be possible to watch television. In reality the relation between members of a family is significant and decisive at meal times. But technical games produce false relations as well as false intelligence even in seeming to make possible a certain initiative. Similarly, video, which seems to enable us to collect programs that please us or to see again movies that we praise, is in truth the sign of a false independence, for it helps to shut us up a little more in a world of images of escape.

I am firmly convinced that the whole system of technical games and amusements and distractions is one of the most dangerous factors for tomorrow's people and society. It leads us into an unreal world, and since we have here a passion or fascination,[7] this unreal world is not the one that is necessary for a day, as in the case of fetes and balls, from

7. I concede that in the past there was fascination with, e.g., roulette or cards. Cf. Dostoyevski's *The Gambler* or Darry Cowl's moving confession in *Le Flambeur* (1986). But these were isolated cases. The whole population was not affected.

which one returned at once to real life. The unreal world here is one of fantasy from which there is no longer any reason to return. It is like an addiction to drugs or gambling. In other words, technical games correspond very well to Pascal's diversions. They divert us radically from any preoccupation with meaning, truth, or values and thus plunge us into the absurd. They also take us out of reality and make us live in a totally falsified world. This is for me the greatest danger that threatens us as a result of technical development. In these conditions we become absurd (etymologically) in relation to reality no less than to truth.

The second basic aspect to which I have referred is that of diversion from public affairs, from society, from politics, from the question of meaning. I am not certain that it is intentional on the part of the authorities, but what is happening is as if there were some policy like that of imperial Rome or Byzantium, namely, to distract people and thus prevent them from thinking, and to do it now with means that are a thousand times more effective and universal. Everything takes place as if the political order of the day were that the people should play and the government would take care of everything. This is a skillful and basic negation of democracy, decentralization, and participation. Do we want democracy? Then we should begin by stopping all television and computer games! This is not to take into account the effects recognized by specialists. Video games, they tell us, accentuate the taste for power, for machismo, for aggression, for manipulation. The machine becomes the child's only friend, invading its personality and shutting it up in a world that has no contact with reality. In the end the young will have only a negative vision of society, a vision of losers. The object of these games is to destroy, and the action in them is accompanied by violent explosions which induce violent behavior. Even nonviolent games like Pac-man keep those who play them in paranoic tension and cause real neurotic stress.[8]

I might not accept this judgment in the case of all games. The essential point as I see it is that of general diversion. People are diverted for nothing (except the pleasure of winning) by a gigantic socio-technical mechanism. People today are perverted, not morally but in intelligence, attention, and scale of values they are perverted by diversion. There is nothing except a constant repetition of games. I would say quite plainly that the greatest threat to Western society today is not communism or Americanism or the economic crisis or drugs or alcoholism or resurgent racism, but our absorption in games and the

8. See "Enquêtes médico-sociales américaines sur les vidéogames," *Libération*, March 23-Nov. 10, 1982.

softening, degradation, disengagement, escapism, and loss of meaning that come in and through games.

2. Sports

Sports are the second great diversion, distraction, entertainment, blinker, deception, illusion, and social conjuring trick for people in the West. Readers will think at once that technique has nothing to do with sports. They will be wrong. Technique plays a double role. First, in the sports themselves, technique transforms their practice by its rigor. Second, technique makes sports the ideal television spectacle and thus changes them. We will simply mention the first aspect in passing, since I studied it in *Technological Society* and it is not so relevant here. Those who engage in sports are constantly breaking records through improved techniques. As they continually seek better results, the techniques become more demanding and extend to the whole of life (regimes, etc.). The result is professionalism. But the greater application of techniques in sports depends on technological discourse. If a match or competition were simply a game (as sports used to be), and if it was simply an individual matter (like golf), there would not be the frantic application of extreme techniques. Technological discourse has transformed sport into an enormous spectacle and fabricated champions, stars, gods of the stadium.

It is the massive diffusion by the media and the power of propaganda over the masses that has transformed sport. It is technological discourse that has overturned it by making it a national and global affair in which little regard is paid to what was once its golden rule: honesty, strictness, fair play.

To see how the public has been captivated by sport we have only to consider its place in the media. On television, on any channel, not a day goes by without several minutes devoted to sports. The newspapers allow it even more space. The remarkable thing is that throughout the year something is taking place in sports every day: football, soccer, tennis, Formula 1 racing, boxing, swimming, cycling, etc. Every day the public has to have its dose of emotion and passion. Matches and games are televised internationally. Ratings show that 50 percent of all viewers watch matches that are televised directly. This is the equivalent of three networks showing films together. In France the price for the right of transmission in 1985 was 500,000 francs (about $100,000) per match, and the annual revenue might be as much as 3,500,000 francs ($700,000).

Naturally, transmission on this scale supports overwhelming advertising. Signs are placed all round sporting arenas. Advertising has given the French a passion for sponsored yacht races in which the sailors carry the names of the sponsoring firms.[9] These sponsorships cost millions, and naturally, advertising carried by the sport brings in as much and more. It would not be possible if television and radio did not constantly infuse sport into the heads of all the French. Broadcasting is indispensable.

Some provincial dailies even devote their first pages to sporting events, with color photographs. The latest match is more important than an international agreement or an assassination attempt. Matches are the most important news. As always, there is reciprocal action, for as the media focus on sporting events, after a while more people are interested and the demand grows. The media can then increase the amount, the intensity, the exaltation of sports even more, for they are now meeting a public demand.

Well and good, but what becomes of sport when it is technicized and made a spectacle in this way? First, we have a transition to total professionalism. Professionalism may not be new, but given the enormous financial stakes it is becoming exclusive. Sport used to be play. People engaged in it when not at work. Little by little professionalism has gained ground. Today the only "true" sport is professional. We thus have a remarkable double phenomenon. On the one side players are bought. The richest club will have the best players. The sums paid to buy them are in the millions. Many clubs go into debt. They thus have to raise money. Some cities will heavily subsidize them, devoting much more money to their clubs than to social prevention or the rehabilitation of prisoners. There is frantic advertising to bring in more spectators. But new stadiums are then needed and the city again helps. The money of taxpayers who are not interested supports this mad expenditure.[10]

It is true that those who are not enthused by the skillful and powerful performances are not wholly normal. Professionalism makes possible technical perfectionism. But it also has some serious consequences. Boys of fourteen with no other abilities are trained exces-

9. The new yachts are also monstrous gadgets using navigational satellites, sophisticated meteorological equipment, a computer displaying the speed of the boat, its course, the true wind speed and apparent wind speed, and a computer to give the most efficient spread of sails in the prevailing wind and sea conditions.

10. I suggested some time ago that taxpayers who are not interested in these sporting spectacles should be allowed to deduct from local taxes the proportion spent on financing the clubs.

sively, for eight hours a day, so that at eighteen they can begin to play in public, but their career will be over when they are about thirty. They will then know nothing else, have no other skills, and not be capable of anything else. During ten or so years of sporting activity they have thus to make a fortune that they can live on for the rest of their lives. For both clubs and players sport has thus become a matter of cash. In the name of technicized sport, in the name of those who thanks to technological discourse have become the gods of the stadium and the fans, money is finally king. Socialism is no better, for it is under the same pressures. There may be less financial competition between clubs but there is more bureaucratic competition!

Another effect of professionalism for sale[11] may be seen in the composition of teams. In Formula 1 races a Renault, Fiat, or Porsche team may be made up of people from any country: Brazilians, Africans, Italians, Portuguese, etc. Originally a city team used to be made up of people from that city, so that one city was really playing another. But what does a match prove today? Simply that the winning club has been able to find the best players on the market by paying the most for them!

If sport is to be pushed at all costs, there need be no hesitation either about extreme brutality. In televised matches that I watch I am always amazed at the brutality of the encounters. We see it even in swimming! It is most striking, however, in boxing, which is no longer an art but merely unleashed violence. We see it in tennis, too. We are amazed by racket strokes that are like sledgehammer strokes. It would seem that blacksmiths are at work. It is not for nothing that tennis champions are given such names as "The Rocket" and "Boom-Boom." But where is the tennis of yesteryear with its grace and suppleness and finesse? Borotra, Lenglen, and Tilden did not play like those who slaughter cattle, holding their rackets like cleavers. This brutality is linked both to the violence of competition and also to transmission for millions of spectators, for brutality is visually much more profitable. How can we be surprised that the brutality of the players affects the public, the thousands of obsessed fanatics, who are no more sporting than I am, but who say that they are because they watch the matches, and who finally, as we see again and again on television, become overexcited, fly into a rage, and smash everything. They are simply contaminated by the gods of the stadium and by the worldwide televising of the spectacle, in which they themselves suddenly become actors.

11. The sale of football players, etc., today reminds us of the auctioning of gladiators, pugilists, and chariot drivers at Rome. There, too, sport had become professional and technical, arousing mad public enthusiasm.

I think that the more techno-sporting propaganda increases, the more outbreaks we will see in the stadiums, just as there were violent conflicts between the great sporting rivals, the Blues and the Greens, at Byzantium. The fans show us the result of complete fascination in diversion.

In contrast to what I have just written, I would recall the way in which a traditional society has been able to integrate and control the most violent of all sports, the bullfight. A fight between a "brute" and a man is the one that could have degenerated most easily into general bestiality. But this has not happened. The barbarous game has been ritualized, the violence given precise forms, the brutality replaced by a controlled elegance, and the collective behavior set within a kind of communal ethic. What was butchery has become a noble art. But today we see the very opposite. My point, however, is that this is principally the fault of technique and technological discourse, which entails the primacy of success and by way of compensation a popular unleashing by fanaticizing.

The fact that sport is now a spectacle comes to expression at every level. I have in mind one astonishing instance of this, namely, the ridiculous gesticulating by the players. When a player scores, for example, he falls on his knees, invokes heaven with frantic gestures, then falls on the necks of team members, and they embrace and congratulate one another while the crowd is roaring. This is more important than any political event. But the gesticulating is only because there are cameras showing it to millions of viewers. In other words, the players all know that it is a show and this is in keeping with the need for absolute diversion. It is the result of the principle: "Play, play, we will take care of the rest."

I want to draw attention to three other results of technological discourse. First, the sports deception. We have as an example the celebration of what are called the "Olympic Games." But the real Olympic Games were something quite different. They were not just a sporting event; they included singing, theater, poetry readings, etc. They were also an occasion for the Greek cities to come together, even in wartime. During the games there was a truce, fighting stopped, Greek unity was restored, and I suppose that diplomatic conversations would use the occasion to solve problems. Our own "Olympic Games" are the very opposite. They are an occasion for sanctions against countries, for divisions, for expressions of conflict. Some countries refuse to go to the USSR, others refuse to go to the USA. This or that accursed country (e.g., South Africa) is excluded. In other words, the games are a mode of combat. They are thus a good demonstration of the tremendous transformation which sport has undergone due to the technicizing of society (not its politicizing, for no world was more political than the Greek). Competitors must be crushed at all cost. This is the pitiless, lawless law of technique and also of sport now that sport has become part of the public's tragic mystique and is exalted by all the media, which heap on sporting figures honors that were previously reserved for generals and which give them the glory of bigger headlines than ever, so that games are plainly no longer viewed as games but as mortal competitions, merciless encounters, a true Manichean battle. Our Olympic Games today, then, are not real Olympic Games of peace and dialogue but games of ferocious rivalry. This is the law of technique and also of modern sport. And technological discourse has to exalt the greatness of sport, which is the greatness of a country as a whole. Sport is so important that the President of France had to postpone a speech that he had to make in order to allow the citizens to watch European Cup matches on television!

A second effect is that the importance of sport is so dominant that a sporting event has to be staged as a show when there is a blank in the daily program. Not a single day must go by without the public having its fill of shows. Monstrosities are thus created like the rally from Paris to Dakar, which was an absurd and insulting waste in countries where famine is rife, showing off Western power in the powerless Third World, an example of sheer vanity. Do not misunderstand me; I am not contesting the courage, endurance, ability, or energy of the participants. What I am saying is that in a world in which courage, intelligence, and endurance may be shown in thousands of necessary ways, it is totally absurd and even odious to waste these qualities on something so imbecile. Why did not the energetic pioneers work for Third World relief, transporting needed provisions to the heart of Africa on routes just as difficult as those to Dakar? Why did they not sail on Greenpeace ships? Last year a television reporter who followed the rally with the important mission of sending back breathtaking pictures that would make millions of viewers gasp was killed in a helicopter and thus earned the right to a deluge of heroic talks. The only sane thing was said by Cavanna under the heading: "Dying as a Fool for Paris-Dakar."

Finally, a good article in *Réforme* on the yacht race to Rhum was interesting because it emphasized how the public was mobilized for this event (500,000 watched the start at St. Malo) and recalled that this sport is really a publicity stunt with sponsors backing the boats, enormous technical research to improve their performance (as in the case of Formula 1 cars, typical waste on the part of a society in disturbing economic imbalance), and technique enjoying an eminent place during the race as satellites and computers were used to make sure that all viewers would not miss the slightest incident. This is a good example of sublime means used to attain ridiculous results, the sport being a game with no particular importance of its own. But why was this excellent article given the title: "Technology in the Service of Fantasy"? This title is worth considering. "Technology" was not in service at all. It dominated and aroused and imposed its own fantasy. In this regard we may follow Castoriadis (*L'Institution imaginaire de la société*). To the degree that fantasy is instituted, if it is hypnotized, fascinated, diverted, and vitiated by shows, denuded of any meaning of its own, robbed of finality, stripped of value, creating nothing, purely passive, fed on merchandise, the whole society which results can be nothing but dust and smoke dispersed as quickly as the exciting but insignificant image of sport. It is ultimately of little importance that sport has become an instrument of big money. "We need to expose and denounce the mystifying mythology of sport, which for the benefit of big investors has become a travesty of circus games."[12]

It is also of no great importance that sport is centralized by the state, cemented by middle-class ideology, and governed by relations to production, so that it is a capitalist substructure.[13] This criticism of leftish dogmatism, which has the merit of destroying idealistic and humanistic talk about the virtues of sport, makes the mistake of relying on analyses of society in 1900. Today big investors, who certainly have a decisive influence, no longer play so radical a part as technique, even in socialist countries. The dominant factor in sport is now technological discourse, the indispensable drug of the average Westerner, sport being merely an occasion. Sport has to bow to the necessity of this discourse no less than to the imperatives of big investors, who profit by it, as do socialist states as well!

3. The Automobile

The automobile is the great symbol of diversion and the associated emptying out of reality and truth.[14] It signifies both being somewhere

---

12. C. Leroy, "Les métaphores du libero," *Le Monde aujourd'hui,* June 1985.

13. See M. Caillat in the magazine *Quel Corps* (a good critique of the actual practice of sports).

14. Cf. B. Charbonneau, *L'Hommauto* (Denoël, 1967); J. Baudrillard, *Le Système des objets* (Denoël, 1968); V. Scardigli, *La Consommation, culture du quotidien* (Paris: PUF, 1983); Sauvy, *Les quatre roues.*

else and being outside oneself (making infamous brutes of fine people who are good parents or spouses). Its very presence makes it possible for me to be elsewhere and outside myself. This is why it has established a truly universal consensus among people in the West. As Scardigli states, it is the product of the happy marriage of science and desire. Its performance as a consumer object coincides with the demands of a system of personalist values (i.e., officially, those of our society). The destruction of traditional society invokes the creation of a space and a time that are structured by and for the vehicle.

The automobile is thus the most perfect symbol of the technical society, and in buying it we are really buying a nexus of social symbols. It combines utility and futility, evasion and fatality, being elsewhere and finally meeting death. The development of sales bears witness to the consensus. The big firms first imposed its use in every country, and now there is agreement that it is the supreme diversion. Nothing can stop it. It was said that the oil crisis would arrest its use. It did not. Sales declined for two years but then rebounded and became bigger than ever. The OECD predicted in 1983 that the number of cars in Europe would double from 1975 to 2000. In France about two million new ones are registered each year, and 2 percent of the homes without cars buy them each year. In surveys the automobile alone is beyond criticism; some people are opposed to television and the computer but none to the car. Thus 93 percent of the upper classes have cars and 84 percent the laborers. More and more families have two cars.

The motorizing of society is a model that is followed everywhere. Third World countries are buying cars. Sales in these countries grow 7 percent each year and many of them have their own automobile factories. The car seems to be a good way of responding to the economic crisis, as it was in 1930. There has never been a strike against the car (in general) or in favor of public transport. Traffic snarls and accidents and parking problems have led to talk of a demotorizing trend, but in fact there is no such trend. People all continue to use their cars more and more to go to work and to go on vacation. Fewer cars can hardly be imagined, since for the last thirty years we have planned and modeled urban and social space in such a way that it is no longer livable without cars. The consensus being there, what can the authorities do but comply with it. Any decisions to the contrary lead to demonstrations and blockages aimed at taxes, fines, impoundings. The principle is always that to reduce the use of automobiles is to mount an attack on freedom. Car owners and motorcyclists are the most important lobby or pressure group in Western countries. The state is totally powerless against them. In the construction of roads and highways they have forced the state to come out decisively and, I believe, irreversibly in favor of the car over public transport. Every obstacle to the automobile has been lifted. The state contributes three-quarters of the funds for research at Renault and Peugeot. Even at times of gasoline shortage the state permits the grotesque aberrations of Formula 1 races. Doubling this state aid, public and private financial institutions basically play the same automobile card.

The great word to justify all this is *freedom*. Again we see diversion. The meaning, experience, and depth of freedom have been diverted (and perverted). Out of freedom we have made the enormous foolishness of being able to take to the road freely in a car at any speed we like. Freedom is thus reduced to escape (from one's neighborhood, from routine, from daily concerns) and to mobility (confounding freedom with going anywhere!). Freedom in solitude: this was imposed on urban dwellers by the destruction of traditional society, but it is now in great demand by car owners, at a speed which isolates them and which creates a "closed sphere of intimacy."[15] As Charbonneau says, the automobile is a product of middle-class freedom. It is no longer a chariot of the gods but is available for every citizen. Unfortunately, when freedom incarnates itself it exacts a price. We can go faster and faster wherever we please. But the roads on which we go are covering the land. And we are all there together on the same roads, though each in his or her own private property (for this is not Russia where people are crammed into trains), and each listening to the same program on his or her own transistor.

Technological discourse tells us that the car is only an instrument of freedom. It helps us to look for work, to enlarge the horizons of our knowledge by a personalized tour, to expand many of our faculties, or to buy the best products at the best price, etc. It is freedom. We can choose our own car. We freely choose to go to the office in it, and on vacation we freely choose our route, though at the same moment that millions of others are making a similar free choice. Since the car means freedom, we are yielding to a basic law, namely, that technological discourse is above all deceptive discourse. As we have seen, the lie is not merely in advertising. Indeed, it is in technological discourse as a whole, which strongly and loudly affirms values (in this case freedom) by the very means which negates these values.

At the same time there are annexed values. The car ensures social integration. Its absence arouses suspicion, rejection by the dom-

---

15. M. Bonnet, "L'automobile quotidienne: mythes et réalités," *L'Automobile et la Mobilité des Français* (Paris: La Documentation française, 1980).

inant society. The right to drive is an initiation rite. The car symbolizes access to adult sexuality and the marvels of modern life. It guarantees our being in touch with progress. It enables drivers to surpass themselves. When they drive, they are no longer conscious of limitations. I would almost say with Scardigli that it gives us access, or enables us to return, to paradise lost. It can compensate for all the frustrations of a society like ours and make us think that we have a part in personal, autonomous development. The day when we finally break loose we put a tiger in the tank, we roar, and we rush upon our prey. But the highway is cluttered, our speed drops to 5 even though it is 120 in our head. Thus the vehicle that was going to take us out of the crowds puts us back in them (B. Charbonneau).

Be that as it may, we live in a kind of splintered space by reason of the new solitude. The traffic flow rules out encounter. Alluring physical communication displaces social communication. Solidarity becomes abstract and impersonal. We no longer help neighbors in distress, but we work several hours a week to pay for the immense mechanism of road safety intended to bring help to unknown people. We no longer communicate with those we pass alongside each day, but an immense web of roads (or telephones) makes "communication" possible. The car both answers and demands this extraordinary dispersal that is called an opening up to the world. It effects (and is required by) the work of industrialization and urbanization. We are in a social void that speed alone can fill. The lofty discourse of technology tells us that speed is our access to paradise. But the reverse side of the picture is already familiar to us. The vaunted autonomy quickly turns into increasing dependence relative to the car's demands. Thus we have the vicious circle that cars make possible the putting of big roads around our cities, and these roads are so practical and economical that they demand the use of cars, etc. This common revolving door is a symbol of the effect of technique on humanity.

But how are we to evaluate the real positive and negative aspects (not those of technological discourse)? Who will add up the increase in noise level, in pollution,[16] in accidents, in the destruction of beautiful country (the French highway network covers 4 percent of the land), in nervous illnesses, etc.? Who will add up the positive benefits, the adding to the country's wealth by exports, the making of isolated regions more accessible, much more democratic access to resorts, etc.? The cost of individual labor offered as a sacrifice to the automobile has

16. For an exhaustive study see "La Pollution automobile," *La Recherche* 149 (Nov. 1983).

been calculated. In developed countries 15 to 20 percent of the gross national product is devoted to transport. In France we work one hour out of seven to pay for transport, or one hour out of five if we have a car. According to EEC statistics, in France one must work 1,317 hours a year for one's car. Some have scoffed at these figures and at those of J.-P. Dupuy and P. d'Iribarne, who in an inquiry into the true speed of cars prove that it is just as fast to go by a bicycle as by a Porsche. It is wrong to mock such statistics, for this is in fact the reality.

But the weight and coherence of technological discourse prevent us from believing it. The ultimate reason for this is that the automobile is a diversion. It is a diversion because it gives us speed, takes us to far places, offers us escape, and grants us apparent freedom. I am told—though we would need employment statistics and opinion polls to prove it—that like the microcomputer the car is used much more for pleasure than for transport. It is a distraction. It displaces us. It is useless and superfluous. It makes us do things we would not do without it. In it we go anywhere to calm our nerves, etc. This is more essential than using it to go to work. It is a diversion because it prevents us from looking at ourselves or meeting our neighbors or being content with one-to-one relations or contributing our personality to everyday life or being responsible at the heart of our community or on Sunday having an ultimate encounter with God. The refusal to do all these things acts as a funnel to send us off in our cars. We have such a passion for this diversion that we may well pay for it with our lives.

It cannot be repeated too often that 1,000 people die each month on the roads in France. But what does death matter to those who are in ecstasy? They are indeed fascinated people. The automobile is an engine of death. As has been said, the car is life, but it will not allow any but a mechanical life. Everywhere it passes, it kills. The human cost of such a hecatomb cannot be estimated: the number of hours lost per year, the premature deaths; every year a million years are lost in the USA. The medical costs in France in 1978 ran to 40 billion francs (Scardigli). Are we sure that the economic gains compensate for this? But for the public, as we know, this is of no importance. The bloody sacrifice, equal to that of the Aztecs, is accepted with complete passivity and even content: the content of drugged and fascinated people, of diverted people. It should also be stressed that we do not take the costs into account because the divine vehicle has two contradictory sides. It is prestigious and yet it is ordinary and therefore reassuring. It is not a coffin but a commonplace means of transport. We do not imagine entrails and blood on its comfortable upholstery. The car does not kill; an accident is to blame. The car does not kill because it kills every day.

If we had just a little awareness and freedom, we would challenge the automobile. But to take note of the harm done by progress is to begin to question the very foundations of our society and to hasten the transition to a different model of social life. Contesting the automobile is going terribly far. It is going against the opinion of fascinated humanity. Protecting the speed of cars is more important than saving people. The first minimal exercise of freedom would be to make cars a secondary accessory which is used only in exceptional circumstances. But modern people, fascinated and diverted, put things the other way round. They want to go out and they are back in their cars. They smash themselves up, thinking to grasp the reflection of happiness.

4. Mechanistic Art

I am not going to take up again a full study of modern art, which I have dealt with at length elsewhere.[17] I will not talk about painting, sculpture, architecture, modern music, poetry, or literature. I will simply look at some aspects of what with amazing lack of awareness has been called art for the last decade. This art completes the panoply of dehumanization and fascination whereby what was for centuries the supreme human activity has now become the most vicious and externally the most grotesque of snares. I call it mechanistic art because as a whole it either depends on technique or is provoked or induced by it. Yet my first point is the rather different one that we see here how far we moderns can be fascinated and made infantile by our environment. I have in mind comic strips. I used to like these when I was a child; indeed, they are childish. I have read some of them with amusement as an adult. It is good to sink back into childhood at times. Naturally, I have never thought of them as art. But lo and behold, they have now become a major art form with their ridiculous characters and their puerile or fantastic stories. There has been a national exhibition of them and a surfeit of articles explaining that this is a major art form, that this is the art of our age, and that a specific originality comes to expression here.

I am simply emphasizing that we have at this point a reinforcing of my thesis that we are becoming infantile. What is pitiable is that this process is taken so seriously. Technique begins to play a part, however, only when we come to what Jack Lang calls the new images—animated

17. See Jacques Ellul, *L'Empire du non-sens.* Cf. also M. Nahas, "Quand l'artiste peintre devient informaticien," *La Recherche* 165 (April 1985).

cartoons, videos, etc., which a strong article in *Le Monde* (Dec. 1983) seriously describes as images resulting from the dialectic between technological challenges, economic stakes, and the defense of an artistic patrimony, the very mark of cultural industries. Nothing more! Perhaps culture has always been an industrial product, but the machine, even though it be the computer, is never more than a machine. Animated drawing is only for programs for young people, for ads, and for instructional films. Nothing in it can pretend to be art. But we have here an economic area. That is what really matters.

Nevertheless, the justification of money does not justify everything. There has to be a spiritual element, art. Hence an enormous machine is set to work to make us view as art the new type of advertising films (weird, startling, exuberant, incoherent). The article in *Le Monde* speaks of brilliant moments. Honestly, I have seen many such clips and films and I regard them as absurd and sometimes even silly. The presence of a Halliday or Isabelle Adjani cannot make them art, nor can the play of light, the phosphorescent colors, the smoke, or the formlessness. I have seen automated dancers in jerky rhythms, dances of cubes, and forms and mixtures of colors. Clips and films and commercials and programs and promotions of this kind have no clear status. But none of them has arrested me and caused me to say that it was beautiful or rich in meaning.

They are all of them absurd and futile. This is what happens when the cleverest techniques give free rein to infantile imaginations which have "power" going for them. I might admire the amazing work of integrated circuits. With the help of computers we can design the smallest circuits in the world. The microchip, we are told, will be the key to industrial production in the 21st century. That may be. But clever though it may be, I do not see in industrial neo-neo-designs the slightest shadow of artistic quality, even though I might imagine that art today is very different from what it has been for three thousand years. What I reject is that art should be a lowering to the basest level of all human specificities.

I must now state as firmly as I can that technological marvels do not suffice for artistic creation and that the techno-sciences are not precursors of the arts. To talk this way is just as ridiculous as to argue about the existence of God. Technical devices may upset people's convictions, give them a sense of relativity, and muddy the frontiers between the certain and the improbable. Seeing that we in the West no longer have any values or certainties, that we are schizophrenic, paranoid, and stupid, one would not have to exert great effort to bring about imbalance and to erase all bench marks, drowning us in information. This is our new ultraconformity. As we have already read and understood for the last thirty years, there is nothing to understand here. This was already true of the new novel. The ultramodern is in fact regressive.

But let us leave aside these tricks which cost fortunes and mobilize an armada of intellectuals and artists at exhibitions. Like music videos, they are a form of entertainment, but they are far less important than music. I am not talking about great music but about the permanent daily music which accompanies young people in all that they do thanks to the Walkman. On buses and trains we have only to look at the faces of those who carry a Walkman to see how serious a matter this is. With their concentrated and inhibited air and attention always directed inward, those who constantly listen to music are abstracted from reality and are strangers to those around them, living on a solitary island and refusing all communication. This device is a new destroyer of human relations and a creator of solitude that can lead only to suicidal attitudes. For a long time now young people have been living with permanent noise, constantly replacing their cassettes. The serious warning here is that a refusal to listen to silence is a refusal to meet oneself or others. Things have grown much worse with the twofold fact of the Walkman, which makes this world of noise incessant, and the change to rock, which is a disaster, especially hard rock.

E. Morin is both nuanced and comprehensive regarding rock. "It is not just musical frenzy but existential frenzy. The premier rock groups feel a rage to live where most get lost. The only ones to save themselves and triumph are those who abandon existential disorder (and drugs) to enter the commercial system, which demands a minimum of regularity of life even if only to honor engagements and respect schedules."[18] (Indeed, these revolutionary geniuses easily enter the star system and the capitalist regime and rapidly make their fortunes!) "At the source of rock is a movement which does not believe in the industrial society. It arises on the fringe. . . . There is in this intense and frenetic music a dionysiac ferment, a panic. . . . Strong stimulation is given to the ferment of rebellion in all adolescents." The thrust of the movement is either to violence or to dandyism. What a system of mass culture has to do in the cultural industry of the song business, however, is to "restrict the dionysiac tendency without destroying it (since it is that which sells), snuffing out the latent element of rebellion, eliminating the explosive tendencies and social explosiveness but keeping the

---

18. E. Morin, *Sociologie* (Paris: Fayard, 1984), though the text to which I refer dates from 1965.

appearance and sound of them." Rock shows are exercises in "socio-logical domestication," in the integration of all groups into a star system, in "the acclimating of rock's original savage force."

But since Morin wrote these words, rock has in fact become diversified and has worsened. It is based on the beat which always characterizes it (the constant repetition of regular sounds combined with syncopated rhythms). Hard rock, according to psychoanalysts, excites sexual instincts with its metallic beat and electric guitars; acid rock orients toward acid (LSD). A medical team in Cleveland has studies the effects of rock: severe traumatisms of the auditory, nervous, and endocrinal systems. It causes breathing changes, glandular secre-tions, contraction of the larynx, and irregular heart rhythms. Some halls are equipped with laser beams, which cause burning of the retina and during dancing result in dizziness, nausea, and hallucinatory phenomena.[19] Above all, there is the intense noise. Hard rock is 20 decibels above what the human ear can tolerate. It assaults every listener by pounding the auditory perception. People do not listen to it; they are submerged in it (with Peter Townshend). Acid rock (Beatles, Rolling Stones) has as its refrain that rock is the source of revolution. Means have been sought to pass on subliminal messages which the hearers do not consciously hear in the incredible noise but which they pick up in the subconscious. Rock groups have been accused of passing on such messages on behalf of drugs. Punk rock does the same on behalf of violence to the self and others. The subliminal message becomes more subtle. Some claim that phrases are added in reverse which we can hear when tapes are played backward. For such messages very low frequencies are used (14 cycles per second) or very high frequencies (17,000 cycles per second). The power of rock lies in the syncopation coupled with the amplified noise. All self-control is lost. All power of reflection and all personal will are inhibited. Rock in its different forms is a true destroyer of personality. When devotees of rock meet in a hall, isolated from one another by the crushing music, exposed to the play of blinding lights, they do what they do without ever looking at one another or speaking a word to one another. The unanimous cry that is most often heard is "Me, me, me." An American study in 1981 showed that 87 percent of American young people spend three to five hours each day listening to rock. The number of hours has risen to seven to nine with the coming of the Walkman. In 1984, 130 million rock records were sold. In general, then, there is a loss of control over the powers of reflection, a permanent diminishing of intelligence, neurosensory over-

19. For details cf. Regimbal, *Le Rock n'Roll* (Geneva: Croisade, 1984).

excitement, a hypnotic and (secondarily) depressive state, and finally serious memory difficulty and loss of neuromuscular coordination.

I want to insist, however, upon the effects of noise. We find this everywhere and it is simply aggravated by rock. It is not really recognized as a serious danger. When it lasts, the different effects are due to the cumulative dose. Recovery from damage demands a longer period than that of exposure. Once 80 decibels are passed,[20] there is a serious alteration of physiology, and around 120 decibels a change in the composition of the blood and an increase of arterial tension, of the cholesterol rate, and of the production of stress hormones. The vibration (i.e., the rhythm of rock) greatly aggravates the effects of noise.[21] Furthermore, it has been verified that excess noise generally weakens the cerebral faculties.[22] But at the same time the organism grows accustomed to noise and ends up by needing it. Noise has now become a veritable drug. We are drowned in urban noise (cars, sirens, engines, cement mixers, bulldozers, planes, helicopters, motorcycles, jackhammers, etc.). And now, under the pretext of music, this pounding by rock has been added. Why do we stress this here? Because this "art" would not exist were it not for technical means. As there would not be the idiotic "sculptures" of compressed vehicles were it not for the presses, there would not be the disintegration of the personality by rock were it not for the technical means that make it possible to produce such an infernal racket (sound systems), the media which tell us with such beatific intensity about the exploits of the rock stars who are their heroes, and finally the technological discourse which extols the marvels that result from the use of all these techniques.

Two things above all show how impotent we are against noise. First, noise is not suppressed but we are given devices to make it tolerable (e.g., soundproof windows, and anti-noise barriers alongside freeways). Second, the measures taken against noise in 1983 are ludicrous: regulation of the noise level of ultra-light planes; a revision of airport noise standards after three years (in 1986 this was forgotten); "aid" (?) to people living near airports; compulsory construction of

20. This was the limit set by France's National Council of Hygiene in 1975. We must not forget that the increase from 80 to 90 is exponential.

21. See H. Laverrière, *Repenser ce bruit dans lequel nous baignons* (La Pensée universelle, 1982).

22. France's National Council on Noise put it this way: "A brain which resounds [*résonne*] does not reason [*raisonne*]" (1982). But the Council has met with a setback (*Le Monde*, Feb. 1986). The amusing thing is that at the very same time the Prefect of Paris was much disturbed about the barking of dogs! Government always provides comic relief.

mufflers that could not be dismantled (but we have not seen them); levels were set for washing machines and contracts were made with pilot towns for soundproof schools and day-care centers. But nothing was done about cement mixers, bulldozers, or motorcycles, and there were no severe penalties for drivers who do not have mufflers. Noise is part of the sacred and inviolable world. As motorcyclists repeat again and again, it is a matter of freedom.

5. Ultimate Idiocies

Disneyland

Undoubtedly, the diversion which is most calculated to make the French infantile and distract them from all serious things is the Eurodisneyland project at Marne-La-Vallée. Investment in this gigantic park of infantile attractions has risen to 10 billion francs (about $2 billion). It is projected that there will be 10 million visitors the first year, which ought to leave a profit of 5 billion francs, but similar calculations were made regarding high-speed trains, and they proved wrong. It is generously estimated that the average visitor will spend 225 francs (about $45) in the park. Thus a husband and wife and two small children will spend about 1,000 francs (about $200) on a visit to Disneyland. As for the positive effects, these are wonderful. Disneyland is an opportunity for the whole region, say M. Cantal-Dupart (the urbanist responsible for Greater Paris) and M. Bayle (editor of the review Urbanisme). Construction over five years will offer 13,300 jobs. Hotels, restaurants, shops, and transport will mean 30,700 permanent jobs. Receipts should amount to 8 billion francs (much of this from foreigners), or 10 percent of all foreign tourism.

Every source of financing has been considered. As always in such projects, there will be a perfect financial balance (the problems will appear much later). Remarkably, too, Eurodisneyland will even things out west of Paris. A wonderful arrangement! Ecologists, of course, do not agree. But the authors cited above, like all serious people, scoff at them. Yet it is necessary to expropriate twenty-five farmers from good agricultural wheat fields. Obviously, the interests of these country folk count for nothing compared with the greatness of a project that is so highly cultural and also so profitable. Furthermore, ecologists are told that Disneyworld in Florida contains many wooded areas and stretches of water, and this ought to reassure them. It remains to be seen how all this will develop.

Others are disturbed about American influence, but they are told that French culture and European history will have a place among the attractions. The techniques used will also be French, as will 90 percent of the workers. The existence of Main Street, the West (cowboys and Indians), New Orleans Square, etc., is of little importance. For my part, I am not so worried about this aspect, for our pseudo-culture is wholly modeled on techno-Americanism. This will be only one more step. I also commend Mr. Giraud, president of the Regional Council of the Ile-de-France, for his honesty in saying that the project is primarily a matter of economics.

Culture is secondary. But I am interested in this statement: "My culture is Mickey."[23] Here at last we hear the truth. The cultural level of educated France is that of cartoons designed for children of eight. Disneyland is not a scandal. It is simply social cretinism pushed to its extreme limit. It is simply the most explosive manifestation of the deculturizing of France and its diversion from serious and basic daily realities. It is the degradation, justified by employment and economics, of what might still remain of the spirit of criticism and reserve.

The Idol

It is good sometimes to close with mockery. This world of the distraction or diversion or perversion of humanity by technology culminates in adoration, veneration, and beatification, in the expression of a properly religious sentiment. This is normal. As Marx showed, alienation leads to religion. Humanity, when diverted, also becomes religious. The star system has been functioning for a long time. In the entertainment world the star becomes the idol in the primary sense, the absolute, transcendent image, a veritable Allah. I have rarely found a more grotesquely exalted paean, however, than in *Le Monde* (Oct. 1984) under the pen of Marc Raturat and with reference—this is interesting— to the first star of music video, Michael Jackson. Those who have not seen Jackson's videos, he wrote, should spend some time in a Tibetan lamasery or amuse themselves planting a flag on the South Pole. Channel 2 (on French TV) talks of the Jackson myth. No gesture of his is unimportant. His squeals transform into nuggets the poor stuff that he sings solo. When he finds a good song, the whole planet pants with him.

The Jackson myth! Plainly there were never any musicians or

23. Article in *Le Monde*, July 10, 1986.

singers before him! What makes him unique and divine, however, is that he is the first music video star. This technique is what has made him without equal, on a rank of his own, truly transcendent. The religion of the star becomes loftier and loftier as the dulled senses of diverted humanity float down with the stream toward they know not what.

CHAPTER XX

Terrorism in the Velvet Glove of Technology

I know that in linking terrorism to technology I will scandalize the technicians who use it peacefully or who invent means to make life better and who have no thoughts of terrorism. I know that others will think that I am exaggerating. But again, we need to understand one another. I am not talking about technical terrorism, although I might have shown that the growth and popularization of technical means has made terrorist acts much easier.[1] My subject, however, is the discourse of technique, or technology in the strict sense, and my point is that this discourse is terrorist. But I must also explain my use of the word *terrorism*. I am not using it here in the literal sense which would relate "terrorist" to those who set off bombs. I am recalling instead the usage around 1968 when a professor might be called a terrorist because his status enabled him to influence his students with lectures which no one could contest, or when teaching the mother tongue was terrorist because, along with the language, images and symbols and judgments were impressed upon the minds of children that they could never cast off, so that they were not allowed to develop freely.

At that time I showed that the analysis was both correct and absurd! It is in this abstract sense of molding the unconscious with no possibility of resistance that I am adopting the word *terrorism* in this context. My point is that the discourse on technique which we encounter everywhere and which is never subjected to criticism[2] is a terrorism which completes the fascination of people in the West and

---

1. See Jacques Ellul, "La démocratisation du Mal," *Sud-Ouest Dimache*, Sept. 1981, p. 5.
2. Occasional denunciations cannot hold their own against the flood of talk in the media.

which places them in a situation of twofold irreversible dependence and therefore subjugation.

This terrorist discourse rests on a picture of tomorrow's society. It is clear, incontestable, and beyond a shadow of a doubt that the society of 2000 will be an entirely computerized society, a communication society, a high-tech society, a society of space colonization, of unlimited energy, of radically transformed production thanks to industrial automation and robotics, a society in which artificial intelligence almost completely replaces human intelligence, and in which material shortages will be largely made up by the creation of new materials (see the enthusiasm of A. Ducrocq, who speaks of the "morning of the elements"), which are just as good as the old and even better. This society will also be one in which the methods of production are so different that it will be impossible to think of work as we know it today, in which communities and transport will be completely changed, in which the main problems of birth, living, and aging will have been solved, and in which nutritional problems will also have been solved thanks to new and inexhaustible foods. It will be a society in which eventually questions of consumption will be pointless because new products and services will make possible new and balanced budgets both public and private.

Am I dreaming? Am I simply sketching another *Brave New World*? Not at all! I am simply reproducing the headings of a most official document which I have quoted before, namely, *Prospective de 2005. Sept Explorations de l'avenir*, the report which a forecasting group drew up for the planning commission and the prime minister.

In the private sector we find exactly the same global view of the year 2000 in such works as those of Ducrocq or Bressand and Distler. We also find it in *Le Monde*. The interesting thing is that these are not "scenarios" (with different parameters, multiple variations, or varying solutions). No, the world will be like this in 2000. We have here a complete and certain forecast. Given what exists today, one cannot contest that what is forecast will come to pass. Money will have given way totally to credit cards. All objects in use will be computerized. All relations will be by the computer route. New materials will make possible structures that are inconceivable today in housing, transport, and also prosthetics. The great difference from *Brave New World* or science fiction, whether films or stories, is that those are works of imagination, so that what they foresee is possible but not certain, and need not be taken seriously. But the reports and works to which I have referred claim irrefutable scientific certainty.

I will not return to what I said earlier about technical unpredictability. What I have noted in reading official or semiofficial texts is a complete failure to mention four phenomena which seem to be very important: the possibility of a nuclear war, the possibility of a general Third World revolt, the possibility of an exponential increase of unemployment, and the possibility of a general financial collapse of the West due to accumulated debt. These four possible disasters are ignored. Furthermore, the report which I quoted deals solely and expressly with France.

I do not want to give the impression of thinking that the experts had no idea at all of the four possibilities I list. I am simply saying that they present a future in which these possibilities have no part, without taking the precaution that all could be disrupted. No, the society of 2005 will be like this. They are thus giving a practical demonstration of my 1950 thesis, violently attacked at the time, that technique is the new fate of our era. Technique is incontestable and inexorable; it is our destiny. When the forecasters portray society as it is then going to be, they make it plain, though without using metaphysical or literary terms, that it cannot be different. An irresistible fate is taking us to that point. This is going much further than the popular slogan that we cannot stop progress. It is closer to the important statement of Professor Bernard on television when, answering the reservations of Professor Testart regarding genetic engineering, he said that we must never impede scientific research.

Our future is clear and irrevocably fixed. This might, of course, be a matter for intellectual debate. One might bring against Ducrocq or Bressand the damage that progress does, or quote critical reports. But this is no longer the situation. The report that I have summarized is an official text which is meant to show the government what it should do and forecast. Above all, we have to see that this is not a mere hypothesis or probability. No, tomorrow's society *is* like this. The main task of the government, then, is on the one hand to prepare young people to enter this society and on the other hand to bring this society into being. It is here that terrorism arises.

This totally technicized, computerized society is inevitable. Thus we have to go with the flow, to make it arrive, to preside at its birth, and to integrate the new generation into this world. We no longer have any choice. There are no options, which would be useless, for we know what the outcome will be. In a different ideological context we, too, must go with the flow of history. But we are well aware that when we do this we are scientifically determined. (Do not forget that Marxism was regarded as the science of sciences; today technologism is.) The ineluctable outcome is dictatorship and terrorism. I am not saying that the governments that choose this as the flow of history will reproduce Soviet terrorism. Not at all! But they will certainly engage in an ideological terrorism.

Three threats are the key to this previsionism. There is the threat of unemployment. If you do not take the path which leads to this society, if you do not prepare to be a technician of one of these leading techniques, you will inevitably be unemployed. Everything in the future will be done by the computer, by telecommunications, by networks and files. Those who are unacquainted with these will necessarily be marginal and totally unemployable. Children are incessantly told that here is their only future. Parents are told that this is the only choice. If their children do not become technicians they will be nothing. Technique is so broad that if the children have training in computers they can move in many different directions. They can even become theologians if they are trained in the biblical and theological use of the computer, as at the Louvain Institute.

There is also the threat to the intelligence. Only those who know how to handle the computer are intelligent in our society. We have only to recall the young computer geniuses to whom we referred earlier. Knowledge of literature, ancient languages, or history counts for nothing. At a pinch there is a place for the humanities (though their stock is constantly reduced). But they have to show that they can help to adapt individuals to the technical world. If they cannot do this, they have no place. Entertainment and diversions can also be admitted, for they prevent people from becoming aware of what our future society demands.

Our world has obviously made great progress as compared with that of the 19th century. Then ideal industrial workers were uneducated people who would do what they were told. Today an appeal is made to every level of intelligence, and that will be even more true in the year 2000. Intelligence, linked to mathematics, is absolutely essential in our society. But it is no longer the intelligence of the humanities, of human beings as such. It is the intelligence that cooperates with the robot and that is modeled on artificial intelligence. The problem is no longer whether the computer is intelligent and a competitor. No, given the intelligence of the computer, the problem is how to educate us so as to correspond to it. We have to learn to pose problems differently in computer language and to solve them by other than classical procedures. We have to think in algorithms (unambiguous rules of thought, transformation that makes possible the passage from one representation to another, execution by a finite number of steps). We have to be shaped by the theory of algorithms, formal grammars, and the complexity of memory (which will make it possible among other things to determine the calculability of an algorithm), by the theory of graphs, and by the semantics of computer language. This is the kind of intelligence that is indispensable in the society which is ineluctably predicted. The interesting thing, however, is that those who are educated this way are completely obtuse when it comes to other forms of intelligence. Happily these no longer count.

During the transitional period, of course, what I am indicating might seem strange and very difficult. Not at all! This terrorism does not terrify; it acclimates. Thus the computer might seem hard to understand and to use. Its achievements are so astonishing that there is a real computer myth. But some have given themselves the task of "demythifying" it. This is being done, however, in a strange way. The formulation is that "the technology in this area of the computer is still in prehistory. . . . The gap between computers which have evolved very fast and business needs which have evolved less fast justifies the need for service companies. We shall continue for another decade or so to demythify the computer. This is for the sake of users, who must not be isolated from suppliers" (*Le Monde*, July 1986). The point is that there has to be a business which helps other businesses to choose the best computer. In other words, demythifying the computer does not mean showing its fragility, mistakes, or dangers, as do Vitalis and Chamoux. On the contrary, it means creating a need for it where the need is not yet felt, encouraging more users, teaching them to buy the model that best suits their needs, proving to them that the computer is not miraculous or mythical but that the only intelligent thing to do is to use it. The computer is really so simple. It can help you a lot. You do not want to die an idiot.

This is the kind of thing that is really being said by an important service company in the demythifying of the computer. There is a true threat to the intelligence, which is global. To be "in," "to be linked"— formulas essentially based on the computer—is a proof of intelligence. These are terrorist formulas. Those who are not "in" or "linked" are treated with sovereign disdain and pity. Advertising also tends to show that for every problem the good technical fairy will have the solution. But we have to be willing. We have to be ready to accept its help and to be comfortable with what it can provide. We must not be defiant or reticent. Terrorism is composed of three steps: psychological preparation, education, and compulsory imposition. We have said a few things about psychological preparation. We now take a look at the other two aspects.

Education or instruction: the idea is simple. Since tomorrow's society has to be thus and cannot be different, we have to prepare young people (and the not so young) to enter it, not to be alienated by what they find in it, to be accustomed to its workings. Around 1930 teaching on automobiles was introduced into the schools. But the automobile was not primarily a working tool. Today we have to prepare children to use the new instruments (e.g., teleconferencing, printers, computer graphics, etc.) as a help in making decisions. Computer science is thus an indispensable subject along with French or mathematics. It can also be integrated with other subjects, helping children to learn history or physics.

This is why I talk of terrorism. *All* children must now learn to use computers. They are shaped by them and adapted to them. This adaptation will one day go so far that orthography itself will have to be changed to fit computers. The computer will mediate all things intellectual and the whole intellectual formation of the child. What we hear repeated a hundred times is always the same. Children must be able to use computers because tomorrow they will be the universal work environment.

What is demanded of children, however, is a transformation of intelligence, for the computer claims to be not only a technique but also a science.[3] It changes our way of imaging things (whether they be physical, economic, linguistic, or biological). It gives us a new way of coding images, words, ideas, language. Everything must go into a code for the machine. It imposes its own language and its own way of putting problems. It produces principles and new concepts. In high schools it is both a science to learn and a tool for the teaching of other sciences. This is computer-assisted instruction. Experiments are being made with it in many high schools.

As the report notes, however, we must begin at the beginning. The computer should not be something supplementary. It is giving rise to a new culture (the famous technical culture) which reconciles theory and technique. It does this so well that when applied everywhere it becomes the central discipline (earlier French and mathematics had been the central disciplines in France, around which all the others revolved). Every field makes use of the computer, and as it will be inevitably in society, so it will be inevitably in education. All methods are good. Children will be taught telematics by telephone, by television, and not just at school. They will have to learn that the computer is the

3. What follows is taken from the report submitted to the President of France by the president of the commission, J.-C. Simon, *L'Éducation et l'Informatisation de la société* (Paris: Fayard, 1981).

irreplaceable tool for the modeling of complex phenomena and that its
concepts are of use in every discipline.

The technical culture becomes increasingly important whether
we like it or not, and schools must give future citizens the culture of
the future. Schools must raise students to the same technical level as
they will find outside. Using a computer can be associated with a game.
It is a pleasure to play around with gadgets instead of being bored by
books. Children must be prepared for added cerebralization (?) after
school by more intelligent (?) cultural activities.

This very complete report then lists the academic sphere in
which the computer can be applied and how it will be done (e.g., the
acquisition of languages, of knowledge, and of modes of reasoning). But
the modern methods of instruction (audiovisual, telecommunications,
computer-assisted instruction) all imply the domination of images over
writing or speech. In evaluating students' work seven criteria are to be
retained: memory, syntax, semantics, induction, deduction, strategy,
and creativity. Each corresponds to a "skilled operator," and students
must be trained to understand the process of intellectual development.
Training in computer science is essential if those who use computers
are not finally to see in them mere push-button tools but an aid to
creativity. The report examines in detail the application to each subject,
the levels of instruction, and computer-assisted instruction in all
advanced countries. It finally shows the advantages and disadvantages
of directed study, in which students are guided by the system following
their answers to multiple choice questions, and nondirected study,
which involves simulation, assisted conception, and programming.

I cannot give all the details. What I want to make plain is that
we have here as terrorist a system of education as there can possibly
be.[4] For young people have no escape, even less so because it is all
presented as a game. What is more ideal than to learn by playing? This
plan of educating all children by computer was envisioned by Giscard
and reinforced by the socialist government. In November 1984 the
Secretary of State, Mr. Carraz, announced that technical instruction
was one of the basic items in the modernization of the schools.
Technical instruction would be obligatory. Computers would be generally available, and schools would be equipped with microcomputers
and software. Mr. Fabius announced in January 1985 a plan, "comput-

---

4. It needs the courage of an anarchist to challenge the terrorism; cf. the
passionate work of Paul Feyerabend, *Against Method* (New York: Schocken, 1978),
which demands among other things the separation of science from the schools—an
excellent point in my view.

ers for everyone," which would put 100,000 microcomputers in primary schools, high schools, and colleges, with 11,000 workshops for computer training. High schools would have semiprofessional materials. The plan would cost 2 billion francs (about $400 million), and 100,000 computer instructors would have to be trained. By January 1986 Mr. Chevènement had put 120,000 computers in educational establishments. A program of "computer culture" had been set up, and it was hoped that computers would lead to more strictness and effort in French writing (cf. the Simon report). Over 110,000 teachers, enthusiastic volunteers, were trained, and it was hoped to open computer classes to the public and especially to parents.

This is in my view one of the main forms of this terrorism and it is based, as we have seen, on the certainty that ours will be a total computer society. Children have to be assimilated to the computer, which becomes the judge of all things with no possibility of critical evaluation by other criteria (e.g., a possible conflict between word and picture) or by a culture founded on other values and made up of other modes of education. If there had been serious reflection (i.e., not based on the assumption that the computer is the key to our society), it might have led to the view that the computer might be one subject of instruction among others, and therefore optional, instead of being made a general teaching aid or something which forces us to adopt new concepts, a new logic, etc., and which in some sense falsifies everything that traditional culture has slowly developed.

The plans and projects of the various governments—for Chirac has naturally endorsed the suggestions of Giscard and the work of Fabius—display a total lack of reflection. For eventually the young people who will be taught a different mode of reasoning and of seeing and interpreting reality might well find themselves twenty years later in a radically different society from that which these futurists predict. What will become of them then? What will they do if there is a crisis or a war? How will they survive, having been trained for an automated, computerized world? For I do not believe at all that those who have been molded in this way will be able to adapt quickly.

That which has enabled human beings to survive, their versatility, is being lost due to the coupling of people and machines. This coupling imposes a true terrorism. Quite apart from disasters, do we not see with what prodigious speed changes have taken place in the three great technical spheres? For example, the speed of change in artificial languages: who would dare state that the knowledge accepted for using computers in 1990 will still be valid for computers in 2010? It might be replied that the principles remain the same. My simple retort would be that there is nothing to prove this. Computers were not the same prior to microprocessors. The vaunted knowledge that is spread abroad today might well be useless for the computer of tomorrow.

But the computer is not the only factor in the modeling of children. Television is another. An important difference is that whereas the subjugation is intentional and calculated in the case of the computer, it is spontaneous and desired in the case of television, and its effect is thus harder to measure. The abuse of the screen by children results in the creation of a more difficult relation to space, for things change on the screen more quickly than the eye can follow, the eye being trained in space to follow real movements. There is thus a false estimation of space which has been confirmed medically again and again. Children who live too much before the set cannot do certain things. Educators have also stated that children who watch too much television are both overexcitable and somnolent at the same time, passing through phases of excitement and debility.

Some say that television opens up children's minds and gives them a smattering of everything. As we have seen, however, we need to know what this culture consists of, for talking about this and that is not a coherent training of the intelligence. Does television really teach children anything? Plainly, it does not (cf. Piveteau, Cazeneuve, Holtz-Bonneau). There are, of course, educational programs. But do children really want to watch these rather than infantile cartoons or films of terror and violence? By what mysterious means will children choose and assimilate good programs and avoid the follies of Mandrake and other science-fiction programs that are full of violence, star wars, etc.? We would have to believe that they have a superhuman wisdom and perspicacity much superior to that of which adults are capable. Furthermore, English psycho-sociologists (cf. the various warnings to "be careful what your children watch," and Brethenoux's thesis) have shown that when certain programs are assiduously watched, behavior can change, but not basic attitudes.[5] This proves that television acts more by structure than by content. Nothing conscious or intelligent can be gained by means of it. Many psychologists—and parents have perhaps found this to be true—argue that television prevents children from engaging in natural activities. They no longer know how to play spontaneously and even less how to invent games on the basis of things which are nothing in themselves but become what they are through the

---

5. See William Belson, *Television Violence and the Adolescent Boy* (England: Saxon House, 1978).

imagination. When no longer fascinated by the screen, children do not know what to do. They are unable to create; they are bored. This is one of the interesting results of terrorism in a velvet glove. But this is no longer the terrorism of discourse about technique. It is the terrorism of technique's own discourse, which is symbolized in televisual images.

This acclimatization, sometimes obligatory and sometimes voluntary, to the most modern technique—and television plays its part by so blinding children with what is done by technical machines that they can no longer distinguish between reality and fiction, everything being a show—must certainly continue after the years in school and college. Adults have to plunge into this bath, too. This is the aim of the grandiose Museum of Science, Technology, and Industry opened at Villette. I will not refer again to the frightful cost. In this context the important thing is the purpose, which is to set visitors in a world full of all that average people today can understand, perceive, and visualize. They must see the most sophisticated tools and machines and also their connections. They must not only see but also experiment with them. There must be interactive presentations in which they can manipulate, move, and converse. (Exceptionally tough equipment is thus needed!) The public must not be merely window-shopper consumers.

To justify the enormous expenditure the same argument is always used: France must be up-to-date. All French people must be given the chance to possess the culture with which to confront the century. They are thus plunged into a universal technical bath which gains its fascination from the extraordinary potential of the machines. (Naturally, there is no place for reflection, for reserve, for a critical spirit!) The obvious aim is to stir many people to become research scientists, technicians, and industrialists.

"Central to Villette is the symbiosis of science and technique, and these are put to work in industry. To bring about the great transformation which the actual crisis both conceals and expresses there is need of more science, more techniques, and more industry. . . . We must help the coming generations to understand the world in which *they will live*."[6] We must thus seduce and integrate them. We must set them on the right track by preadaptation to what is thought to be the future. This exposition obviously presupposes unending progress, final evolution, and the ultimate possibility. Incessantly, then, it puts visitors in the future world, for the world in which they actually live is not that of highest progress. This is why this museum is part of the terrorism in a velvet glove which is never seen to be anything but a diligent

6. See P. Delouvrier, "Pour l'apprentissage du futur," *Le Monde*, April 1984.

servant. It is a true instrument of the thirst for progress and the incitement never to question anything.

At times, however, this terrorism can cease to be benevolent and gentle. It can become constraining and incontestable. We see this when people are practically prohibited from producing by their own means their own electricity. Électricité de France is not a state monopoly but it acts as though it is. There is no formal law against setting up a windmill or tapping a stream on one's own property. But there are many administrative obstacles and the charges are prohibitive. A more clear-cut example is the decision to put all homes on cable and to make whole groups take Minitel. The telephone company graciously makes a gift of this, but we can be sure that it will recover the cost later. The government, leaning on capital, industry, and more or less captive markets, is following a strategy very much determined by the spread of the computer. But in telematics it is plainly using its monopoly in telecommunications. Great Britain took the lead with Prestel, but France has adopted a more extreme policy with the free distribution of Minitel, first creating an electronic directory (cf. the Vélizy experiment in 1980-81).

There has been no real debate about these moves in spite of the promise of a big parliamentary debate in 1982. The same is true of the large-scale authoritarian distribution of videotex. After 1982 no debate was possible because the social actors were too much engaged in integrating videotex to take time to discuss choices and problems.

I talk about terrorism, then, for two reasons. First, great transformations are imposed on human and social relations without consulting the interested parties. (We grew used to this, of course, with nuclear power stations.) In our so-called democracy decisions are made to change society without considering the opinions of the "captive sovereign." The second reason, however, is more startling. A decision has been made to eliminate the old means. Thus the telephone directories are to be replaced by those of Minitel and videotex. This device will tell us all we want to know. We are not free, then, to reject it. We have no other means of finding information. This is a wholly terrorist measure. And we have to admit that it will grow worse as equipment is perfected. We will then be forced into futuristic technicization.

Our final task is to ask who are the agents of this technical or technocratic terrorism. First, of course, there are the political powers acting either by constraint or by massive propaganda. The government that acts in this way thinks that it is adopting important policies. Governing is foreseeing. Since society will necessarily be this computerized, automated society, the virtue of government is to preadapt institutions and people to what will happen just as surely as the sun will rise tomorrow. What we said earlier about advertising and publicity is important here, but we need not repeat it. Second, there are the technocrats and the technostructure. Their position is clear: the more society becomes technicized (the atom, the computer, satellites, genetic engineering), the more indispensable they become, the more power they have, the more important they are, the more money they make, the more difficult they are to uproot. Their propagation of all techniques and their crushing of nontechnicians and nonspecialists by their science and authority is an expression of both their self-interest and the strengthening of their situation. They cannot act in any other way. They are forced to reject increasingly what remains of democracy. We talked about this, too, in depicting the transition from the present system to an aristocracy.

But the first two groups who are responsible for this terrorism find a guarantor or opening quite ready in the public. I will state again an older law that I laid down in my book on propaganda, namely, that propaganda cannot succeed without the complicity of those at whom it is aimed. This is how things are today. There is complicity on the part of the public. Being badly informed—that goes without saying— the public is full of admiration for all that modern means can do and blinded by their obvious achievements. There can be no discussion with the man who walked on the moon or with the robot; we are struck dumb by such marvels. Because they are spectacular, all the techniques have become obvious. Evidence of this kind cannot be questioned. This is why the terrorism can be in a velvet glove. Its rests on advance evidence.

There remain two other groups that bear responsibility for this terrorism, namely, intellectuals and the churches. Their situations are similar in many ways. First, if they do not basically engage in this technicization, they are afraid of appearing reactionary and thus calling down upon them the scorn and derision of the parties of progress. Then they have the duty of forming the culture of their day, but how can they do this unless they integrate technique into it? Then teachers, aided and equipped thanks to the daily papers which give them ample crops of technical innovations, adopt what they think will be the best way of adapting children to tomorrow's society (e.g., history as the history of techniques and economics, geography as the geography of natural resources and economic changes, and similar inimitably silly fads). If teaching Racine or Roman history was a horrible terrorism in 1968, what are we to say of this type of teaching? The first had at least the merit of putting children in the concrete reality of their society and thus giving them the chance to adopt a critical standpoint. The second plunges them underwater so that they learn nothing and want nothing but science and techniques. This type of education is double terrorism.

To me, however, the churches seem worst of all. Whether we take the World Council of Churches or the papacy, they have become the privileged agent of technological enthusiasm. They are in a panic lest they should be thought to be behind the times, obscurantist, out of things. To show their good faith and broad-mindedness, they defer. Should they have the audacity to confess to reservations and to raise the question of truth, they are put in their place and told that they have nothing to teach anyone. Reference is naturally made to Galileo (wrongly, for Galileo was not prosecuted, as is said, on the purely astronomical issue). At a pinch it may be conceded that the church can express an opinion on the moral plane, for example, in matters of in vitro fertilization or the freezing of embryos or surrogate motherhood. But never must it meddle in basic matters. Happily, the clergy are only too willing to sound the trumpet to the glory of technique, signalling, as has been said, "the end of the era of suspicion."

Two recent works point the way. First, Michel Boullet shows that the hierarchy is increasingly open to artificial "communication."[7] His path is simple. As the influence of the churches diminishes, they need to become aware of the importance of the media, to invest in this sphere, and to train professionals. (Billy Graham made this discovery half a century ago, but the European exodus from the churches has continued!) Boullet makes the remarkable judgment that fear of the media is not Christian. I am not sure what is Christian, but in any event the effect of the media on their audience, and children in particular, is certainly unlikely to involve the transmission of even a grain of truth. Pierre Babin, a former collaborator of McLuhan, goes much further.[8] The actual status of the church, he thinks, has changed in the communication society. The audiovisual irruption has produced a spiritual renewal (a greater miracle than any at Lourdes!) and is transforming catechesis (as is obvious, though without making children more open to the truth). This is a theological debate into which I do not wish to enter here. I am simply saying that the church's spokespersons now favor broadcasting. A technological wager: The official church has not yet taken up the challenge but it can hardly fail to do so given its traditional sociological conformism and the example of the World Council.

7. Michel Boullet, *Le Choc des médias* (Paris: Desclée, 1986).
8. Pierre Babin, *L'Ère de la communication: réflexion chrétienne* (Paris: Centurion, 1986).

But before going into that, let us examine the vigorous thesis of Babin. Television, he says, has replaced the crucifix in the home and the imagination. The faith of the new generation is also marked more by the emotional and symbolic force of the audiovisual media than by intellectual adherence to a doctrine or dogma. These statements certainly merit reflection. Since when has faith in the Lord Jesus Christ been intellectual adherence to a dogma? Babin's "Christian reflection" also seems to me to be strangely subject to the primacy of fact (i.e., to that which is the basis of all technological procedure). This is the actual situation, and the church and revelation must be adapted to it. Babin calmly states that pleasure, beauty, and symbol will tomorrow be the privileged paths of faith and the knowledge of God rather than learning. What pleasure? What beauty? Works of modern art inspiring faith in God by their "beauty"? As for symbol, we have already studied the fact that the modern world is opposed to symbols.

The work of Olivier Rabut is another example of conformity to the technical society.[9] The procedure may seem to differ but it is really the same. Christianity must be modernized, that is, adapted to our technical society. The modern world *demands* a change of mentality on the part of Christians. Christian modes of thinking and even essential beliefs must change. Christian doctrine must be liquidated because it does not stand up to the pressures of science and technique. But vital Christian spirituality will survive. What does this spirituality consist of? Are not shamanism and voodoo just as spiritual? We never break free from the vague happiness of those who appeal to "living" rather than thinking (which is by no means new). The odd thing is the primacy of belief in modernity and progress, which have reason to call all things into question, including Christianity. This is again a belief in fact as the ultimate value to which everything must bow. No attempt is made to reverse things and to let Christian faith, clearly explained in an intellectual manner, judge these facts, these life-styles, these instances of pseudo-progress, this way of being modern. Authors of this type never dream of such a thing.

I will conclude my discussion of Roman Catholic writers with a series of articles by Michel Albert entitled "La Bonne Nouvelle cachée dans le développement économique."[10] Here—another surprise— there is perfect agreement between the gospel and economic growth. Thus far, Albert says, the gospel has been located in countries of famine, poverty, and "predation." It has recommended a sharing of wealth,

9. Olivier Rabut, *Peut-on moderniser le christianisme?* (Paris: Cerf, 1986).
10. Four articles in *France catholique,* Nov.-Dec. 1985.

condemned the accumulation of riches, and opposed powerlessness (i.e., being on the side of the lambs in societies divided between lambs and wolves). But that was all tied to the existing economic situation. It has changed today. Thanks to research and development we are entering an age of economic plenty. The wealthier we become, the more we can help others to become wealthy. The search for wealth is a good thing. Economic development is giving us a new gospel. All economic creation becomes a matter of human relations, that is, of religion. We are entering a world of creation in which individuals and groups are not enemies any longer but must have relations of trust. This is a fiduciary world based on a recognition of others and confidence in them. The rate of trust will grow with that of the gross national product. What are being created are no longer relations of competition but of partnership.[11] The new gospel is that of wealth, trust, and a decline of power.

We will now take a brief look at the Christian naivete of the World Council of Churches (WCC) in its attitude toward technique. I have in mind the Boston conference of four hundred scientists and theologians on the theme "Faith and Science in an Unjust World" (July 1979).[12] We note that already in the title there is emphasis on an "unjust world." In accordance with the dominant WCC trend, the world (i.e., the Western world!) is judged to be unjust. This is in line with the great predominance of African and Latin American churches in the WCC. Thus the problem of science and revelation is not really studied, but the whole problem is first set in the context of the injustice of the Western world. That said, after reading the two big volumes we are completely disappointed, provided we have any knowledge at all of the issue. The approach is mostly descriptive. We are simply told what is. No effort is made to evaluate, from the perspective of revelation, the scientific and technical developments. At best there is only juxtaposition: Science says this and theology that, with no interaction, and usually a justification of science as such, with perhaps some ethical limitations. The only thing that is condemned is the break between technique and humanity. Prometheus is good, but his deviation into Faust is reprehen-

---

11. This is all incredibly naive. Older societies were not always societies of penury. Relations of trust have existed before and were a hundred times stronger than they are today. Power is not any the less in our societies. Partnership is no more important now than it was eight centuries ago. The most glaring mistake, however, is the equation of the gospel with economic growth!

12. *Faith and Science in an Unjust World*, 2 vols. (Philadelphia: Fortress, 1980). In this attempt at Christian reflection we are also surprised to discover science in Islam and Buddhism, and Arab, Indonesian, African perspectives, etc.

sible. The promotion of science is a moral duty. It has an essential role in the building of a better world (Hombury Brown). In no sense is it an enemy of faith. Its modernity brings emancipation and postulates autonomy. There are also "audacious" formulations like the following: "Love and a just, participatory, ecologically responsible society are public criteria for the verification or falsification of science."

In the whole work what is said about science is astoundingly trite and for the most part presents techno-scientific realities that are totally outdated. There is no sense of the real situation and the reasoning is simplistic (e.g., that redistributing income will increase productivity and consequently the returns). We also find that the churches are beginning to take note of the lessons taught by Christians who live in socialist countries. There we find the ideal for a proper application of science. The dominating impression in both the scientific and the theological discussion is one of banality. What is true, I would say, is banal, and what is novel is erroneous. Thus it is said that everyone in the West now sees the need for a limitation of growth. Increasing the production of material goods is the necessary condition for transition from the realm of necessity to that of freedom (C. T. Kurien). The ethical sphere is just as badly treated as the scientific.

The whole work is a plea for science and technique so long as they are set in a socialist world. The work has been partly published in French with some added articles by Swiss theologians which are much better.[13] In these additions we find a much more refined sense of the problem but also great timidity. The conformist talk of the WCC is more relevant in the present context than these articles. The question that I have to ask is why the churches have so little judgment and so little critical spirit in a matter which concerns not only dogma but the conception of humanity as a whole and even the possibility of a revelation that is beyond the reach of science. I think that all the churches' reactions stem from the fear of not being modern, of not being up to date or "with it." It is much more important for them to preserve contact with their contemporaries than with God, to talk as society does than to listen to God's Word. They are thus victimized by the terrorism of opinion and communication as regards technique. To escape their own panic the churches become in their turn the cassette recording terrorism in the velvet glove of technology. This ensures that they will not be judged.

13. *Science sans conscience; foi, science et avenir de l'homme* (Geneva: Labor et Fides, 1980). This edition contains sixteen conference reports and articles by P. Gisel, J.-L. Blondel, M. Faessner, et al.

"This society is inevitable and we are thus preparing young people to enter it, to find a place and a job in it." This is the terrorist argument, as we have said. What seems not to be considered is that this society is not inevitable. By preparing people to enter it, by giving them no other aim than to be competent in it, by creating among them a frantic need to work on technique, by soaking them in the knowledge and coherent practices of this society, we are making it increasingly probable. What will finally make it inevitable is neither the development of science and technique nor economic needs but the shaping of people who can do nothing else and will not be comfortable in any other society. What makes techno-science inevitable is the belief that it is, the pseudo-predictive boasting, and the assuring of people that it is in process of realization.

Last Words

1. Inventing Humanity

This is the fine title of an excellent book,[1] and it has been taken up for some time by the press. Inventing humanity! A. Jacquard borrowed the phrase from Sartre and his use of it to describe the ideal objective of history. He follows the Marxist line: As we have thus far known only humanity's prehistory, modern humanity is only a preliminary sketch of what humanity is to become. For Marx, however, this will be the result of a transformation of the economic and social milieu. But today things have changed. Obviously, the tremendous power that humanity has now been given implies for it a transformation. We are condemned to invent humanity, says Jacquard, because of the technical change which has followed the first "invention of humanity" at the time of the determinative breakthrough made by *Homo sapiens*. This is also the view of those who talk about "science with a conscience."[2]

If we regard Bergson's idea of the supplement of the soul as totally ridiculous and idealistic, then modern orientations seem to be extraordinarily disturbing or blind. It is not at all clear what is meant by these reassuring formulas. What are the options and possibilities? We may set aside at once cybernetic or computer-man, who has electrodes implanted in the brain and who does exactly as commanded. A second suggestion is more tempting since we are already on the way to it by genetic engineering. This is the way of artificial insemination, of preservation of embryos in vitro, of the production of clones, of the endless reproduction of the same models, of the accurate detection of

1. A. Jacquard, *Inventer l'homme* (Editions Complexe, 1984).
2. Cf. E. Morin, *Science avec conscience*.

defects in the embryo, of the insemination of women with the sperm of unknown men, of the preservation of the sperm of great men, etc.

It is now fully possible to make the kind of people we want.[3] But this is where Jacquard shows wisdom, and we must read his book as a warning as well as a magnificent introduction. His judicious thesis is that although we are technically capable of doing anything by genetic engineering, there is no scientific indication of where we ought to go. It is very feeble to think merely that we should make hundreds of Einsteins.[4] Jacquard is in search of a path, a certainty, and he comes up against the limitations of human judgment, which has not kept pace with human knowledge. No matter what answers one may give, there is always, he thinks, a question which makes all the other questions and answers ludicrous.

Our knowledge is not on the same plane as our uncertainties, which are not of the same order or in the same sphere of comprehension as phenomena. The great venture of genetic engineering comes up against a giant obstacle. No one can answer the very simple question: What kind of person do we want to create? One that is above all intelligent, or religious, or muscular, or perfectly balanced physiologically, or altruistic, or egotistic, or fully integrated into the collective, or sensitive to beauty and artistic, or endowed with the critical judgment that makes for autonomous individuality, or conformist, or individual? We have to choose, for we cannot have them all. We cannot have a person that is both rigorously rational and highly spiritual.

When human types are presented to us in novels, we see clearly the uncertainty regarding the ideal model, and in futurist films or science fiction there is even horror at the model that is coming, since in spite of omnipotence it is either an evil genius or a stupid hero. There is no other option unless we are prepared for the silliness of *E.T.* Here, then, is the proof that no scientists, psychologists, sociologists, moralists, or philosophers are able to tell us what is the ideal model of

---

3. See A. Maillet, *Nous sommes tous des cobayes (sur l'expérimentation biologique et clinique sur l'homme)* (Editions Jeune Afrique, 1980); J.-P. Changeux, *L'Homme neuronal* (Paris: Fayard, 1983); Vance Packard, *The People Shapers* (Boston: Little, Brown and Co., 1977). I might have devoted this whole book to genetic engineering, for it is the most flagrant example of our unreason. But I am less familiar with this field than the others. Cf. the excellent article by T. Deutsch, "Les manipulations génétiques," *La Recherche* 110 (April 1980), in which he demonstrates the madness of biologists. Cf. also J. Testart, *L'Oeuf transparent* (Flammarion, 1986).

4. The case of Einstein is interesting because (1) he was a mathematical genius and a good man, (2) he could not foresee the results of what he was doing, and above all (3) his general political and religious ideas in his books are rather puerile.

humanity that we must reproduce with our technical methods. We know how. But what?[5] And even if we knew what, another small detail is being overlooked. This ideal person that we invent and produce by genetic engineering will have no freedom but will simply be the model that it is programmed to be. But freedom is a big thing. Can a person be ideal without it?

Face to face with these basic problems, I will never cease repeating the old joke of J. Rostand who, when working on the embryos of toads, said in 1960 that he could now create a toad with two heads or five paws, but he had not yet managed to make a supertoad. This is vital: a toad better adapted to live the life of a toad than one that is reproduced normally. I do not believe that the person created by genetic engineering will be a superman. A human being born naturally is a human being. Let genetic engineering rectify some of nature's mistakes and make possible the avoidance of certain psychological or physiological tragedies, but let it stop there. It is not its task to invent a new humanity in spite of the paeans of victory that regularly sound forth.

What other ways are there? Transition to an ideal socialist society is supposed to produce the new humanity. Aragon's novel on Communist humanity makes this claim. There is no need to insist that we do not know what this humanity is. Yet it is often thought that changing society or the social environment will produce the new humanity. The new humanity will no longer be planned and directed but will be the long-term or short-term result of the influence of the environment. In an interesting way we find here the antithesis of nature and culture, or of the natural and the artificial, or of the spontaneous and the planned. We might proceed to make the new humanity directly by technical means. Or we might wait for a change in the natural environment by social neo-Darwinism—the appearance of a mutant.

Very curiously, Scardigli combines the two. On the one hand he sees a humanist trend in the consumer society which "is rehabilitating the person."[6] Each has to believe that the future will be better than the present and act accordingly. The consumer society is thus reintroducing humanity as the principle and goal of economic activity. It is in humanity's name that there is innovation. Finite material needs are giving place to desire, and an infinity of objects is responding to an

5. See S. J. Gould, *The Mismeasure of Man* (New York: Norton, 1981); A. Jacquard, *Endangered by Science?* tr. Mary Moriarty (New York: Columbia University Press, 1985); cf. also "Les manipulations génétiques: des risques encore mal évalués," *La Recherche* 107 (Jan. 1980).

6. V. Scardigli, *La Consommation, culture du quotidien* (Paris: PUF, 1983), pp. 56-57.

infinity of desires. That may well be. But what "humanity"? We know the universal consumer: hypertension, cholesterol, obesity. "Desire" tells us nothing. Is this really the ideal humanity that we want? On the other hand, the same Scardigli tells us that the new humanity can be "produced," and he recalls the progress in medicine, quoting such banalities as the fact that the invention of the pill gives women a freedom that corresponds to personalist (?) values, that there are now convenient ways to eliminate pain, that cosmetic surgery can give us all beauty, etc. He also makes astonishing statements, for example, that the isolation of the sick proves that we regard them as autonomous persons, or that medicine today denies any participation of those around the patient in either the outbreak or the cure of sickness; patients are the persons responsible for their own treatment.[7]

This is in contradiction with two points made by analysts of the medical establishment, namely, that those around the patient play a vital part in healing and that patients participate less and less in the treatment that physicians prescribe. In other words, some people believe that the global evolution of society is leading to a positive transformation of humanity in which we will be on the same level as technical devices.

In reality all these are misleading illusions. The computer will not enable us to invent humanity nor will the coupling of people and computers (e.g., artificial intelligence, for it is not a matter of intelligence). Again, the more wide-ranging coupling of humanity and machine does not enable us to predict a human transformation. This transformation might be in any direction. As for genetic manipulations or operations on the brain, we have already stated what we think of these. We always run up against the same obstacle. We can invent humanity, but we have not the slightest idea of what humanity to invent.

Socrates did not have the same idea of humanity as Buddha, and neither of them thought in terms of *homo economicus*. As for obtaining by slow spiritual education a wise and moderate person who will use technical power for the common good, will this really come about when machines of increasing power are put in the hands of this and all persons? Is not the tempting model closer to home than Prometheus? The superman of Nietzsche seems the most probable. As regards artificial intervention, we have to remember that all the attempts had serious negative and harmful effects, not because technical and scientific progress is inadequate but because such effects are inherent in

7. Ibid., pp. 180ff.

every enterprise of this kind. As we have already seen, there can be no pure progress. Perverse effects are inherent in the existence and development of science and technique. Thus inventing humanity as though it were a technical and scientific activity will have destructive effects for humanity, or may well create monsters. We must have no illusions about this. As for inventing humanity spiritually, ethically, and educationally, I wish it could be done, but it should have been done during the last 2,500 years of effort. The sorry conclusion to which we are led is that in fact humanity has already been invented. It is part of the great design of which we spoke at the outset. This great design is the complete integration of humanity into the technical system.

2. The Great Design

The great design has three panels; it is a magnificent triptych. On the central panel is humanity perfectly adapted to the requirements of the smooth functioning of sciences and techniques. People are trained for this from their youth. Their main mission is to promote it. On the left panel is fascinated humanity: fascinated by the marvels of science and technique and by the ever-growing opportunities of our life. On the right panel is diverted humanity: games and distractions of all kinds, gadgets, etc. People here are diverted from seeing reality. They constantly flutter around the many brilliant lamps and possibilities of escape. If we close the side panels, on the central panel we have the representation of a perfectly balanced, happy, and fulfilled humanity, never protesting, knowing no trouble, calmed by hypnotics, *mens sana in corpore sano*, kept healthy by jogging and other kinds of exercise.

The great design is, above all, that there should be no conflicts: not within an individual, not with neighboring groups, not with corporations where one works, not with political authorities. We have not yet arrived at this point, but when we consider how ardently people in the West enter into false conflicts (e.g., electoral), we have to think that we are close. At root our society is not at all demanding, contrary to the impression one might have!

What, then, is required of people today? Essentially four things. Their first and chief duty is to work well, painstakingly, and punctually. The second is not to be bothered about collective matters, not to become involved, not to meddle, to leave things to those who are qualified to see to them: politicians to govern, the churches to dispense tranquillity, doctors and hospitals to see to the sick and elderly. Each one has a sphere—play, play, and we will take care of the rest. The third thing is to be a good consumer, to have good wages and to spend them, consumption being an absolute duty, the only imperative duty, for if people do not consume the pace will slow down, money will not circulate, and there will not be enough work. The final thing is to follow the opinions propagated by the media, to adopt the information and themes for reflection that are proposed, and not to seek further afield, since the information provided is sufficient, and occasionally a scapegoat will be found, a terrible enemy, though not too close or powerful, on which the crowd can vent its anger and show its independence of spirit. These, then, are the four duties of people today.

There are, of course, some serious obstacles: the unemployed, Third World famine, terrorism. But we are assured that on the one hand the black marks will soon be erased, and on the other hand that those who cannot be reached are lost in a confusing mist, tragic and distant figures. In sum, the four requirements of the great design, which are well on the way to being met, will make it impossible for people to have an individual view of their own lives or of the reality of the world in which they live. Amid all the extolling of the world of communication and information, the great choice which is being made is that of ignorance. Such is the great design.

Seldom is this admitted in principle but it often is in fact. J.-J. Salomon quotes an amusing passage from World Health Organization report which argues that from the standpoint of mental health, the most satisfactory solution to the problem of the future peaceful use of nuclear power is to raise up a generation that will have learned to accommodate itself for the most part to ignorance and uncertainty. If we do not know the risks, we will not worry, and this is best for everybody. This is mental health. Here is the balanced humanity of our closed triptych. We have to live with the conviction, says Scardigli, that as medicine can solve all problems, so what is damaged here can be repaired elsewhere. This combination explains our blindness. If we were to take note of the harm done by progress, we would begin to challenge the very foundations of our society. But this is unthinkable. Everything is in place for us to live in blissful ignorance. We are diverted and distracted. As I have said, but need to repeat, the concern to focus our energies of dissatisfaction, of protest, of questioning on false targets is one of the major tasks of the numerous communication systems.

The choice of ignorance agreed upon between those who work the machine and have an interest in this ignorance and those who are part of the machine and have an interest in their own peace of mind or mental health has the remarkable effect of completely erasing responsibility from our society. Everyone has become irresponsible. I am not referring now to the irresponsibility resulting from our insurance system. This is a simple matter and is of no particular interest. I have in mind two great blocks of irresponsibility in our society. There is that of the decision makers and that of the "untouchables." The decision makers—the politicians, administrators, and technicians—are fundamentally irresponsible. There are two main reasons for this. First, the events for which they ought to be responsible are mostly far too complex, and second, the process of decision itself is similarly far too complex.

The first point is plain. Who can be responsible for the enormous waste that we have seen? Politicians have made decisions, administrators have drawn up memoranda, and technicians and experts have made their own specific contributions. But among them, who is really responsible? The President who sets the general plan in motion? The cabinet members who in their own areas refine it? Who is responsible for a nuclear accident? The magnificent response in the case of Chernobyl is that human error on the part of a worker was responsible. A wretched fellow turned the wrong handle. Fortunately, he is dead. Take any matter and this is how it goes. But our present-day operations are a hundred times too complicated for a single person, or even ten, to be responsible. In the great decisions that have now to be made there is only one in which the responsibility is clear: that of a head of state pressing the button for nuclear war. But this brings us back to simple cases.

Where do we still see responsibility? When one policeman kills a criminal; when one surgeon botches an operation. These are precise and limited incidents. They are not complex; they concern only one person. Elsewhere there is mitigation. Who can foresee the landslide that causes a dam to shift and finally to crack? Who can calculate exactly the trajectory of the second, discarded engine of a rocket? Who can envision the consequences of depositing barrels of dioxin in a quarry, etc.?

Every day I would say that hundreds of complex matters (not all of them equally grave) can have serious negative effects. But responsibility is mitigated in the two ways that we have indicated. The actual event is complex, for it is only one possible result of the enormous complexity of the decisions that lead to the unfortunate incident. The global phenomenon is intrinsically complex. It has not been made such. All that we attempt with our leading techniques (e.g., space, computers, lasers, and atoms) is the result of a series of detailed operations and interwoven micro-decisions. In no operation does there seem to be one clear responsibility.

To this situation there corresponds the inevitable complexity of

administrative machinery and industrial organization. Where are we
to seek those who are to blame, those who are responsible, in the tangle
of decisions and micro-executions? In his *Nomenklatura* Voslensky has
given us an incredible example of administrative complication in the
decisions of the supposedly all-powerful Soviet Central Committee. No
one can be held responsible for anything. Those in power are not
responsible because their fault or responsibility can never be pinpointed.

Then there are the "untouchables." In our society we have two
great classes of untouchables. There are the scientists, whose absolute
autonomy is guaranteed by the omnipotence of sacrosanct science.
Who would ever dare accuse a scientist? Then there is the political
class.[8] Though its members are divided into warring parties, they are
all essentially at one in defense of their class status. So long as there is
a political class in so-called democratic countries, there can never be
true politics or true democracy.

There can, of course, be political scandals. These are usually
financial, but for the most part they are on a small scale and involve
only individuals who have not kept the rules. Beyond that there is no
responsibility for political decisions made by a group in which each is
defended by all.[9] When by chance action is taken, one need not fear
that the guilty party's career will be ruined, for there are always enough
political friends to provide an honorable position.

As for scientists, they are universally untouchable, for they carry
with them our future. This is true even though it may be realized that
in many ways science participates in the bluff of technology. I will take
up only one aspect. (The deifying of science is not my present theme
beyond what I have already said on this score.) I refer to scientific fraud.
We are familiar with financial fraud but there is also scientific fraud.[10]

8. I have often denounced the total irresponsibility of this class resulting
from the fact that politics is both a living and a career. Some simple decisions would
put an end to it: no plurality, no mandate; the impossibility of having more than one
reelection; the exclusion of elections when one has fulfilled four times a representative mandate. Thus simply and assuredly one could achieve true representation and true democracy.

9. Whenever sanctions are taken against a political group for their decisions,
it is always in the course of a "revolution." In 1917 and 1933 politicians as a whole
were condemned by the victors. But unless they are killed they will reappear with
glory a few years later, as happened in the case of those who were condemned in
France in 1940.

10. See M. Blanc, G. Chapoutier, and A. Dawchin, "Les fraudes scientifiques," *La Recherche* 113 (July 1980); P. Tuillier, "Le scandale du British Museum,"
La Recherche 125 (Sept. 1981); cf. also 106 (Dec. 1979).

One of the most eminent of British psychologists, Cyril Burt, made up facts and experiments that he reported regarding intelligence and heredity. Chemistry has also had its frauds, for example, when a laboratory made up the results of its work on plutonium. Then there was the famous Piltdown skull, or the less familiar scandal of experiments relating to evolutionary continuity and their interpretation. More serious was the complete fraud of F. Moewvs in molecular biology. In biology, too, P. Kammerer achieved considerable renown by proving the heredity of acquired characteristics. Unfortunately, it was discovered that one of the proofs in his laboratory had been forged in India ink. Fraud of this kind is common even in serious laboratories. We might mention finally Schubert's tremendous fraud when he announced that he had a complete cure for plutonium poisoning on the basis of falsified experiments (1978).

The other type of fraud is not in scientific work itself but in the technical conditions of its execution. Some experiments ought not to be carried out because they are so dangerous. Early research on encephalitis involved many experiments whose author himself admitted in 1977 that he never performed them. He was so sure of his ideas that he regarded the experiments as useless, though he later wrote them up! Often the scientists concerned may be world famous, like Schubert with plutonium. Thus even in sacred science, the pure search for truth, there are schemes, fraud, and falsehood. Science is not just "impure" in terms of its political or social consequences. It is also impure in terms of the irresponsibility of scientists who are above suspicion. We should note, however, that the immunity from responsibility which politicians and scientists enjoy is possible only in a general climate of irresponsibility.

I ought to have recalled that the word *irresponsible* has a double sense. The mentally deficient are not responsible for their acts. But that is not the meaning here. Those who commit a crime, or attack public morals or their neighbors, or violate the truth but cannot be questioned on the matter, are irresponsible because their social status puts them outside the norms and relieves them of any investigation. There is also, however, a third sense, that of evasion. People may be held responsible but deep down inside they mock at responsibility and try in every way to escape their obligations. They seek protection. This is the point of insurance. But it is not just the financial world that engenders and reinforces the inner feeling of not being responsible for anything. The general climate of irresponsibility makes possible the irresponsibility of politicians, scientists, and technicians.

In opposition I would refer to the old example of the ship's captain. If there was an accident, whether the captain was personally responsible or not, he would assume responsibility and go down with his ship even though most of the passengers and crew might be saved. Similarly, in the 19th century when a businessman or banker failed, whether he himself made the unfortunate speculations or not, he would commit suicide. Absurd? Yes, like everything that relates to human dignity. But it solved nothing? It did not put his affairs back in order? And does general irresponsibility settle our economic and political affairs? This was a matter of honor, the exact opposite of irresponsibility. But no one today knows what honor is.

I want to quote at some length from Olivier Merlin's remarkable article in *Le Monde* called "Béjart et les robots."[11] Its theme is the "Mass for the Future."

> Whether the younger generations like it or not, the delightful times that await our grandchildren will in all probability have restored to honor the bestiality of former ages, the domination of strong males, and heroic songs on the computer. Such French privileges as charm, compliments, and trifles will be for ever excluded from these implacable relations in which any reference to culture is regarded as a crime. The essential merit of Béjart's mass is that it gives us a wild vision of the new world whose programmed gesticulations will one day regale humanity. . . . The pure rhythmics swarm with ideas that test whether the nerves are in a messianic (?) state. Drive this point home: the prehistoric plantigrade descending from the trees will one day call the tune and *homo sapiens* or the supreme thinker of Teilhard de Chardin will be viewed as a little joke. The second part of the mass, which would have been enough alone and which Béjart calls "the conventional world," takes on direct intensity from the fact that four robots move out in their glass cages. They are terrifying, these robots, unbearably ugly. . . . In the night of disheveled relations a spark of altruism will seem to be born from the depths of their mechanized being. . . . Yet I do not mitigate in the least my profound disapproval of the musical nexus. Béjart has always shown a liking for horrifying Hindu or Far Eastern rhythms, and in trying to astound the middle class he astounds himself.

It seems to me that Merlin's text is important when we consider that this is supposed to be a mass *(ite, missa est)*, that its author is regarded as France's greatest modern artist in music and ballet, and that art has the role of being an accurate and absolute reflection of a society at its

11. Olivier Merlin, "Béjart et les robots," *Le Monde*, Feb. 1984.

height. I think that the work expresses very well the meaning of the ideal shape of the finished product, the bluff of technology.

Is this a closed situation? Is there no way out? Is collective spiritual and material suicide the only result that is incontestably held out to us by the actual bluff of technology? Having stated that we can foresee nothing with certainty, I can hardly fall into falsely prophetic prediction in looking at the logical consequences of the bluff. But I can give a warning. If we have any chance of emerging from this ideologico-material vice, of finding an exit from this terrible swamp that is ours, above all things we must avoid the mistake of thinking that we are free. If we launch out into the skies convinced that we have infinite resources and that in the last resort we are free to choose our destiny, to choose between good and evil, to choose among the many possibilities that our thousands of technical gadgets make available, to invent an antidote to all that we have seen, to colonize space in order to make a fresh beginning, etc.; if we imagine all the many possibilities that are open to us in our sovereign freedom; if we believe all that, then we are truly lost, for the only way to find a narrow passage in this enormous world of deceptions (expressing real forces) as I have attempted to describe it is to have enough awareness and self-criticism to see that for a century we have been descending step by step the ladder of absolute necessity, of destiny, of fate.

Following Hegel, Marx, and Kierkegaard, I have often said that we show our freedom by recognizing our nonfreedom. But this is no longer a philosophical or theoretical matter of the mind. It is no longer a matter of debate between the servile and the free will. Our back is to the wall. We must not cheat or think that we can extricate ourselves by talk. Seeing the Hydra head of trickery and the Gorgon face of hi-tech, the only thing we can do is set them at a critical distance, for it is by being able to criticize that we show our freedom. This is the only freedom that we still have if we have at least the courage to grasp it. Nothing is more certain.

Are we then shut up, blocked, and chained by the inevitability of the technical system which is making us march like obedient automatons thanks to its bluff? Yes, we are radically determined. We are caught up continuously in the system if we think even the least little bit that we can master the machinery, prepare for the year 2000, and plan everything. Yet not really, for the system does not stop growing, and thus far we have no examples of growth that does not reach the point of imbalance and rupture. For the last twenty years balance and cohesion have been increasingly difficult to maintain. Not really, for as we have seen, the gigantic bluff is self-contradictory and it leaves a margin of chaos, it covers gaps without filling them, it gives evidence of mistakes, and it has to multiply deceptions to veil the absence of feedback in the system. Even without nuclear war or an exceptional crisis, we may thus expect enormous global disorder which will be the expression of all the contradictions and disarray. This must be made to cost as little as possible. To achieve that, we must meet two conditions. We must be prepared to reveal the fracture lines and to discover that everything depends on the qualities of individuals. Finally, not really, if we know how little room there is to maneuver and therefore, not by one's high position or by power, but always after the model of development from a source and by the sole aptitude for astonishment, we profit from the existence of little cracks of freedom, and install in them a trembling freedom which is not attributed to or mediated by machines or politics, but which is truly effective, so that we may truly invent the new thing for which humanity is waiting.

BIBLIOGRAPHY

I do not pretend to offer here a complete bibliography on techniques since 1977 (see *Technological System* for earlier works). I cite only those books which allow the reader to find the basic elements of my analyses. I have also limited the references to certain authors who remain fundamental, like Max Weber, Bertrand de Jouvenel, Georges Friedmann, and Bernard Charbonneau.

Ader, M. *Le Choc informatique.* Denoel, 1984.

Aizcorbe, R. *Fin de Milenio.* Buenos Aires: Occitania, 1985.

"Ambiguïtés du progrès." Special number of *Lumière et Vie.* Lyon, 1981.

Amery, C. *La Fin de la providence.* Paris: Seuil, 1977.

Amir, S., et al. *La crise, quelle crise?* Maspero, 1982.

Atlan, H. *Entre le cristal et la fumée.* Paris: Seuil, 1979.

Aznar, G. *Tous à mi-temps.* Paris: Seuil, 1981.

Beillerot, J. *La Société pédagogique.* Paris: PUF, 1982.

Berleur, J., et al. *Une société informatisée—Pourquoi? Pour qui? Comment?* Presses universitaires de Namur, 1982.

Birou, A., and Henry, P. *Pour un autre développement.* Paris: PUF, 1977.

Blanc, M., et al. *L'État de sciences et des techniques.* Maspero, 1984.

Boucher, W. *Study of the Future: An Agenda for Research.* National Science Foundation, 1977.

Bourguinat, H. *Les Vertiges de la finance internationale.* Economica, 1987.

Bressand, A., and Distler, C. *Le Prochain Monde.* Paris: Seuil, 1986.

Brethenoux, D. "Étude de la réception télévisuelle: Sémiologie T.V., réception T.V." Dissertation, University of Bordeaux, 1985.

413

Brzezinski, Z. *Between Two Ages: America's Role in the Technetronic Age.* New York: Viking, 1970.

Bussy, J.-C. *La Forêt de l'An demain, La maison rustique,* 1980.

Caballero, F. *Essai sur la notion juridique de nuisance.* Pichon, 1981.

Camilleri, J. A. *Civilization in Crisis.* Cambridge: Cambridge University Press, 1977.

Castelli, E. *Il tempo inqualificabile.* Cedam, 1975.

Castoriadis, C. *Crossroads in the Labyrinth.* Cambridge: MIT, 1986.

————. *The Imaginary Institution of Society.* Tr. K. Blamey. Cambridge: MIT, 1987.

Cazeneuve, J. *L'Homme téléspectateur.* Denoel, 1983.

Centre d'études et de recherches de biologie, R. I., océanographie médicale. *Les Pollutions chimiques de la mer.* C.E.R.B.O.M., 1972.

Cérézuelle, D. "Le Mythe de la technique." Dissertation, University of Dijon, 1976.

C.F.D.T. *Les Dégâts du progrès.* Paris: Seuil, 1977.

Chamoux, J.-P. *Menaces sur l'ordinateur.* Paris: Seuil, 1986.

Charbonneau, B. *Le Chaos et le Système.* Anthropos, 1973.

Chesneaux, J. *De la modernité.* Maspero, 1983.

Chevalier, J. M. *L'Économie industrielle en question.* Paris: Calmann-Lévy, 1977.

C.N.R.S. *L'acquisition des techniques par les pays non initiateurs.* 1973.

Coriat, B. *Science, Technique et Capital.* Paris: Seuil, 1976.

Cosmao, V. *Changing the World.* Tr. J. Drury. New York: Orbis, 1984.

Derian, J. C., and Staropoli. *La Technologie incontrôlée?* Paris: PUF, 1975.

Ducrocq, A. *1985-2000, le Futur aujourd'hui.* Plon, 1984.

Dumouchel, P., and Dupuy, J.-P. *L'Enfer des choses.* Paris: Seuil, 1979.

————, eds. *L'Auto-organisation.* Colloque de Cerisy. Paris: Seuil, 1983.

Dupuy, J.-P. *Ordres et désordres.* Paris: Seuil, 1982.

Durandin, G. *Les Mensonges en propagande et en publicité.* Paris: PUF, 1982.

Durbin, P., et al. *Research in Philosophy and Technology.* Greenwich, CT: JAI Press, 1982.

"Écologie, l'écologie, enjeu politique," *Le Monde* 1978.

Éducation 2000. No. 19, "Informatique au present." 1981.

Ekeland, I. *Mathematics and the Unexpected.* Chicago: University of Chicago Press, 1988.

Ellul, J. *Changer de révolution: L'Inéluctable Prolétariat.* Paris: Seuil, 1982.

————. *L'Empire du non-sens: L'Art et la société technicienne.* Paris: PUF, 1980.

———. *The Humiliation of the Word*. Tr. J. Hanks. Grand Rapids: Eerdmans, 1985.

———. *The Political Illusion*. Tr. K. Kellen. New York: Knopf, 1967.

———. *Propaganda: The Formation of Men's Attitudes*. Tr. K. Kellen and J. Lerner. New York: Knopf, 1965.

———. *The Technological Society*. Tr. J. Wilkinson. New York: Knopf, 1964.

———. *The Technological System*. Tr. J. Neugroschel. New York: Continuum, 1980.

———. *What I Believe*. Tr. G. W. Bromiley. Grand Rapids: Eerdmans, 1989.

Encyclopédie de l'Écologie, Le présent en question. Paris: Larousse, 1977.

Feyerabend, Paul. *Against Method: Outline of an Anarchistic Theory of Knowledge*. New York: Schocken, 1978.

Galbraith, J. K. *The Affluent Society*. 2nd ed. Boston: Houghton Mifflin, 1970.

———. *The Nature of Mass Poverty*. Cambridge: Harvard University Press, 1979.

Gaudin, T. *A l'écoute des silences*. 1980.

———. *Les Dieux intérieurs*. Éditions Cohérence, 1984.

Gellibert, J. "Le Choix de la biomasse comme énergie. Révélateur des mythes et des conflits de la société technicienne." Dissertation, University of Bordeaux, 1986.

Geze, F., et al. *L'État du monde 1984*. La Découverte, 1985.

Giarini, O., and Loubergé, H. *La Civilisation technicienne à la dérive*. Dunod, 1979.

Gille, B. *Histoire des techniques*. "La Pléiade." Paris: Gallimard, 1978.

Girod, P. *La Réparation du dommage écologique*. Pichon, 1974.

Goldschmidt, B. *The Atomic Complex: A Worldwide Political History of Nuclear Energy*. American Nuclear Society, 1982.

Goodman P. *Growing Up Absurd*. New York: Vintage, 1962.

Granstedt, I. *L'Impasse industrielle*. Paris: Seuil, 1980.

Grant, G. *Technology and Empire*. Toronto: Anansi, 1969.

Groetendieck, et al. *Pourquoi la mathématique?* "10-18." 1974.

Guinness, O. *The Dust of Death*. Downers Grove, IL: Intervarsity Press, 1973.

Hartung, H. *Les Enfants de la promesse*. Paris: Fayard, 1972.

Henry, M. *La Barbarie*. Paris: Grasset, 1987.

Holtz-Bonneau, F. *L'Image et l'Ordinateur*. Paris: Aubier, 1986.

———. *Déjouer la publicité*. Éditions Ouvrières, 1976.

Hottois, G. *Le Signe et la Technique*. Paris: Aubier, 1985.

Illich, I. *Tools for Conviviality.* New York: Harper & Row, 1973.

Iribarne, P. d'. *Le Gaspillage et le Désir.* Paris: Fayard, 1976.

Jacquard, A. *Endangered by Science?* Tr. M. Moriarty. New York: Columbia University Press, 1985.

———, ed. *Les Scientifiques parlent.* Paris: Hachette, 1987.

Jalee, P. *The Third World in World Economy.* Tr. M. Klopper. New York: Monthly Review Press, 1969.

Jamous, H., and Gremion, G. *L'Ordinateur au pouvoir.* Paris: Seuil, 1978.

Janicaud, D. *La Puissance du rationnel.* Paris: Gallimard, 1985.

Janicaud, D., et al. *Les Pouvoirs de la science.* Vrin, 1987.

Jézéquel, J.-P., Ledos, J.-J., and Regnier, P. *Le Gâchis audiovisuel.* Éditions Ouvrières, 1987.

Jouvenel, B. de. *La Civilisation de puissance.* Paris: Fayard, 1976.

Kahn, H., and Wiener, A. J. *The Year 2000.* New York: Macmillan, 1967.

Kerorguen, Y., ed. "Technopolis—L'explosion des cités scientifiques." *Autrement* 1985.

Koch, C., and Senghaas. *Kritischestudien zur Politikwissenschaft. Texte zur Technokratiediskussion.* Europäische, 1971.

Kolm, S.-C. *Philosophie de l'économie.* Paris: Seuil, 1986.

Lagadec, P. *La Civilisation du risque.* Paris: Seuil, 1981.

———. *Major Technological Risks.* New York: Pergamon, 1982.

Launay, B. de. *Le Poker nucléaire.* Syros, 1983.

Laverrière, H. *Repenser ce bruit dans lequel nous baignons.* La Pensée universelle, 1982.

Lavigne, J.-C. *Impasses énergétiques: défis au développement.* Éditions Ouvrières, 1983.

Lefevre, B. *Audiovisuel et Télématique dans la cité.* Paris: La Documentation française, 1979.

Longacre, D. J. *Living More with Less.* Scottdale, PA: Herald Press, 1980.

Lussato, B. *Le Défi informatique.* Paris: Fayard, 1981.

Marchand, M., and Ancelin, C., eds. *Télématique, Promenade dans les usages.* Paris: La Documentation française, 1984.

Mattelart, A., and Schmucler, H. *Communication and Information Technologies: Freedom of Choice for Latin America?* Tr. David Buxton. Norwood, NJ: Ablex, 1985.

Mayz-Vallenillea, E. *Ratio Tecnica.* Mont Avila, Caracas, 1985.

Melot, N. *Qui a peur des années 1980?* Monaco: Rocher, 1981.

Mendel, G. *54 millions d'individus sans appartenance.* Laffont, 1983.

Michalet, C.-A., *Le Capitalisme mondial.* Paris: PUF, 1977.

Mirabail, M., et al. *Les Cinquante mots clés de la télématique.* Privat, 1981.

Montmollin, M. de. *Le Taylorisme à visage humain.* Paris: PUF, 1981.

Morgenstern, O. *On the Accuracy of Economic Observations.* 2nd ed. Princeton: Princeton University Press, repr. 1972.

Morin, E. *Pour sortir du XXᵉ siècle.* Nathan, 1981.

————. *Sociologie.* Paris: Fayard, 1984.

Morin, J. *L'Excellence technologique.* Publi Union, 1985.

Negri, A. *I tripodi di efesto, Civiltà tecnologica e liberazione dell'uomo.* Sugar Cie Editions, 1986.

Neirynck, J. *Le Huitième Jour de la Création. Introduction à l'entropologie.* Presses polytechniques romanes, 1986.

Nora, S., and Minc, A. *L'Informatisation de la société.* Paris: La Documentation française, 1978.

Nordon, D. *Les Mathématiques pures n'existent pas.* Actes Sud, 1981.

O.C.D.E. *L'État de l'environnement.* 1985.

Packard, V., *The People Shapers.* Boston: Little, Brown and Co., 1977.

Partant, F. *Que la crise s'aggrave.* Solin, 1978.

————. *La Fin du développement, Naissance d'une alternative.* Paris: Maspero, 1982.

Pelissolo, J.-C. *La Biotechnologie demain?* D.G.R.S.T., 1980.

Pinaud, C. *Entre nous, les téléphones.* INSEP, 1985.

Piveteau, J. *L'Extase de la télévision.* INSEP, 1984.

Politik und Wissenschaft. Verlag Buch, 1971.

Poquet, G. *Économies de matières premières.* Futuribles, 1977.

Prospective 2005. Sept Explorations de l'avenir. C.N.R.S. Report. 2 vols. 1986.

Puel, G. *Pourquoi la pauvreté.* Éditions Tiers Monde, 1986.

Puymege, G. de, et al. *Autour de "l'Avenir est notre affaire,"* Fondation Veillon, 1984.

Rapport et annexes de la Commission "Informatique et Libertés." Paris: La Documentation française, 1974.

Ravignan, F. de, *La Faim, pourquoi?* Syros, 1983.

————. *Naître à la solidarité.* Paris: Desclée, 1981.

Regimbal, J. P., et al. *Le Rock n'Roll.* Geneva: Croisade, 1983.

Rifkin, J. *Entropy: A New World View.* New York: Viking, 1981.

Rodes, M. *La Question écologique.* 1978.

Ronze, B. *L'Homme de quantité.* Paris: Gallimard, 1977.

Roqueplo, P. *Penser la technique.* Paris: Seuil, 1983.

Rostand, J. *Peut-on modifier l'homme?* Paris: Gallimard, 1956.

Rougemont, D. de. *The Future Within Us.* New York: Pergamon, 1983.

Salomon, J.-J. *Prométhée empêtré.* Paris: Pergamon, 1981.

Salomon, J.-J., and Schmeder, G. *Les Enjeux du changement technologique.* Economica, 1986.

Salomon, J.-J., and Lebeau, A. *L'Écrivain public et l'ordinateur* Paris: Hachette, 1988.

Scardigli, V. *La Consommation, culture du quotidien.* Paris: PUF, 1983.

Schumacher, E. F., and Gillingham, P. N. *Good Work.* New York: Harper & Row, 1978.

Schuurman, E. *Reflections on the Technological Society.* Toronto: Wedge Publication Foundation, 1976.

Shumann, J. B., and Rosenau, D. *The Kondratieff Wave.* New York: Dell, 1974.

Simon, J.-C. *L'Éducation et l'Informatisation de la société.* Paris: Fayard, 1981.

Stanley, M. *The Technological Conscience.* New York: Free Press, 1978.

Susskind, C. *Understanding Technology.* Baltimore: Johns Hopkins University Press, 1973.

Technisches Zeitalter. "Technik." V. Schilling, 1965.

Ternisien, J. A. *La Pollution et ses Effets.* Paris: PUF, 1986.

————. *La Lutte contres les pollutions.* Paris: PUF, 1968.

Touscoz, J., et al. *Transferts de technologie.* Paris: PUF, 1978.

UNESCO, *La Culture, la société et l'économie dans un monde nouveau.* La Baconière, 1976.

Vincent, B. *Paul Goodman et la Reconquête du présent.* Paris: Seuil, 1976.

Vitalis, A. *Informatique, Pouvoir et Libertés.* Economica, 1981.

Voyenne, B. *L'Information aujourd'hui.* Colin, 1980.

Weber, M. *Essais sur la théorie de la science.* Plon, 1965.

Weizenbaum, J. *Computer Power and Human Reason.* San Francisco: Freeman, 1976.

Welger, C., et al. "Informatique, matin, midi et soir," *Autrement* 1982.

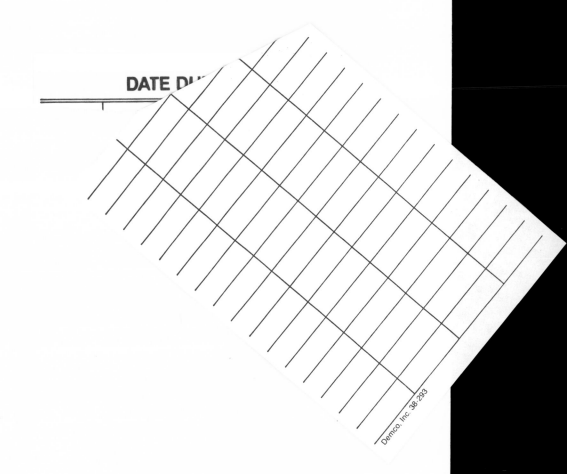

DATE DU

Demco, Inc. 38-293

/303.483E47TB>C1/